Alcohol, Science and Society Revisited

Alcohol, Science and Society Revisited

Edith Lisansky Gomberg,
Helene Raskin White and
John A. Carpenter,
Editors

Ann Arbor   *The University of Michigan Press*
New Brunswick, New Jersey   *Rutgers Center of Alcohol Studies*

Library of Congress Cataloging in Publication Data
Main entry under title:

Alcohol, science and society revisited.

   Includes bibliographical references and indexes.
   1. Alcoholism—Addresses, essays, lectures.
2. Alcoholism—Research—Addresses, essays, lectures.
3. Alcoholism—Treatment—Research—Addresses, essays,
lectures.   I. Gomberg, Edith S., 1920–   .   II. White,
Helene Raskin, 1949–   .   III. Carpenter, John A.
[DNLM: 1. Alcoholism. 2. Social problems. WM 274
A35257]
HV5047.A37 1982   362.2'92   82-4839
ISBN 0-472-10024-6   AACR2
ISBN 0-472-08028-8 (pbk.)

Foreword

In the beginning were . . . Haggard and Jellinek. When these two men—
in background and personality as different as can be imagined—got
together to participate in a common venture to explore alcohol problems, a new era dawned for interdisciplinary scientific endeavor.
The history of this uncommon association of two outstanding men
has been documented many times, especially by Mark Keller in numerous writings, so I shall not repeat it here. Enough to note that
soon after E. M. Jellinek came to Yale University to join H. W. Haggard, the historical first Summer School of Alcohol Studies was
launched. An immediate success, it became an annual event—continued to this day at Rutgers University.

    The design of the school was innovative in that it linked sci-
entific with social perspectives and research in the field of alcohol
problems. Participants at the school, faculty as well as students, came
from widely diverse fields, ranging from biochemists and sociologists to old-time advocates of prohibition and leaders in the then
young Alcoholics Anonymous movement.

    The success of the school inspired the idea of publishing in
book form a selection of the lectures during one session, together
with the discussions and questions that followed each lecture. Thus
the original *Alcohol, Science and Society* was born. The editors of
that book could not foresee that they were publishing a work destined to become a basic text for a new multifaceted discipline—
alcohology. *Alcohol, Science and Society* had an enduring success,
as evidenced by its many reprintings. But it is now almost forty years
old. So much has happened since then in the field of alcohology that
the need to update the contents and to incorporate the knowledge
accumulated since then is obvious. And so, the intention of the editors of the present volume to undertake this task will be welcomed
throughout the world concerned with alcohol problems.

    An examination of the table of contents reveals the wide scope
of this undertaking. It shows also that the editors have not slavishly
followed the outline of the original book. While some of the topics
are appropriately the same, the new knowledge acquired during the
elapsed years and the ensuing possibly controversial opinions make
for an entirely new undertaking. In addition, many of the chapters

deal with new topics that had not yet surfaced in the 1940s. For not only has knowledge advanced, the concepts themselves have changed. The approaches to alcohol problems which were ground breaking in the 1940s have undergone revision—some subtle, some drastic—and the contributors to this volume have taken this into consideration and have not avoided controversy.

This volume represents the much-needed all-encompassing survey of the current state of knowledge and opinion—and uncertainty and conflict—about the major issues around alcohol, drinking, drunkenness, and alcoholism. Fortunately, most of the contributing authors were sensitive to the fact that their essay would be read and needed to be understood by the other contributors and by readers from disciplines and professions other than their own. They have mostly avoided esoteric jargon and written in English. The contents of this volume should thus be understandable to the broad audience that needs to be concerned with these issues.

In summarizing the current knowledge in their particular areas of expertise the authors in this volume have not avoided the exposition and advocacy of their own views and perspectives on approaches to the issues surrounding alcohol. The resultant diversity of viewpoints among the various disciplines constitutes a specially useful feature of this revisit to *Alcohol, Science and Society*. Not only alcohologists but all students of alcohol problems and anyone concerned with possible solutions to the problems in this multidimensional field will find in this book the fundamental knowledge now essential for rational approaches.

The designers and editors of this book have succeeded in making this volume a basic textbook of alcohology.

Vera Efron
Stratton, Vermont

Preface

The idea for this volume originated in a deceptively simple question: in the flurry of research and treatment activity of the last 35 years, just how much have we learned and how much progress have we made? In 1944, the lectures given at the Yale University Summer School of Alcohol Studies were recorded, and 29 of the lectures were published in a volume entitled *Alcohol, Science and Society*. The volume encompassed or at least reflected the available information of the time as well as the "ideas and beliefs" of a number of professional persons: biostatisticians, physicians, physiologists, psychologists, sociologists, ministers, and legal authorities. It is important to note that most of the lectures concerned themselves with alcohol and drinking per se: blood alcohol concentration, alcohol and metabolism, alcohol and aggressive behavior, alcohol and economics, alcohol and traffic, the Temperance Movement, etc. Only 11 of the 29 lectures dealt with alcoholism per se, and of these, only 6 dealt with intervention and treatment of alcoholics.

The Summer School of Alcohol Studies became the model for a large number of similar schools and institutes over the years, sometimes university-sponsored and sometimes state-sponsored or hosted by an organization. But present-day summer schools and institutes bear little resemblance to the original, and Professor E. M. Jellinek, first director of the School, would find his caveat unheeded. He wrote, in the introduction of the 1945 volume:

> Those who are interested primarily in the question of what to do with the alcoholic may find it a bore to have to listen to such lectures as, let us say, the philosophy of the temperance movement or the economic aspects of alcohol in modern society. On the other hand, those who are interested in alcoholism primarily as a national manifestation may be impatient with the lectures which are devoted to the individual problems of the alcoholic and to the treatment of the alcoholic. But I can assure you that unless those interested in the public care and therapy of the alcoholic learn about the economic and social involvements which have led their patients to the need for treatment, efforts at rehabilitation will be frustrated. And those who are interested in alcoholism as a national phenomenon will never be successful until they have realized the importance of those individual problems whose aggregate forms the national manifestation *(pp. 6–7)*.

The gulf between therapists and social scientists has not diminished. And since the majority of those who attend the summer schools and institutes are clinicians, the emphasis has shifted toward treatment issues and skill training. That does not in any way invalidate Professor Jellinek's view that therapists and social scientists need to know about each other's work and findings and methods.

In 1979, we three editors came together at the Rutgers Center of Alcohol Studies (formerly the Yale Center of Alcohol Studies) and discovered that each had cherished the same hope, i.e., to put together a collection of "lectures" by the elder statesmen of alcohol studies. These "lectures" would discuss the progress in treatment and research, the new frontiers, the blind alleys, and even the possible directions for the future. They would take, as a point of departure, the original 1945 volume and trace events from there to the present.

This daydream achieved realization but with limits. One limit was a shortage of elder statesmen (or, to bring it up to the present, elder statespeople). Professor Jellinek died in 1963. It was anticipated that Dr. Giorgio Lolli would contribute to the volume but he died in Rome while the book was in planning. Dr. Selden D. Bacon retired. We considered asking the 1945 lecturers to bring their areas of expertise up to date, but few had pursued careers from then to now in alcohol studies.

The limitations of reality being what they were, we listed the topics which were major present concerns in alcohol studies and sought out senior and younger contributors who knew an area of alcohol studies well and who were willing to bring some historical perspective to bear. Half of the contributors are indeed seniors who have spent virtually all of their professional lives in alcohol-related studies. Contributors like Editor Emeritus Keller, and Drs. Lemert, Cahalan, Straus, Lester, Maxwell, Gusfield, and Rubington have been known for some time for their alcohol-related research and writing. The remaining contributors were drawn from the next generation of researchers and teachers but all of them have proven track records in research and teaching in alcohol studies.

It was not possible, or even desirable, to try and parallel the original 29 chapters. Some concerns do lend themselves to parallel: Professor Jellinek wrote about heredity in 1945, and Dr. Goodwin writes on the same question, one in which there is much new evidence and interest. Dr. Jolliffe wrote about the diseases of alcoholism in 1945, and Dr. Lieber addresses himself to similar medical consequences, brought up to date. Some issues of alcohol and psychological factors are brought up to date by Drs. Pandina and Gomberg.

It is a reflection of changed emphasis that the present volume contains much more discussion about alcoholism per se than does the original volume. The Yale (now Rutgers) Center of Alcohol Studies was a multidisciplinary research center dedicated to both basic and applied research on alcohol, and the 1943 Summer School of Alcohol Studies reflected those interests. Another shift in emphasis is the lessened interest in the Temperance Movement; although historians continue to write about temperance and the experiment with Prohibition, not all contemporary readers would know what was meant by "wet" and "dry" propaganda. There is more current interest in the 1980s in epidemiology, cross-cultural study, study of "special populations," the etiology of alcoholism, animal models of alcoholism and the therapies like behavioral modification and family treatments. But many of the older issues do remain: we are still struggling with definitions of alcoholism, we continue in various disciplines to study the effects of alcohol, and concern with issues of policy and social control are very much with us although the federal role and public attitudes have changed much over the last 35 years, coming full circle, as many of our contributors note.

Each contributor was asked to prepare not an all-inclusive review of the literature but a "lecture" in which the 1945 state of knowledge was the baseline, the contributor considering the question: what have we learned, what has progress been in this particular area of alcohol studies since 1945? This required not only a historical perspective but a finely honed discriminatory sense to separate the more important from the less important findings in the avalanche of published materials since 1945. Inevitably, contributors varied in their interpretation of these instructions, but the present volume does represent a progress report, a state of art summation. And, inevitably, there is some repetition. Different contributors offer different perspectives or interpretations of the importance of the same event to the field of alcohol studies, yet all of these are necessary to help clarify where we are.

There remains a task for an historian: writing a history of the movement which included both alcoholism therapy and alcohol-related research since the 1940s. Such a history would include popular attitudes and behaviors relating to alcohol, the growth of Alcoholics Anonymous, the out-of-the-closet phenomenon of celebrity alcoholism, the development of cadres of professional and paraprofessional caretakers, the so-called alcoholism lobby and the alcoholic beverage industry lobby, the changing fashions in treatment, and the shifting roles of community, state and federal governments in treatment and research.

This book should be read as though each of the contributors had been invited to a School of Alcohol Studies to sum up the status quo in the early 1980s. Since the *Journal of Studies on Alcohol, Inc.* published the original *Alcohol, Science and Society* in 1945, it was decided to make the new volume a joint venture of the University of Michigan Press and the Rutgers Center of Alcohol Studies, of which the journal is a part. It has been a most felicitous cooperation. This book owes much to many people. Professor Alfred S. Sussman, Dean of the Horace H. Rackham School of Graduate Studies at the University of Michigan, has encouraged this venture from the beginning. At the Center of Alcohol Studies, we thank Timothy G. Coffey for his help in the early stages of planning and organization, and Marian Parra and Carol Miller for their clerical assistance. Finally, we express gratitude to the people who contributed their time and expertise in writing for this volume.

E.L.G.
H.R.W.
J.A.C.

Contents

Contributors

MARK KELLER, Professor Emeritus, Center of Alcohol Studies, Rutgers University.

MARY K. ROACH, PH.D., Former Associate Editor, *Journal of Studies on Alcohol*; Rutgers University.

ROBERT J. PANDINA, PH.D., Associate Professor and Acting Director, Center of Alcohol Studies, Rutgers University.

DWIGHT B. HEATH, PH.D., Professor, Department of Anthropology, Brown University.

EDWIN M. LEMERT, PH.D., Professor, Department of Sociology, University of California, Davis.

DON CAHALAN, PH.D., Professor of Public Health (Emeritus) and former Director, Social Research Group, School of Public Health, University of California, Berkeley.

ROBERT STRAUS, PH.D., Professor and Chairman, Department of Behavioral Medicine, College of Medicine, University of Kentucky.

DAVID LESTER, PH.D., Professor, Center of Alcohol Studies, Rutgers University.

DONALD W. GOODWIN, M.D., Professor of Psychiatry, University of Kansas Medical Center.

EDWARD MAJCHROWICZ, PH.D., Senior Research Scientist, Laboratory of Preclinical Studies, National Institute on Alcohol Abuse and Alcoholism.

EDITH LISANSKY GOMBERG, PH.D., Professor, School of Social Work, Institute for Social Research, Institute of Gerontology, The University of Michigan.

HELENE RASKIN WHITE, PH.D., Assistant Professor, Center of Alcohol Studies, Rutgers University.

CHARLES S. LIEBER, M.D., Chief, Section of Liver Disease and Nutrition, Alcohol Research and Treatment Center, Bronx V.A. Medical Center and Mt. Sinai School of Medicine (CUNY).

A. ARTHUR SUGERMAN, M.D., MED. D. SC., Associate Psychiatrist and Director of Medical Student Training, Carrier Foundation.

PETER E. NATHAN, PH.D., Professor and Director, Alcohol Behavior Research Laboratory, Graduate School of Applied and Professional Psychology, Rutgers University.

MILTON A. MAXWELL, PH.D., Professor Emeritus, Sociology, Center of Alcohol Studies and former Director of the Summer School of Alcohol Studies, Rutgers University.

PETER STEINGLASS, M.D., Associate Professor of Psychiatry and Behavioral Sciences, Center for Family Research, George Washington University School of Medicine.

EARL RUBINGTON, PH.D., Professor, Department of Sociology and Anthropology, Northeastern University.

REV. DAVID C. HANCOCK, President, Prevention of Alcohol Problems, Minneapolis, Minnesota.

ROBIN ROOM, PH.D., Scientific Director, Alcohol Research Group, Institute of Epidemiology and Behavioral Medicine, Institutes of Medical Sciences, Berkeley, California.

JAY LEWIS, Editor and Publisher, The Alcoholism Report, Washington, D.C.

JOSEPH R. GUSFIELD, PH.D., Professor, Department of Sociology, University of California, San Diego.

1

Alcohol, Science and Society: Hindsight and Forecast

Mark Keller

We knew a lot about alcohol when we published *Alcohol, Science and Society* in 1945.

We know a lot more now: We know now that we now know less than we thought we knew then. Some of our certainties have been clouded by un.

Before comparing knowledges of then and now, it is useful to review the historical setting. An understanding of the background will facilitate the evaluation of what has been learned that's important and grant perspective on what yet needs to be learned.

Alcohol, Science and Society was a collectanea of lectures delivered at the second annual Yale Summer School of Alcohol Studies, in 1944; the School then consisted almost entirely of lectures. In the "discussion period" following each lecture the students asked questions; the lecturer answered. Marty Mann established a record: she asked one lecturer a series of nine questions. The average student asked 1.1 questions. A portion of the "discussions" was published with each lecture. E. M. Jellinek, as "chairman" at most of the discussions, occasionally helped the lecturers by contributing a comment.

Of the 24 lecturers, 6 were from the staff of the Yale Laboratory of Applied Physiology which sponsored the Summer School, a staff already then harboring a majority of non-biologists but not yet formally constituted as an interdisciplinary Center of Alcohol Studies; 3 were borrowed from other departments of Yale University; and 15 were distinguished authorities from other universities and institutions.

The volume was designed to serve as a source book on a range of topics wide enough to cover most of the recognized alcohol-related problems: the metabolism of alcohol; its physiological and psychological effects; its functions in primitive and complex societies; contemporary drinking patterns; economic aspects; alcoholism and heredity, personality, family, aggression, disease, poverty; the Temperance Movement, the churches, control laws, propaganda; treatment of alcoholism by medicine, religion, social work, penology, Alcoholics Anonymous.

The texts of the lectures and discussions were first mildly edited by E. M. Jellinek, then more firmly by Howard W. Haggard, and finally copy-edited by me. For unexplained reasons Haggard and Jellinek did not agree about title-page credit, so the book was sent to press without an indication of editorship. (It should have been "Edited by Howard W. Haggard and E. M. Jellinek.") It was a successful book, for it contained in readable language the most reliable knowledge of the time, and it sold and sold and sold; after 9 years and five printings, it was still in demand. Moreover, it was making money for a publications division of the Yale Center of Alcohol Studies which consistently operated with a deficit. So instead of letting it go out of print I wrote a new "Preface to the Sixth Printing" in 1954, updating much of the contents, and we sold some additional thousands of copies before letting it go out of print.[1]

How did all that knowledge come to be centered in the Yale Laboratory of Applied Physiology and in *Alcohol, Science and Society*?

What exactly did we think we knew?

It started when National Prohibition ended. By coincidence the vitamin age in nutrition was ready to burgeon in the early 1930s. Physicians in Cincinnati (Blankenhorn and Spies), in Boston (Maurice Strauss), and in New York (Norman Jolliffe) demonstrated that some of the classical diseases of alcoholism could be cured even while the patients continued their drinking, provided therapeutic doses of essential nutrients (especially B vitamins) were administered simultaneously. Other "alcoholic" diseases were suspected of being caused by malnutrition rather than, as hitherto thought, by the direct action of that demonic substance alcohol. There was no lack of "chronic alcoholic" patients in the hospitals, mostly products of the 13-year Prohibition Era, as subjects for study. And one day, in New York's Bellevue Hospital, Norman Jolliffe, who for years had been studying polyneuropathy, pellagra, Wernicke's encephalopathy, Korsakoff's psychosis, cirrhosis of the liver, scurvy, ariboflavinosis, anemia, and beriberi heart in several thousand alcoholics, made a historically significant observation: The alcoholics whom he

---

[1]It was newly reprinted by Greenwood Press, Westport, Conn., in 1972 and 1977.

had cured a year ago and two years ago kept coming back with the same or new disorders. He said, they keep on drinking the same way and developing more symptoms and severer diseases. It isn't enough to cure those diseases. The question is, Why are they drinking like that? That's what we should be studying!

Jolliffe had discovered alcoholism as Columbus discovered America. It had been there all the time, but it needed a long sail after a lot of thought to stumble on it.

A Research Council on Problems of Alcohol with a distinguished scientific advisory committee was formed. A review of the biological literature on the effects of alcohol on man was undertaken. For its conduct, Jolliffe enlisted E. M. Jellinek, and thus brought him into alcohol studies. As offshoots of that review (*1*) Dr. Howard W. Haggard founded the *Quarterly Journal of Studies on Alcohol*; (*2*) a now-classic volume, *Alcohol Addiction and Chronic Alcoholism*, edited by Jellinek on behalf of the Research Council's scientific advisory committee, was published;[2] (*3*) the rudimentary *Classified Abstract Archive of the Alcohol Literature* and the *International Bibliography of Studies on Alcohol* (the first an ingenious Jellinekian invention, the second a joint Jellinek–Keller idea) were moved to Yale to be fostered and developed conjointly with the *Journal*; (*4*) Jellinek was supported by Haggard in enlisting an inter-disciplinary staff that was to become the Center of Alcohol Studies.

The Research Council could not attract adequate financial support; it went out of existence in 1949. Studies and activities related to alcohol became centered at the Yale Laboratory of Applied Physiology and did flourish there. Sociology, psychology, anthropology, law, documentation, education, economics, penology, medicine, were added to its physiology and biochemistry. The Summer School of Alcohol Studies was started. A National Committee for Education was formed (to become the National Council on Alcoholism). The Yale Plan Clinics were launched. It became necessary to publish *Alcohol, Science and Society*, and *Lay Supplements* to the *Journal*, and other books, so that a documentation–publications division was created. And, behold, within the Yale Laboratory of Applied Physiology a formal Center of Alcohol Studies with five divisions was born—and the child was bigger than the parent. When historical

[2]It was Volume I of "The Effects of Alcohol on the Individual" (1). As there was yet no Center of Alcohol Studies and so no documentation–publications division, and Jellinek and I had meantime moved to Yale at Haggard's invitation, we let the Yale University Press publish it. But the projected Volumes II and III were never produced, partly because we had become too busy with other developments leading to the creation of a Center of Alcohol Studies.

changes at Yale caused the Laboratory to be discontinued and the Center transferred to Rutgers University, the roles were appropriately redistributed: the Laboratories became a division of the Center.

The story has moved ahead of itself: the developers and purveyors of the knowledge are already far ahead of the knowledge itself. We must go back to the mid-1940s when *Alcohol, Science and Society* was published.

We did believe in "The Problems of Alcohol"—that was the title of Lecture 2, by E. M. Jellinek. The title, illustrated by Vera Efron in a Lay Supplement of the same title as a ramifying vine (2), is a significant statement. It parted from older conceptions of "the alcohol problem" which had filled the historical problem-oriented literature dominated by the temperance–antialcohol movement. We now were invited to consider not that alcohol was THE problem but that there were many problems in which alcohol was involved in a variety of ways. This viewpoint, now widely understood, is still not universally accepted. Many people still believe in "the alcohol problem" and, again, in "the" solution—to suppress, or at least limit, the availability of alcohol.

Several critical knowledge issues that were dealt with in *Alcohol, Science and Society* are still major issues but the knowledge taught now differs substantially.

In the biological area the outstanding recent acquisitions relate to the question of direct damage by alcohol to human tissues. New knowledge of the role of microelements of nutrition gained in the 1920s and 1930s was reflected in the teaching, in *Alcohol, Science and Society*, that most, but not necessarily all, of the commonly known "alcoholic" diseases were due to malnutrition rather than the direct action of alcohol, as formerly supposed. The role of vitamin B_1 deficiency in causing polyneuropathy, beriberi heart and Wernicke's encephalopathy was demonstrated conclusively, as well as the role of nicotinic acid deficiency in pellagra. Nutritional defects were suspected of involvement in liver cirrhosis, Korsakoff's psychosis, and delirium tremens, but never conclusively demonstrated to this day.

In recent years the elaborate years-long experiment with baboons by Rubin, Lieber and their colleagues, as well as other experiments by the same group, has led to a reverse conclusion (3–5). Some of the primates maintained over many years with a liquid diet which, though half the calories were supplied by alcohol, provided adequate amounts of vitamins and other nutrients, developed liver cirrhosis. The experimenters concluded that their work proved that the cirrhosis resulted from direct damage by alcohol, and this claim has been generally accepted, though some skepticism survives

(6). The same experimenters likewise claim to have demonstrated direct causation by alcohol of cardiomyopathy and direct damage by alcohol to other organs.

Nevertheless it is too soon to pronounce the authorities of *Alcohol, Science, and Society* in error on this score. The recent experiments may have claimed more than is justified; they need confirmation and the reinforcement of other evidence. Although they provided all the known necessary nutrients in the liquid diet, the experimenters have not demonstrated that all those nutrients, including those that may protect the liver against cirrhosis, or prevent damage to other organs, were actually absorbed. Indeed, they have noted that absorption studies were not performed (4). Moreover, they themselves have shown in other experiments the alcohol-induced malabsorption of essential nutrients (5, 7); nor have they explained how contact of some tissues with the tiny concentrations of alcohol that reach them—less than ½ of 1%—can cause the pathologies attributed to direct damage by alcohol. Perhaps the next generation of physiologists–biochemists will resolve the question.

Science has its temporal fashions. A period in which all work points to nothing but damages by alcohol is followed by one in which the researchers of the previous generation are shown to have overlooked important factors in their experiments and their conclusions are challenged and reversed. Then a new generation of scientists reverses its predecessors. That's where we seem to be now. In the first third of this century alcohol did it all. In the middle third it wasn't alcohol; malnutrition did it (though alcohol helped). Now it is alcohol again (but malnutrition helps). It is hard to resist predicting the next revision.

The current reversal of ideas about liver cirrhosis, cardiomyopathy, and other organ damage is overshadowed by the dramatic exposition of the fetal alcohol syndrome. Observation of inferior physical and mental development in children of alcoholics (anyhow, drunkards) is as old as history; once upon a time this was blamed on intoxication of parents during conception. By *Alcohol, Science and Society* times the birth of defective fetuses (as well as smaller litters and more frequent abortions) had been demonstrated in alcoholized animals. But by then it was logical to attribute those findings to malnutrition, for the early experiments had been performed without thought to the nutritional protection of the animals. The defects in human children were likewise logically attributed to the commonly observed malnutrition of the alcoholic mothers, especially as similar defects were seen in children of malnourished nonalcoholic mothers.

What is new in the recent reports of the fetal alcohol syndrome

is the possibility that alcohol may be a teratogen, a direct cause of
fetal mal-development if introduced in sufficient quantity especially
in the early months of pregnancy (8, 9). The finely designed experiments that would resolve the question of whether alcohol is a teratogen, and if so, under what circumstances and quantities of ingestion, have yet to be carried out. A decisive advance in knowledge is
awaiting achievement.

Is alcoholism inherited? The authorities of *Alcohol, Science
and Society* were sure it was not. Wherever one could point to the
more frequent occurrence of alcoholism among children of alcoholics it was possible to point out also the circumstances, such as poverty or familial mental disorder, that could account for the pathetic
coincidence. And, then, there was the study, started by Barbara Burks
and just concluded at the Yale Laboratory of Applied Physiology by
psychologist Anne Roe, of children of alcoholic parents who were
adopted and raised by nonalcoholics. Examined as adults, not one
showed signs of any problem with alcohol. The monograph detailing
the study (10) and Roe's lecture in *Alcohol, Science and Society*
decidedly supported the environmentalist side in the nature–nurture
debate about the etiology of disorders such as alcoholism.

A reversal now predominates! Recent studies of adoptees and
separated twins strongly suggest that alcoholism does indeed run in
families (11). Geneticists do not accept this evidence as conclusive
(12); strict geneticists will not conclude from this sort of evidence
that "alcoholism is inherited." But most people, including hard scientists, including some geneticists, will accept that there is a likelihood or at least a possibility that some people inherit genes that
render them more susceptible to developing alcoholism; or that some
people inherit genes that render them relatively immune to developing alcoholism; or that both kinds of genes may exist. Surely this
field is awaiting further cultivation. It is not unimportant to establish
whether there is a genetically more-at-risk population. As the knowledge stands today it is not unreasonable to assume that children of
alcoholics constitute such an at-risk population. It is also possible
that children born to mothers who drink heavily during pregnancy
are adapted to alcohol in utero. Are they born with an alcoholism-
prone constitution?

In psychology a grand reversal has occurred. The authorities
in *Alcohol, Science and Society*, relying on their own experiments
and on the exhaustive critical review of the biological–psychological
literature by Jellinek and McFarland (13), proclaimed with assur-
ance that the action of alcohol in the central nervous system was
always as a depressant. The apparent contradiction, that people often

seemed stimulated after drinking, being more active, more talkative, more sociable, more aggressive, was explained away by a brilliant pilpulistic postulate: People are commonly inhibited; by depressing the inhibitory function of the cortex, alcohol serves as a disinhibitor. The result is like a release of the brakes—people behave as if stimulated because their inhibitions have been depressed by the alcohol.

Already in the 1950s ingenious experiments with exposed nerves challenged the notion of alcohol always acting as depressant (14). Manifestly, at relatively low concentrations, alcohol stimulates, while at higher concentrations it does indeed depress nervous activity. The revised teaching is that the action of alcohol is biphasic (15). This allows a simpler understanding of such phenomena, observed in experiments by psychologists, as that at low concentrations of alcohol in the organism (as achieved after one or two ounces of whisky in a man of average size) people are more fluent (16), perform better on certain tasks of memorizing (17), and—in subjects of high intelligence—perform better at solving difficult problems in logic (18). The same subjects perform relatively poorly at higher alcohol concentrations. On the other hand, some recent investigators have reported that even social drinkers perform less efficiently at mental tasks than abstainers (19). The latter findings are in line with the alcohol-is-to-blame revision.

Incidentally, the designation of these effects—whether of improved or deteriorated performance—as psychological seems questionable. The effects take place in the brain. They are neurological. What they have to do with a mysterious psyche that no one has ever seen is a mystery. This may be a problem in semantics. In the use of terminology there has been no progress since *Alcohol, Science and Society*, but that is not a problem of the alcohol field alone.

Crime and aggression in relation to alcoholism received a sophisticated psychoanalytic interpretation in *Alcohol, Science and Society*, along with some unusually good statistics from a study in Sing Sing Prison. "Inebriates" were reported to constitute only about 25% of the felon population rather than the 60% or so usually claimed. Nevertheless, since 25% of the general male population are not alcoholics, it is obvious that alcoholics were—and are—grossly overrepresented in the penitentiary. From this it was concluded that alcoholism plays a large role in criminality. That conclusion is thought to be confirmed by all recent statistics on the prevalence of alcoholics among incarcerated felons (20, 21).

One skeptical voice has been heard to suggest that the statistics are misinterpreted and misleading (22). It seems possible, the skeptic hypothesizes, that the excess of alcoholics in prisons is due not to their more frequent commission of felonies but to the propensity of alcoholics to perform inefficiently when they engage in crimes, just as when they indulge in honest endeavors. Their ineffectiveness results in less successful concealment, less successful escape; and when detected, less successful defense, with resulting more frequent conviction. Moreover, alcoholics sometimes need a respite from the struggle with life, a sabbath from alcohol; they sometimes seek institutionalization, welcome even imprisonment. Robert Straus's Mr. Moore strikingly exemplifies the pattern (23). Conceivably alcoholics sometimes cooperate with the prosecution, perhaps unconsciously, to get themselves sentenced to prison. In other words, alcoholics are the ones whom the police most often catch, the prosecutors must easily convict, and the judges most regularly sentence to prison. If alcoholics are four times as likely to be caught, thrice as likely to be convicted and twice as likely to be sentenced to prison, that could amply account for the increased presence of alcoholics in prisons, without any excess of crimes being committed by them. To what extent the relative inefficiency of alcoholics as criminals, and perhaps also their need for periodic institutionalization, accounts for their excess presence in the nation's penitentiaries has apparently not yet come under investigation. But reports connecting alcoholism with child abuse, wife beating and other acts of violence are a growing phenomenon.

One more consideration casts doubt on the assumption that alcohol has a one-way effect in relation to crime and aggression: If moderate alcoholization stimulates, and permits or promotes aggressive behavior, heavy alcoholization depresses and should logically inhibit aggressive behavior. How many planned crimes or aggressions were not carried out because the planner drank, as alcoholics do, beyond the moderate amount that evokes Dutch courage, and got too drunk to do anything but continue inbibing? Relevant to this is the not-to-be-forgotten Pittman and Gordon study of Skid Row alcoholics (24): Many of these men had started out in their youth on a criminal career, as their early arrest records showed, but gradually, as alcoholism became their dominant lifeway, they had abandoned crime, settling down to being arrested only for drunkenness.

Social scientists with a skeptical outlook on common assumptions could take a fresh look at the supposed relation of alcohol and alcoholism to crime and aggression.

Passing a law is a favorite American way of dealing with public problems. The national Congress, fifty state legislatures, a proliferating host of regulatory bureaucracies, thousands of local legislative bodies, are legislating and regulating ceaselessly. Even on

Sunday. Not even a justice of the Supreme Court knows all the laws he must obey. And it is possible that there are more laws about alcohol than on any other object. Whether any of these laws is effective in mitigating alcohol-related problems is matter for speculation. Perhaps abolition of all the laws about alcohol would not make much difference.

Of course the laws against drunken driving empower the police to arrest drunken drivers, and the courts to punish them. Everybody agrees that that's good. But it does not seem that these laws prevent much drunken driving.

How hopefully the director of the Traffic Division of the National Safety Council reported in *Alcohol, Science and Society* on the beginning of the use of chemical tests for intoxication. There was only one portable automatic testing instrument in existence at that time—the recent invention of Professor Leon A. Greenberg in the Yale Laboratory of Applied Physiology. But already chemical testing was spreading, and police and courts were increasingly accepting the procedure and the results. As soon as enough of these portable instant testing machines were available to catch the breath of suspected drivers, surely the drunken-driving problem would be mostly solved.

The war ended. Chemical test laws and practices were established just about everywhere. Improved testing instruments were devised. And it all made no difference. In the 1920s about 25,000 persons were killed annually in motor vehicle accidents. In the 1970s about 50,000. That does not reflect an increase. The population is bigger, there are more cars, more highways, more drivers and more driving. But drunken-driving accidents and fatalities have not been anywhere near abolished by the grand advance of automatic and on-the-spot chemical testing and all the related legislation, including lowering the blood alcohol concentration at which a legal presumption of fault is made.

What's needed is the wisdom to recognize that lawmaking is not the same as behaviormaking. What's needed is to discover how to change people's attitudes toward drinking, driving, and driving after drinking. Everybody knows that. But everybody is busy adding more laws. In this the fashion has not changed.

In social as well as biological issues the pendulum keeps swinging from the view that alcohol-is-to-blame to it's-not-alcohol-that's-to-blame to alcohol-is-to-blame. *Alcohol, Science and Society* reflected the beginning of a reaction from an alcohol-is-to-blame era, marked by the extreme attempt through the 18th Amendment to rid society of alcohol altogether. So we then groped through an it's-not-alcohol-that's-to-blame period. Nutrition was blamed for diseases, and for alcoholism itself; personality and environment were blamed for the psychological and social problems, and for alcoholism itself. Now we seem to be entering a renewed alcohol-is-to-blame swing. Alcohol is being blamed for direct damage to vital tissues—the liver, the heart, the brain, the testicles, the muscles, the pancreas, the intestines, the fetus (25). Alcohol is blamed for abused wives and children. Alcohol is blamed for inferior mental performance even in social drinkers. Alcohol is blamed for causing alcoholism—alcohol addiction.

That critical conclusion has led some social scientists to propose that the solution is the suppression of alcohol. Of course they do not put it in such crude terms. Anyhow not prohibition. The latest findings, based on the impressive lognormal curvature of the distribution of alcohol consumption in some studied places, indicate that alcohol problems, presumably alcoholism itself, are proportional to the average volume of alcohol consumed by a population (26). Hence, reduce the average volume of alcohol consumed by the average member of the population, and we will reduce the incidence of problems, no doubt alcoholism too. Of course the necessary reduction of consumption is to be effected by passing more laws, laws to make it at least inconvenient and expensive to obtain alcoholic beverages.

The question yet to be researched is, how does the proportionality work? Is it only that the more alcohol consumed, the more problems; or is it also that the more heavy drinkers or alcoholics there are in the population, the more alcohol is consumed as well as the more problems there are? Is it possible that making it inconvenient and expensive to get alcoholic beverages will affect the consumption of the moderate non-problem-causing drinkers, while the heavy drinkers and alcoholics, being needy, will get theirs somehow anyhow? The questions ought to be resolved by some more research before legislators are persuaded to adopt the interpretations of the believers in the distribution-of-consumption hypothesis.

The economic costs of inebriety were duly summed up in *Alcohol, Science and Society* and came to a grand total of $778,903,000 for the year 1940. It was impressive: nine digits—approaching a billion dollars! This topic was hardly pursued systematically again until 30 years later the National Institute on Alcohol Abuse and Alcoholism hired a private consulting firm for the task. Based on that effort the Institute's second Report to Congress (27) in 1974 set the annual economic cost of alcohol-related problems at 25.3 billion dollars. Its third Report (25) in 1978 revised the estimate upward to

42.75 billion. Even allowing for the reduced value of the dollar, and for the increase in population, and for the inclusion of more elements of cost in the later estimates, the 1940 price seems modest.

The costs of alcohol-related problems are based on such measurable items as hospitalization for alcoholic diseases, imprisonment for drunkenness, support of dependents, absence from work, spoilage at work, damages for accidents, and so forth. The account can hardly be complete, staggering as it seems. Since the cost is attributed to inebriety or alcohol misuse, the estimates have to be accepted to the extent that the analyses seem reasonable. There is a recent tendency, however, to begin again, as in the days of what Bacon has called the Classic Temperance Movement (28), to attribute these costs to "alcohol use," even in academic circles and among professionals working in the alcohol field. It is a manifestation of the revisionist blame-it-on-alcohol movement.

The economic effect of a product or effort is usually gauged in terms of cost–benefit. With respect to alcohol, only cost has ever been examined. No one has attempted to balance the alcohol account by estimating the value of the benefit derived by the vast majority of drinkers from drinking. That there are such benefits (aside from the increased life span of drinkers), that there must be such value, has been rationally hypothesized. But this value has never been calculated in economic terms. Not that anyone knows how to do it, any more than anyone knows how to cost the human suffering caused by alcohol misuse. But the existence of the benefits ought not to be ignored when proposals to inconvenience and deprive non-harm-causing drinkers are considered in the revisionist period.

Alcoholism was one of the troubles for which alcohol was unblamed—absolved—in *Alcohol, Science and Society*. The cause was not in the bottle but in the man. The reasoning was sound; it cannot be faulted today: For probably 95% of drinkers never develop alcoholism. There must be a difference in the people who do develop alcoholism; perhaps also in their circumstances.

And yet the etiology of alcoholism—here defined as alcohol addiction—is not known. Even how alcohol intoxicates, and that process may be involved in the development of the addiction—is not yet known. But good theories have emerged in the past third of a century. Perhaps the work on membrane reaction to alcohol leads to understanding of the process of intoxication. Perhaps the theories of the learning psychologists will yet explain the process of addiction. But by no means can these processes be seen as "absolving" alcohol. It is alcohol that intoxicates. It is alcohol to which the cells of the central nervous system adapt and become tolerant. It is alcohol that those cells then "learn" to need imperiously, irresistibly. In that sense, one may say, alcohol causes alcohol addiction. But that conclusion may not, after all, go far enough. In the last analysis there will be the question, Why do some individuals, but not most, drink enough to learn to be addicted? This question needs answering, and probably can only be resolved by such longitudinal research as has only now been undertaken at the Rutgers Center of Alcohol Studies.

Fully 20% of *Alcohol, Science and Society* was devoted to treatment of alcoholism by doctors, the clergy, the social workers, lay therapists, Alcoholics Anonymous. All the treatments discussed assumed that the goal was the patient's total permanent abstinence— that an alcoholic should never drink again, could never safely drink again. Back of this view was not merely the passionate opinion of the just becoming widely known Alcoholics Anonymous but the experience of generations of professional therapists. Recently the experiments of several behavioral psychologists have suggested that at least some alcoholics can achieve moderated drinking or even controlled drinking (29, 30) which may prudently be designated "social."

The issue has been confused by widely publicized reports, based on defective sampling, unscientific interview methods, inappropriate analyses and amateurish interpretation, of vast numbers of vaguely diagnosed "alcoholics" becoming "normal drinkers" after relatively limited amounts of various poorly defined treatments (31). In fact, the total recovery of some reliably diagnosed alcoholic patients, with ability to drink moderately for long years, has been reported in the literature a number of times (32–34). There is no reason to doubt that some alcoholics—true alcoholics, alcohol addicts—can recover completely and become controlled drinkers. That it is no easy achievement is suggested by the commonly repeated failures of such efforts. Yet the newer experiences of the behavioral psychologists (35) and the several older reports suggest that it is not impossible, as some had inferred. But whether it is possible for all, or for some, or for very few—for whom and under what circumstances and by what means—remains to be learned. It requires the carefully controlled approach of scientific investigation with humanely selected strictly diagnosed patients, and prolonged sophisticated follow-up.

Finally, in *Alcohol, Science and Society* the question whether alcoholism is a disease was hardly raised. Expressed and unexpressed, alcoholism as disease was confidently assumed. The fact that of five lectures on treatment only one was medical did not suggest that alcoholism might not be a disease but only that it includes elements, behaviors, symptoms, that could be dealt with, under an umbrella of comprehensive medicine, by clergymen, social workers, lay therapists, and Alcoholics Anonymous, equally with psychologists, psychiatrists and other physicians. Nowadays, not consciously but in spirit connected, it is a part of the revisionist movement to dismiss the disease concept of alcoholism (36). The attack on it is sometimes naively misdirected: People who have had some trouble on account of alcohol misuse are recklessly labeled alcoholics, or lumped by unqualified diagnosticians with alcoholics, and the fact that they cannot be regarded as having a disease is then exhibited as evidence that "alcoholism" is not a disease. There are, nevertheless, more substantial arguments about the nature of alcoholism. In fact, it comes down to two questions of definition: What is a disease? And, assuming a strict definition of alcoholism as alcohol addiction, Is drug addiction a disease? It may be that the definitive answer to these questions can come only from the neurosciences. First they will have to resolve the fundamental question of addiction—what happens in the central nervous system when a person becomes intoxicated, repeatedly intoxicated, tolerant (or adapted) and addicted. It will then be possible positively to characterize the self-injurious addictions, including alcoholism.

Summary and Conclusion

This hindsighted reflection on *Alcohol, Science and Society* has not attempted to discuss all the topics in that book but has focused on those that seem specially pertinent today.

A fashion pendulum in the consideration of alcohol-related problems swings periodically from alcohol-is-to-blame to it's-not-alcohol-that's-to-blame. The thrust of social and scientific action is influenced accordingly. *Alcohol, Science and Society* reflected a swing from the former to the latter viewpoint. Currently the trend is back to blaming alcohol and renewed efforts to suppress it. Sensitivity to history favors the prediction of a reverse swing in due course.

Moving through the biological to the so-called psychological to the social realms, these are the outstanding specific questions, either considered in *Alcohol, Science and Society* or risen subsequently, that want intensive pursuit in the immediate future:

Does heavy drinking directly damage the liver, the heart, the pancreas, the testicles, the muscles, the brain? Is alcohol a fetal teratogen, and if so, under what circumstances? Does heavy drinking by pregnant women render their children susceptible to alcoholism?

What is the relation between heredity and alcoholism? How does alcohol intoxicate? How is alcoholism (alcohol addiction) reflected in the central nervous system? Under what circumstances does alcohol intoxication or alcoholism cause or inhibit crime and violence? Which sort of laws effectively reduce which alcohol-related problems? Does moderate drinking, not involving intoxication and other troubles, deteriorate or improve brain function? Does moderate drinking affect the life span?

The preceding reflections and questions lead only to the classical conclusion: We need more research.

More-sophisticated research.

REFERENCES

1. JELLINEK, E. M., ed. Effects of alcohol on the individual; a critical exposition of current knowledge. Vol. 1. Alcohol addiction and chronic alcoholism. New Haven; Yale University Press; 1942.
2. [JELLINEK, E. M.] The problems of alcohol. (Lay Supplement No. 1.) New Haven; Quarterly Journal of Studies on Alcohol; 1941.
3. RUBIN, E. and LIEBER, C. S. Experimental alcoholic hepatitis: a new primate model. Science 182: 712–713, 1973.
4. RUBIN, E. and LIEBER, C. S. Fatty liver, alcoholic hepatitis and cirrhosis produced by alcohol in primates. New Engl. J. Med. 290: 128–135, 1974.
5. BARAONA, E., PIROLA, R. C. and LIEBER, C. S. Small intestine damage and changes in cell population produced by ethanol ingestion in the rat. Gastroenterology 66: 226–234, 1974.
6. HARTROFT, W. S. On the etiology of alcoholic liver cirrhosis. Pp. 189–197. In: KHANNA, J. M., ISRAEL, Y. and KALANT, H., eds. Alcoholic liver pathology. (International Symposia on Alcohol and Drug Addiction Series.) Toronto; Alcoholism and Drug Addiction Research Foundation; 1975.
7. LINDENBAUM, J. and LIEBER, C. S. Alcohol-induced malabsorption of vitamin B_{12} in man. Nature, Lond. 224: 806, 1969.
8. STREISSGUTH, A. P. Maternal alcoholism and the outcome of pregnancy; a review of the fetal alcohol syndrome. Pp. 251–274. In: GREENBLATT, M. and SCHUCKIT, M. A., eds. Alcoholism problems in women and children. New York; Grune & Stratton; 1976.
9. HOLLSTEDT, C., OLSSON, O. and RYDBERG, U. The effect of alcohol on the developing organism; genetical, teratological and physiological aspects. Med. Biol., Hels. 55: 1–14, 1977.
10. ROE, A. and BURKS, B. Adult adjustment of foster children of alcoholic and psychotic parentage and the influence of the foster home. (Memoirs of the Section on Alcohol Studies, Yale University, No. 3.) New Haven; Quarterly Journal of Studies on Alcohol; 1945.

11. GOODWIN. D. W. Alcoholism and heredity; a review and hypothesis. Arch. gen. Psychiat. **36:** 57–61, 1979.

12. WINOKUR, G., TANNA, V., ELSTON, R. and GO, R. Lack of association of genetic traits with alcoholism; C3, Ss and ABO systems. J. Stud. Alc. **37:** 1313–1315, 1976.

13. JELLINEK, E. M. and McFARLAND, R. A. Analysis of psychological experiments on the effects of alcohol. Quart. J. Stud. Alc. **1:** 272–371, 1940.

14. GRENELL, R. G. Alcohols and activity of cerebral neurons. Quart. J. Stud. Alc. **20:** 421–427, 1959.

15. KELLER, M. Alcohol consumption. Encyc. Britan. **1:** 437–450, 1974.

16. HARTOCOLLIS, P. and JOHNSON, D. M. Differential effects of alcohol on verbal fluency. Quart. J. Stud. Alc. **17:** 183–189, 1956.

17. CARPENTER, J. A. and ROSS, B. M. Effect of alcohol on short-term memory. Quart. J. Stud. Alc. **26:** 561–579, 1965.

18. CARPENTER, J. A., MOORE, O. K., SNYDER, C. R. and LISANSKY, E. S. Alcohol and higher-order problem solving. Quart. J. Stud. Alc. **22:** 183–222, 1961.

19. PARKER, E. S. and NOBLE, E. P. Alcohol consumption and cognitive functioning in social drinkers. J. Stud. Alc. **38:** 1224–1232, 1977.

20. GERSON, L. W. Alcohol consumption and the incidence of violent crime. J. Stud. Alc. **40:** 307–312, 1978.

21. HANCOCK, D. N. Alcohol and crime. Pp. 264–270. In: EDWARDS, G. and GRANT, M., eds. Alcoholism; new knowledge and new responses. London; Croom Helm; 1977.

22. KELLER, M. Discussion. J. Stud. Alc., Suppl. No. 8, pp. 315–316, 1979.

23. STRAUS, R. Escape from custody; a study of alcoholism and institutional dependency as reflected in the life record of a homeless man. New York; Harper & Row; 1974.

24. PITTMAN, D. J. and GORDON, C. W. Revolving door; a study of the chronic police case inebriate. (Monographs of the Yale Center of Alcohol Studies, No. 2.) New Haven; 1958.

25. U.S. NATIONAL INSTITUTE ON ALCOHOL ABUSE AND ALCOHOLISM. Alcohol and health; third special report to the U.S. Congress from the Secretary of Health, Education, and Welfare, June 1978; technical support document. Ernest P. Noble, editor. (DHEW Publ. No. ADM 79-832.) Washington, D.C.; U.S. Govt Print. Off.; 1978.

26. BRUUN, K., EDWARDS, G., LUMIO, M., MÄKELÄ, K., PAN, L., POPHAM, R. E., ROOM, R., SCHMIDT, W., SKOG, O.-J., SULKUNEN, P. and ÖSTERBERG, E. Alcohol control policies in public health perspective. (Finnish Foundation for Alcohol Studies, Vol. 25.) Helsinki; 1975.

27. U.S. NATIONAL INSTITUTE ON ALCOHOL ABUSE AND ALCOHOLISM. Second special report to the U.S. Congress on alcohol and health, June 1974. Mark Keller, editor. (DHEW Publ. No. ADM 75-212.) Washington, D.C.; U.S. Govt Print. Off.; 1974.

28. BACON, S. D. The classic temperance movement of the U.S.A.; impact today on attitudes, action and research. Brit. J. Addict. **62:** 5–18, 1967.

29. CADDY, G. R. and LOVIBOND, S. H. Self-regulation and discriminated aversive conditioning in the modification of alcoholics' drinking behavior. Behav. Ther., N.Y. **7:** 223–230, 1976.

30. SOBELL, M. B. Alternatives to abstinence; evidence, issues and some proposals. Pp. 177–209. In: NATHAN, P. E., MARLATT, G. A. and LØBERG, T., eds. Alcoholism; new directions in behavioral research and treatment. (NATO Conference Series. III. Human factors, Vol. 7.) New York; Plenum; 1978.

31. ARMOR, D. J., POLICH, J. M. and STAMBUL, H. B. Alcoholism and treatment. (Report No. R-1739-NIAAA.) Santa Monica; Rand Corporation; 1976.

32. SHEA, J. E. Psychoanalytic therapy and alcoholism. Quart. J. Stud. Alc. **15:** 595–605, 1954.

33. DAVIES, D. L. Normal drinking in recovered alcohol addicts. Quart. J. Stud. Alc. **23:** 94–104, 1962.

34. KENDALL, R. E. Normal drinking by former alcohol addicts. Quart. J. Stud. Alc. **26:** 247–257, 1965.

35. NATHAN, P. E., MARLATT, G. A. and LØBERG, T., eds. Alcoholism; new directions in behavioral research and treatment. (NATO Conference Series. III. Human Factors, Vol. 7.) New York; Plenum; 1978.

36. ROIZEN, R. Naming names; a note on drinker self-characterizations. Drinking & Drug. Pract. Surveyor, Berkeley, No. 9, pp. 18–20, 1974.

2

The Biochemical and Physiological Effects of Alcohol

Mary K. Roach

In the Introduction to *Alcohol, Science and Society*, E. M. Jellinek (1, *p.* 7) humorously justified the lectures on alcohol's physical effects, saying "it is only through the physiological properties of alcohol that alcoholism becomes a problem . . . if it had the properties of milk or water we would not be gathered here." Such justification is no longer required. Few today would argue against a biochemical or physiological basis for alcohol intoxication, tolerance and physical dependence, although these responses may be influenced by social and psychological factors. However, Jellinek continued, "physiological investigations have not yielded the answer as to why some men crave alcohol to excess." This remains true; in the ensuing 35 years, biochemists and physiologists have not discovered why some people become alcoholics. But we do know a great deal more of what the body does to alcohol and what alcohol does to the body.

By "alcohol" is meant ethyl alcohol or ethanol,[1] the major "active ingredient" in all alcoholic beverages. Alcohol affects almost every tissue in the body in some way, and to itemize all its actions in such a brief review as this would not be feasible. Consequently, I will first examine the absorption and metabolism of alcohol and then focus on two especially sensitive tissues—liver and brain—to illustrate the ways in which alcohol exerts its effects. Much of the work described here is derived from animal studies, and applicability to the human situation is usually assumed if not proven. The biochemistry and pharmacology of alcohol have been reviewed frequently in the past decade (e.g., 2–11), and interested readers are referred to those sources for more detailed information.

---

[1]Both terms will be used in this article: "ethanol" when a specific molecular action is discussed, "alcohol" when more general effects are described.

The Blood Alcohol Curve

The duration and severity of the pharmacological effects of alcohol are determined not only by the total amount consumed but also by the rates of its absorption and elimination, which may be either accelerated or slowed under various conditions. Absorption and elimination are best depicted as a graph of blood alcohol concentration (BAC) against time (Figure 1), which may be divided into stages as follows: (1) a rapidly rising initial phase where absorption is the dominant activity; (2) a peak or a plateau of varying widths, as alcohol is distributed to the tissues and where absorption and elimination may proceed at approximately equal rates; (3) a relatively linear descent, as elimination becomes the major activity; and (4) an exponential decline when BAC becomes less than about 0.01% (in humans).

Absorption

One obvious influence on the rate of absorption is the route of administration. In experiments with laboratory animals, alcohol may be given in a variety of ways—orally, either in drinking fluids or through an inserted stomach tube; by injection, e.g., into a vein (i.v.), into the abdominal cavity (i.p.) or under the skin; or by inhalation

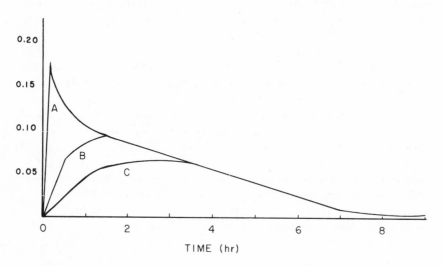

FIGURE 1.—*Diagrams of Expected Blood Alcohol Curves after Three Modes of Alcohol Administration.* A=i.p. injection, B=orally in fasted person, C=orally in fed person.

of ethanol vapor—and each method has its special applications and disadvantages. Absorption and distribution are most rapid after i.v. or i.p. injection, slowest after ingestion. Of course, human beings generally prefer their alcohol by mouth, although other routes (e.g., i.v.) may be used medically or experimentally.

Ethanol's small polar molecule (CH_3CH_2OH) is infinitely soluble in water and readily crosses cell membranes by passive diffusion, i.e., it moves spontaneously across a concentration gradient, from a region of higher to lower concentration. Thus, ethanol enters the blood stream without molecular change, by absorption across the mucosal surface of the gastrointestinal tract (GIT), most rapidly from the small intestine, less so from the stomach and large intestine. The speed of absorption depends on the concentration, but the relationship is biphasic—alcohol at 15 to 30% concentration is absorbed more rapidly than either lesser or greater concentrations. Lesser concentrations are absorbed more slowly because of the smaller concentration gradient, while greater concentrations irritate the gastric mucosa and also slow motility, thus retarding absorption.

Numerous other factors influence absorption, including the type of alcoholic beverage (e.g., absorption of pure ethanol is most rapid, beer slowest), the presence of food in the stomach (absorption is slowed) and the rate of blood flow to the GIT (increased circulation enhances the concentration gradient and speeds absorption).

Any of these variables might alter the shape of the rising portion of the curves in Figure 1. For example, curve A shows the rapid absorption occurring after an i.p. injection; curve B is typical of alcohol absorption in a fasted person and curve C shows delayed absorption, with a broad plateau, which might be due to food in the stomach. Thus absorption might be complete within minutes or prolonged for hours with obvious consequences for the maximum BAC attained and the extent of pharmacological effect.

Distribution

Once alcohol enters the general circulation it is distributed throughout the body, with those tissues having a greater blood supply and capillarization receiving alcohol more rapidly than tissues with less blood flow. Thus, the brain, lungs, kidneys and liver equilibrate with blood alcohol within minutes, whereas some tissues, such as resting skeletal muscle, may require an hour or more. At equilibrium, however, the alcohol is dispersed uniformly throughout the body water, the amount in the tissues varying only with their relative water content.

Elimination

Alcohol's removal from the body begins before absorption is complete, as soon as the tissue concentrations are sufficient to initiate removal mechanisms. Small amounts of alcohol are excreted unchanged in urine and sweat and exhaled in expired air; the quantities depend on the BAC and may be increased by such conditions as diuresis (frequent urination), profuse sweating and hyperventilation. However, over 90% of ingested alcohol is eliminated by metabolism, i.e., biochemical conversion of the ethanol molecule through several steps to carbon dioxide and water. The metabolic reactions in this conversion will be discussed in detail below; for now I shall consider only over-all elimination and the factors that influence it.

The rate of alcohol elimination is usually calculated from the slope of the descending arm of the BAC curve (Figure 1), generally accepted as linear during the period after absorption and distribution are complete but before the curve becomes exponential at low BACs. This linearity implies that the elimination rate is independent of alcohol concentration, and, although there is recent controversy about the validity of this assumption, for all practical purposes we can assume that the elimination rate after a moderate dose is constant. This constancy can be explained if the alcohol-metabolizing system is fully active and cannot proceed at a faster rate under existing conditions (i.e., the system is "saturated"). As long as the alcohol concentration remains above saturating levels, elimination will proceed at the maximum rate. When the BAC falls below 0.01–0.02% in humans, the system is no longer saturated, the metabolic rate becomes concentration dependent and an exponential decay curve is obtained, as in the lower portion of Figure 1. Accurate alcohol elimination rates can thus be determined only over the linear region of the BAC curve, excluding both the extreme upper and lower portions. (The rapid decline immediately after the peak of curve A, Figure 1, is due to equilibration of blood alcohol with tissues. After an i.p. injection, ethanol is absorbed more rapidly than it is distributed, and blood levels temporarily exceed tissue levels. Once a blood–tissue equilibrium is attained, the rate of decline slows and becomes linear.)

In 1932 Widmark (12) formulated an equation, still used, that interrelates the amount of alcohol ingested, the BAC, body weight and the alcohol elimination rate. A careful study of the Widmark formula is recommended for anyone measuring BACs or alcohol elimination rates.[2] The formula has several useful applications, including

[2]See Kalant (13) for a review.

the estimation of the expected BAC from a given dose and, conversely, the estimation, from measured BACs, of the amount consumed. Thus with Widmark calculations, we can estimate that for a 70-kg man of average leaness a dose of 40 ml (1.4 oz) of pure ethanol would result in a BAC of about 0.05%; 75 ml (2.5 oz), 0.1%; and 100 ml (3.4 oz), 0.15% about 1 hr after drinking. The actual value attained, of course, would depend on the rate of absorption as influenced by the various factors discussed above.

The rate of alcohol elimination in humans is on the order of 0.1 g per kg of body weight per hr (a decline in BAC of about 0.015% per hour), but more than twofold variations among individuals have been noted. Rates have been reported to differ among racial groups, but some of the results should be reexamined, as they may have failed to take into account factors, such as variation in the proportion of body fat, which influence the measurement of elimination rates.

Attempts to alter alcohol elimination by administering drugs or through other means have met with mixed success. Elimination may be slowed by fasting, by liver disease and by the enzyme-inhibitor pyrazole. The simple sugar fructose is, however, one of few agents which enhances the disappearance of alcohol in vivo. Another is alcohol itself: prolonged alcohol ingestion may elevate the elimination rate, and a twofold increase is seen in some drinking alcoholics. This ability of alcohol to stimulate its own metabolism might be considered as an example of metabolic tolerance (i.e., with enhanced metabolism, a larger quantity of alcohol is required to produce the same pharmacological effect).

Alcohol Metabolism

As stated earlier, the major mechanism for alcohol removal is by metabolism to carbon dioxide and water. This begins by enzymatic oxidation in a stepwise manner, first to acetaldehyde and then to acetate, occurring primarily in the liver. More than 75% of the acetate formed is transported from the liver to other tissues where oxidation to carbon dioxide is completed. The first step, oxidation to acetaldehyde, occurs much more slowly than the subsequent steps and is generally considered to be rate limiting for the over-all process.

Ethanol oxidation in vivo was considered to be catalyzed exclusively by the enzyme alcohol dehydrogenase (ADH) until 1968, when Lieber and DeCarli (14) reported a new mechanism that formed acetaldehyde from ethanol. This reaction, which they called the microsomal ethanol-oxidizing system (MEOS), clearly differed from ADH but in several ways resembled another ethanol-oxidizing enzyme system—the hydrogen peroxide-dependent catalase reaction. Over the last decade these three ethanol-oxidizing reactions have been intensively investigated by those attempting to differentiate and to quantitate their function in alcohol metabolism.

Alcohol Dehydrogenase

As one of the earliest enzymes to be isolated and crystallized in purified form, ADH has been thoroughly examined. It is found in many different organisms—from yeast to man—and in most mammals 80 to 90% of its activity is in the liver. Lower activities are found in other tissues, especially stomach and intestine, but also kidney and lung. Extremely low levels of activity have been reported in brain tissue, and its importance there is doubtful.

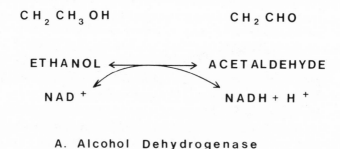

A. Alcohol Dehydrogenase

B. MEOS

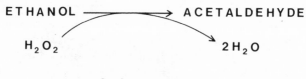

C. Catalase

FIGURE 2.—*Enzymatic Reactions of Ethanol Oxidation.*

The zinc-containing enzyme protein is soluble, i.e., not associated with cell membranes. Reaction requires the coenzyme, nicotinamide adenine dinucleotide (NAD), which accepts hydrogen and electrons from ethanol, forming the reduced coenzyme NADH (Figure 2a). ADH also catalyzes the oxidation of a variety of other alcohol compounds, including methanol ("wood alcohol"), isopropanol ("rubbing alcohol") and ethylene glycol ("antifreeze"). It is effectively and specifically inhibited by pyrazole and its derivatives, extremely useful agents in alcohol research.

Some scientists have asked why animals, including humans, have such active ADH when ethanol is not a natural part of their diet. One suggestion is that the enzyme is required to dispose of ethanol formed naturally by intestinal bacteria. On the other hand, the fact that some steriod hormone products are good substrates for ADH has led other investigators to propose these compounds as the natural, endogenous substrates. Such concerns are of little more than theoretical interest, however, since for a large part of the human population, the principal ADH substrate is beverage alcohol.

Although we often speak of ADH as though it were a single enzyme, there is actually a family of similar enzymes, with slightly different molecular forms and ethanol-oxidizing capacities (sometimes called isoenzymes or "isozymes"), which catalyzes the NAD-dependent oxidation of ethanol. Different isozymes are found not only among various plant and animal species, but also within the same species and even within the same individual. Human ADH isozymes were first reported by von Wartburg and co-workers (15) in 1965 when they described "atypical" ADH, three to six times more active than is "normal" ADH. The frequency of occurrence of atypical ADH is only 10 to 20% in most White populations, but it has been reported in 85% of a Japanese population. While there has been speculation that this enzyme difference may contribute to the increased alcohol sensitivity frequently found in Orientals, greater ADH activity in vitro does not always correlate with increased ethanol metabolism in vivo.

Prolonged alcohol intake has been reported by some researchers to increase ADH activity and by others to cause no change. Even when ADH activity was increased there was not always a corresponding increase in the alcohol elimination rate, suggesting that other factors may control the rate of alcohol metabolism in vivo.

Microsomal Ethanol-Oxidizing System

The MEOS has several properties in common with the microsomal drug-metabolizing enzymes or "mixed-function oxidases" of the liver. Both enzyme systems are associated with the membranes of the smooth endoplasmic reticulum ("microsomes") and require molecular oxygen and reduced nicotinamide adenine dinucleotide phosphate (NADPH) for reaction (Figure 2b). Another similarity with the mixed function oxidases is that MEOS is inducible, i.e., its activity can be increased by repeated administration of a range of compounds including certain drugs (e.g., phenobarbital), insecticides (DDT) and carcinogens (benzpyrene). Lieber and DeCarli (14) showed that the activity of both MEOS and the mixed-function oxidases also increased with repeated ethanol doses and concluded that the MEOS was responsible for the enhanced alcohol elimination often seen after prolonged ingestion. More recently, they have proposed that MEOS is more active when the alcohol concentration is high (over 20 mM) and that MEOS is responsible for increased elimination at high BACs.

Other workers (e.g., 16, 17) found that the NADPH requirement of the MEOS could be replaced by hydrogen peroxide-generating systems (e.g., enzyme : substrate combinations such as the ratio of glucose oxidase : glucose) and suggested that the supposed MEOS activity might actually be due to catalase, an enzyme known to oxidize ethanol in a peroxide-dependent reaction. Catalase is an abundant contaminant of microsomal preparations. These workers believed that the peroxide-generating system substituted for a microsomal reaction that forms hydrogen peroxide from NADPH and molecular oxygen. Other evidence, such as inhibitor studies, supported the similarity of MEOS and catalase.

In the last decade considerable research effort has been given to attempts to resolve this conflict. While the majority of the activity originally attributed to MEOS is undoubtedly due to catalase, efforts to separate the enzymes by physical means (e.g., DEAE-cellulose column chromatography) have indicated that microsomes may contain a small amount of ethanol-oxidizing activity, even in the apparent absence of catalase. This result has been challenged by others, however, and no doubt the MEOS–catalase debate will continue.

Catalase

Although catalase's ethanol-oxidizing ability has been known for decades, it was generally believed that the enzyme was unimportant in ethanol metabolism in vivo. The controversy over MEOS renewed interest in catalase, however, and recent studies have led at least one group (17) to propose that the enzyme contributes significantly to hepatic ethanol oxidation when ethanol concentration is high (over 80 mM).

In the intact liver cell, catalase is localized chiefly in the peroxisomes, but with tissue homogenization it is dispersed, leading to contamination of microsomes, as mentioned above. Catalase's chief physiological role seems to be the rapid destruction of hydrogen peroxide, a toxic product of several cellular reactions. In the presence of a continuous supply of hydrogen peroxide, the enzyme also catalyzes the oxidation of a variety of substrates, including ethanol (Figure 2c) and methanol, and both substrate oxidation and peroxide destruction may occur simultaneously. The kinetics of these combined processes are quite complex, and failure to appreciate this complexity may have confounded the MEOS–catalase debate.

Summary

These then are the three enzymes currently known or proposed to oxidize ethanol to acetaldehyde in vitro. Their respective contributions in vivo have not been proved. Of ADH's considerable in vivo activity there is no doubt, but the significance of the other enzymes remains in question. Although it has been claimed that the alternative pathways might metabolize 50% of ingested alcohol, a more realistic estimate is that no more than 10 to 20% of ethanol oxidation in vivo could be accomplished by enzymes other than ADH (e.g., 17).

Aldehyde Dehydrogenase and Acetaldehyde Metabolism

The aldehyde dehydrogenases, the major family of enzymes responsible for the second step of alcohol metabolism (the conversion of acetaldehyde to acetate), are a heterogeneous group of enzymes found in many body tissues. Liver activity is very high, and acetaldehyde formed there is rapidly oxidized, with very little escaping to the blood under normal circumstances. There is a lack of concensus about blood acetaldehyde levels after alcohol administration, and reports have ranged from 0.01 to more than 0.1 mg per dl, with the lower values thought to be more accurate because of improved analytical methods. Blood acetaldehyde levels are relatively independent of alcohol dose when the dose is above the level required to saturate the alcohol-oxidizing system. This would seem to argue against a major role for acetaldehyde in intoxication, as most of alcohol's effects are dose dependent.

   The "anti-alcohol" drug disulfiram (Antabuse) inhibits aldehyde dehydrogenase and, if taken with alcohol, increases blood acetaldehyde two to sixfold. However, the actions of acetaldehyde in producing the symptoms of the disulfiram–ethanol reaction are not clear.

As with ADH, aldehyde dehydrogenase activity requires the coenzyme NAD, which is reduced to NADH as acetaldehyde is oxidized to acetate.

Alcohol and the Liver

The liver plays a central role in the body's metabolism by transforming nutrients and recycling or removing wastes, thus maintaining a stable internal environment and furnishing other tissues with a steady supply of the substances essential for their function. By disturbing this normal activity, alcohol metabolism in the liver may profoundly affect the entire body. To understand the many ramifications of this simple metabolic transformation, it is necessary to discuss some basic biochemistry relative to alcohol metabolism.

As ethanol is oxidized through acetaldehyde to acetate, hydrogen and electrons are transferred enzymatically to the coenzyme NAD, yielding the chemically reduced form, NADH. Thus, for each molecule of ethanol metabolized to acetate, two molecules of NADH are produced. The amount of NADH in the liver thus increases rapidly, and soon the ratio of NADH to NAD may be two to five times the prealcohol level. For ethanol oxidation to continue, the NADH must be transformed back to NAD, since the liver's content of this coenzyme is limited. There are several reversible reactions catalyzed by other dehydrogenase enzymes which can carry out this transformation. NADH may also be reoxidized in another type of reaction mechanism which is coupled to the production of cellular energy and is the major oxygen-consuming process in the cell. It is generally accepted that it is this process—the energy-coupled reoxidation of NADH—that controls the rate of ethanol metabolism, and not the amount of alcohol dehydrogenase present in the liver.

If the rate of NADH reoxidation is increased, alcohol metabolism is increased. Unfortunately, there are few agents which will markedly accelerate NADH oxidation in vivo without adversely affecting other metabolism. Consequently, stimulating ethanol oxidation to speed detoxication has not been especially successful clinically although fructose, mentioned above, is moderately effective. Fructose enhances the reoxidation of NADH, and studies have reported ethanol metabolism to be doubled in humans intravenously infused with fructose.

Because the process of NADH oxidation is coupled to oxygen consumption, it follows that if the liver's oxygen use were increased, NADH turnover and, subsequently, ethanol oxidation would be increased. In a series of publications since 1970, Israel and colleagues (reviewed in 19) have shown that prolonged alcohol consumption will increase oxygen uptake by the liver between 50 and 100%, a condition they characterize as a "hypermetabolic state." As a consequence, there is a proportionate increase in ethanol metabolism. The increased oxygen use is seemingly mediated by ethanol-induced homonal changes and is diminished after either thyroidectomy or adrenalectomy. This phenomenon would seem to offer a plausible explanation of the increased alcohol metabolism frequently seen after prolonged drinking by alcoholics, and in animal studies as well. These authors believe that increased oxygen demand by the liver makes it more vulnerable to conditions that reduce oxygen availability and that the relative hypoxia in the liver may be important in the production of alcoholic liver disease. As with most alcohol studies, there are some who question this interpretation, and only further research will resolve the differences.

In considering other alterations that occur in the liver following ethanol ingestion, it is important to understand that, for the most part, it is not the ethanol molecule per se, but the increase in NADH resulting from ethanol oxidation that disrupts liver activity. A few examples will illustrate how this comes about. The extent to which these effects occur depends on the dose of alcohol and the duration of drinking—some change may occur even after low doses (BAC = 0.05%), but the more extreme effects require significantly higher doses.

Lactic Acid Metabolism

Excess lactic acid from muscle and other tissues is transported to the liver where, in the absence of ethanol, it is oxidized to pyruvic acid by the enzyme lactic dehydrogenase, simultaneously converting NAD to NADH.

$$\text{Lactic acid} + NAD^+ \rightleftarrows \text{Pyruvic acid} + NADH + H^+$$

This reaction is reversible, however, and during ethanol oxidation the reverse direction (i.e., to form lactic acid) is favored because of the elevated NADH : NAD ratio. As a consequence, lactic acid accumulates in the liver and blood; the resulting hyperlacticemia is seen especially in those who exercise after drinking.

Apparently as a secondary effect of the lactic acid increase, the blood level of uric acid may also rise. This occurs because the normal excretion of uric acid is blocked by the excess lactic acid, the two acids apparently competing for the same transport sites in the kidney. One clinical manifestation of hyperuricemia might be an increased propensity for gouty attacks. In this review, however, our special interest in this effect is as an example of the way one action— hepatic alcohol oxidation—may work through several intermediates to produce a remote event, in this case, increased blood uric acid.

Glucose Synthesis

Among the most important functions of the liver is the maintenance of a constant blood glucose level for use by tissues, such as brain, for which glucose is the sole energy substrate. One way the liver accomplishes this is by the synthesis of glucose (gluconeogenesis) from smaller carbon compounds which enter the gluconeogenic pathway via NAD-linked oxidation. The conversion of lactic to pyruvic acid, described above, is one such entry reaction; there are several others. During the ethanol-generated increase in NADH relative to NAD, most of the reactions favor the reverse direction: glucose precursors become unavailable and glucose synthesis declines. If other sources of blood glucose (e.g., glycogen) are depleted, as after extended fasting, alcohol ingestion may result in severe hypoglycemia because of this suppression of gluconeogenesis.[3] In fact, numerous cases of alcoholic hypoglycemic coma and death have been reported since the condition was first described in 1941.

Lipid Metabolism

For many years, prolonged alcohol consumption has been known to produce deposits of fat, i.e., triglycerides, in the liver, but the generation of an alcoholic fatty liver is complex and, even now, not fully understood. Many aspects, such as the role of malnutrition, are still debated after decades of research, but one contributing factor appears to be the increased levels of NADH relative to NAD resulting from ethanol oxidation.

Key intermediates in lipid metabolism are compounds called "free fatty acids" which may either be stored in adipose tissue as triglycerides, synthesized in the liver from the excess dietary carbohydrate or derived from dietary fat. Normally these compounds are metabolized in the liver through a multistep reaction sequence,

---

[3]In well-fed persons, heavy alcohol ingestion may result in a transient elevation of blood glucose, thought to be due to the stress-induced release of epinephrine (adrenaline), which stimulates the liver to break down the storage form of glucose (glycogen) and release it into the blood. In this case, alcohol is acting as a stressor, causing epinephrine release.

producing NADH and, ultimately, cellular energy. However, after alcohol ingestion the energy needs of the liver cell are met by the NADH from ethanol and acetaldehyde oxidation, and fatty acid oxidation is depressed. This, coupled with a stress-induced increase in fatty acid release from adipose storage depots, results in their accumulation in the liver during alcohol intoxication.

The excess fatty acids are converted to triglyercides in a reaction sequence requiring a glycerol compound, α-glycerophosphate, also increased in concentration because of ethanol oxidation. Thus alcohol ingestion seems to initiate a number of events which promote the synthesis of hepatic triglycerides: increased release of fatty acids from adipose, decreased fatty acid oxidation and increased synthesis of α-glycerophosphate, the latter two arising from the changed NADH : NAD ratio. No doubt other factors are involved as well. In the absence of alcohol, most triglycerides synthesized in the liver are exported as lipoprotein complexes to be stored in adipose tissue. With alcohol present, however, the triglycerides accumulate in the liver, either because they are synthesized at a rate that exceeds the liver's export capacity or because of some breakdown in the export process. Regardless of the reason, they remain in the liver, and with continued drinking, form fatty deposits. The relationship between fatty liver and other forms of alcoholic liver disease is still debated, but some have suggested that fatty liver may be a first step in the progression to alcoholic liver cirrhosis (discussed by Lieber in this volume).

Other Actions

The effects described above are cited to illustrate some of the ways liver metabolism may be deranged by alcohol oxidation and the shift in NADH relative to NAD. This is not to say that all of alcohol's actions in the liver result from the NADH increase. Alcohol has other effects there, but they have not been so well characterized. Certainly the most pronounced biochemical effects seem to arise from the increased NADH : NAD ratio and the subsequent metabolic derangement.

These effects occur while alcohol is being metabolized. When alcohol is no longer present after a short period of drinking, the NADH : NAD balance is restored and liver metabolism proceeds normally. During prolonged alcohol consumption, however, the liver may undergo longer-lasting changes, some of which may occur in an effort to counteract the continued presence of alcohol. Such changes, sometimes characterized as "adaptive," are subject to various interpretations, as in the case of the "hypermetabolic state," discussed above.

Is the metabolic derangement seen in the liver present in other tissues after alcohol ingestion? It is unlikely, because the capacity for ethanol oxidation with the concomitant increase in NADH : NAD is limited or negligible in most other tissues. Thus, in nonhepatic tissues, especially in brain and other nerve tissue, where alcohol dehydrogenase activity is weak or absent, alcohol must work in other ways.

Alcohol and the Brain

Without question the brain is the most complex organ of the body—biochemically, anatomically and functionally—and presents a proportionately difficult challenge to investigators. Many early studies of alcohol's effects on this tissue should be considered with caution in light of present understanding of neuronal systems; current and future studies should have more validity as alcohol researchers continue to acquire knowledge and apply techniques of increasing sophistication in this difficult field.

Alcohol is not oxidized in the brain in any significant degree. This conclusion is supported by (1) the negligible ADH activity of brain; (2) the lack of radioactively labeled carbon dioxide production from radioactive ethanol added to brain tissue; and (3) the failure to find a consistent increase in NADH relative to NAD in the brain after ethanol administration. Thus, the profound metabolic changes that are observed in liver would not be similarly produced in brain. Nevertheless, there is a marked disturbance of brain metabolism after alcohol, and the central nervous system (CNS) depression,[4] tolerance[5] and physical dependence[6] that occur undoubtedly result from alcohol-induced alterations in the physiologic or metabolic function of brain.

While we do not know for certain how alcohol produces its CNS effects, increasing evidence supports the view that it acts pri-

[4]Depression of the CNS is here defined as a reduction in the functional activity of the nervous system produced by alcohol or other sedative or anesthetic dugs.

[5]Tolerance is here defined as a decrease in the magnitude of a response to alcohol or other drug, as a result of prolonged exposure. An increase in dose will restore the level of response. The tolerance mechanisms may be metabolic (increased elimination), pharmacologic (nervous system adaptation) or behavioral (learned reaction). Tolerance as discussed in this section is pharmacologic.

[6]Physical dependence on alcohol is here defined as an altered state of the CNS produced by prolonged exposure to alcohol. The altered state is thought to be an adaptive change to counteract alcohol's continued presence. When alcohol is withdrawn, the resulting neuronal hyperexcitability may be called the "withdrawal syndrome."

marily at the membrane level. Nerve cell membranes have specialized functions not found in other cells, i.e., electrical excitability and the transmission of nerve impulses are membrane-dependent phenomena, and thus, by disrupting membrane activity, alcohol could alter these fundamental functions of nerve tissue.

Membranes are complex organizations of lipids (mainly phospholipids and cholesterol) and proteins (including certain enzymes). Both the composition and physical state of the lipids influence membrane function, and it is here that alcohol is presumed to have its major effect, either by altering lipid–protein interactions or by changing lipid organization within the membrane. Although the link between such membrane changes and alcohol-induced depression or other CNS effects have not been identified, ethanol resembles a number of other general anesthetics in its action on membranes. Moreover, there is a direct correlation between the lipid solubilities of the general anesthetics, including ethanol and other short-chain alcohols, and their anesthetic potency, further supporting an association between the ability to depress the CNS and an interaction with membrane lipids.

At concentrations as high as 4 to 5% in vitro, ethanol has been shown to expand membranes, increasing their volume and disrupting the lipid–protein associations within the membrane. Even at low-to-moderate concentrations, comparable to mildly intoxicating levels in vivo, ethanol has been found to increase membrane "fluidity." This term describes the organization of lipid molecules within a membrane; increased fluidity means there is increased disorder among the lipids, with consequences for a variety of membrane functions. Presumably as a result of the disorganization of membrane structure, ethanol interferes with a number of membrane-dependent activities, including transport processes, ion permeabilities and activities of membrane-bound enzymes. These effects are produced by a direct physical action of the ethanol molecule itself and do not require its metabolic transformation.

The observations by Chin and Goldstein (20) of the effect of ethanol on membrane fluidity are among the most stimulating of recent studies of alcohol and the CNS. These authors discuss the possibility that ethanol and other anesthetic agents may depress the CNS through disordering membrane lipids and support this concept by citing other evidence that elevated pressure, which decreases fluidity and thus increases membrane order, can reverse the anesthetic effect. Another significant observation by these authors (21) was that membranes from alcohol-dependent animals resisted the fluidizing effects of ethanol. Such membranes were found to have an increased cholesterol content, and other workers have reported additional differences in the lipid composition of membranes from brains of alcohol-dependent animals. These changes in membrane composition and organization during the prolonged presence of ethanol may represent an adaptive change as neuronal membranes attempt to counteract the continued effects of ethanol, and, as such, may be a mechanism in the development of tolerance or dependence. Further research will be necessary to test the validity of these intriguing hypotheses.

Even if we accept that ethanol's primary action in the brain is at the membrane level, we still have no conclusive evidence to indicate which of the several membrane functions, altered secondarily, leads to CNS depression or the adaptive changes of tolerance and dependence. Several possibilities have been proposed including an inhibition of sodium conductance, a decrease in sodium and potassium transport and an increase in calcium binding. Whether any or all of these membrane functions are directly involved in ethanol's neuronal effects remains to be seen.

Another approach to the study of ethanol's effects on the nervous system has been to examine the metabolism of the neurotransmitter[7] substances of the brain. Most frequently studied have been two biogenic amines, norepinephrine and dopamine, but the results have been somewhat contradictory. In early studies in the 1960s, brain levels of norepinephrine were variously reported to be decreased, increased and unchanged after ethanol administration. Analytical techniques have improved since then, and our understanding of transmitter activity has changed. We now know that measuring steady-state concentrations of neurotransmitters is not as meaningful as determining the rate of metabolic turnover, which reflects the degree of neuronal activity. In one study (22), biogenic amines in rat brains were shown to respond to ethanol in a biphasic manner. Norepinephrine turnover was initially accelerated with no change in dopamine turnover, but, at 5 hr after ethanol, both norepinephrine and dopamine turnover were reduced. In physically dependent rats during withdrawal, norepinephrine turnover was accelerated and that of dopamine decreased. While these results indicate changes in the activity of noradrenergic and dopaminergic neurons during alcohol administration and withdrawal, one questions whether this is the cause of alcohol's CNS effects or merely a

[7]Neurotransmitters are compounds which are synthesized and stored in nerve terminals and are released into the synapse when a nerve is stimulated. After release, they may interact with receptors on the postsynaptic cell, thus "transmitting" the nerve impulse.

consequence. In many such instances, it seems difficult to determine whether an observed effect causes—or is caused by—changes in neuronal activity. The distinction is important if alcohol's mechanism of action is to be understood.

The metabolism of other neurotransmitters has been studied, with equally confusing and contradictory results. The inhibitory neurotransmitter γ-aminobutyric acid has been reported to increase and to remain unchanged. Acetylcholine seems to react in a biphasic manner, initially increasing and then decreasing. An explanation for all these results will require further study, but it seems clear that variations in the time after ethanol administration at which observations are made will influence the results obtained and may explain some of the discrepant data.

Another area of neurotransmitter research that has generated both interest and controversy during the last decade deals with the chemical condensation between amine neurotransmitters and aldehydes to produce pharmacologically active substances. The premise is that dopamine or other amines combine either with acetaldehyde from ethanol oxidation or with intermediate aldehyde derivatives of the amines themselves to form compounds known collectively as tetrahydroisoquinolines (TIQs). Among the more controversial hypotheses was that of Davis and Walsh (23) that dopamine and its aldehyde derivative condensed to form the TIQ tetrahydropapaveroline, which happens also to be an intermediate in the plant biosynthesis of morphine alkaloids. These authors speculated that such a condensation product might mediate some of the pharmacological actions of alcohol and that alcohol dependence might actually be a physical dependence on the morphine-like compounds. Convincing support for this hypothesis has not been forthcoming, however, and one persuasive argument against it is that there is little or no similarity between the physical dependence produced by alcohol and that produced by morphine. Opinion regarding the significance of the other aldehyde condensation products should be withheld until further data are available. Although TIQs have been shown to have pharmacological activity, only very recently has evidence been obtained that such condensation occurs in vivo in humans: urine of recently drinking alcoholics was found to have measurable levels of salsolinol, the condensation product of dopamine and acetaldehyde. That such condensation occurs in brain tissue in vivo under normal conditions has not been shown.

Indirect evidence against a significant role for the TIQs in alcohol's CNS effects can be drawn from studies (e.g., 24) showing that states of intoxication, tolerance and physical dependence indistinguishable from and cross-tolerant with those produced by ethanol result from the administration of tertiary butyl alcohol (*t*-butanol). This four-carbon alcohol cannot be oxidized to an aldehyde and thus cannot form TIQS. That it is as pharmacologically active as ethanol seems a good indication that the TIQS and in fact acetaldehyde itself are relatively unimportant in alcohol's CNS actions. The amine–aldehyde condensation hypothesis continues to hold the interest of a number of researchers, however, and these investigations will no doubt continue.

The results with *t*-butanol may also be offered as further support for the membrane-level action of ethanol. Smaller doses are required of *t*-butanol than of ethanol to produce comparable pharmacological effects. That is, of course, partly due to the fact that *t*-butanol is cleared from the blood more slowly than ethanol. More significant, however, is that *t*-butanol is four-to-five times more lipid soluble than ethanol and thus has greater membrane-disrupting capacity. The use of *t*-butanol in laboratory animal studies should help to clarify many of the questions remaining about ethanol's CNS effects.

Other Effects

This review has stressed two main mechanisms by which alcohol seems to work its effects: in liver, through its oxidation and the subsequent shift in NADH relative to NAD, and in brain, apparently by disturbing membrane structure and function. The emphasis on brain membranes is not meant to imply that ethanol has no effects on the cell membranes of liver and other tissues; ethanol can be shown to influence most membranes to an extent compatible with their lipid composition and functional characteristics. Certainly, ethanol's effects on membranes of the hepatic endoplasmic reticulum (microsomes) are well known. However, because of the unique properties of neuronal membranes, ethanol's actions there have special significance.

Neither should the emphasis on only two mechanisms be interpreted to mean that ethanol works in no other way, but other activities have not been so thoroughly studied. For example, ethanol has been shown to suppress some aspects of hepatic protein synthesis, but the mechanism of this action has not been detailed.

There is another activity of ethanol that contributes to its many physical effects at BACS higher than 0.1%. This action is reflected in changes in the endocrine system that increase or decrease the output of a variety of hormones in response to the stress-induced activation of the hypothalamic–pituitary–adrenal system (e.g., 25). Increased hormonal levels then affect target organs, resulting in such changes as the transient hyperglycemia and fatty acid release mentioned earlier. Still to be resolved is whether these hormonal effects result from a direct stimulation of the hypothalamus or pituitary by ethanol or, secondarily, from an action of ethanol on other CNS functions.

Conclusion

Research into the biological aspects of alcohol seems rife with conflict and controversy—two or more opinions can be produced on almost any topic. Some diversity exists in all research endeavors, but alcohol studies employ such an abundance of experimental variables that perhaps it is remarkable that research groups ever agree. Much of the conflict can probably be attributed to differences in ethanol dose, route of administration, time of observations, the species or strain of animals used, and age, sex or nutritional status of subjects. These are the obvious variables—factors that are usually controlled and recorded. Also likely to contribute to discrepant results, but not usually noted, are the uncontrolled variables, such as stressful handling of laboratory animals, failure to maintain body temperature during alcohol intoxication and differences in the preservation of sample material after collection.

Despite, or perhaps because of, the controversy, biological research with alcohol has proliferated in the decades since the first *Alcohol, Science and Society*. Perhaps before the third edition, some of the conflicts will be resolved.

REFERENCES

1. JELLINEK, E. M. Introduction to the curriculum. Pp 1–12. In: Alcohol, science and society; twenty-nine lectures with discussions as given at the Yale Summer School of Alcohol Studies. New Haven; Yale University Center of Alcohol Studies; 1945.
2. WALLGREN, H. and BARRY, H., 3d. Actions of alcohol. 2 vol. Amsterdam; Elsevier; 1970.
3. KISSIN, B. and BEGLEITER, H., eds. The biology of alcoholism. Vol. 1. Biochemistry. New York; Plenum; 1971.
4. KISSIN, B. and BEGLEITER, H., eds. The biology of alcoholism. Vol. 2. Physiology and behavior. New York; Plenum; 1972.
5. ROACH, M. K., MCISAAC, W. M. and CREAVEN, P. J., eds. Biological aspects of alcohol. (Advances in Mental Science, Vol. 3.) Austin; University of Texas Press; 1971.
6. HAWKINS, R. D. and KALANT, H. The metabolism of ethanol and its metabolic effects. Pharmacol. Rev. **24:** 67–157, 1972.

7. GROSS, M. M., ed. Alcohol intoxication and withdrawal; experimental studies. (Advances in Experimental Medicine and Biology, Vol. 35.) New York; Plenum; 1973.

8. MAJCHROWICZ, E., ed. Biochemical pharmacology of ethanol. (Advances in Experimental Medicine and Biology, Vol. 56.) New York; Plenum; 1975.

9. LIEBER, C. S., ed. Metabolic aspects of alcoholism. Baltimore; University Park Press; 1977.

10. GROSS, M. M., ed. Alcohol intoxication and withdrawal—IIIa; biological aspects of ethanol. (Advances in Experimental Medicine and Biology, Vol. 85A.) New York; Plenum; 1977.

11. MAJCHROWICZ, E. and NOBLE, E. P., eds. Biochemistry and pharmacology of ethanol. 2 vol. New York; Plenum; 1979.

12. WIDMARK, E. M. P. Die theoretischen Grundlagen und die praktische Verwendbarkeit der gerichtlich-medizinischen Alkoholbestimmung. (Fortschr. naturw. Forsch., Heft 11.) Berlin; Urban & Schwarzenberg; 1932.

13. KALANT, H. Absorption, diffusion, distribution and elimination of ethanol; effects on biological membranes. Pp. 1–62. In: KISSIN, B. and BEGLEITER, H., eds. The biology of alcoholism. Vol. 1. Biochemistry. New York; Plenum; 1971.

14. LIEBER, C. S. and DECARLI, L. M. Ethanol oxidation by hepatic microsomes; adaptive increase after ethanol feeding. Science 162: 917–918, 1968.

15. WARTBURG, J.-P. VON, PAPENBERG, J. and AEBI, H. An atypical human alcohol dehydrogenase. Canad. J. Biochem. 43: 889–898, 1965.

16. ROACH, M. K., REESE, W. N., JR. and CREAVEN, P. J. Ethanol oxidation in the microsomal fraction of rat liver. Biochem. biophys. Res. Commun. 36: 596–602, 1969.

17. THURMAN, R. G., MCKENNA, W. R., BRENTZEL, H. J., JR. and HEESE, S. Significant pathways of hepatic ethanol metabolism. Fed. Proc. 34: 2075–2081, 1975.

18. ROGNSTAD, R. and GRUNNET, N. Enzymatic pathways of ethanol metabolism. Pp. 65–85. In: MAJCHROWICZ, E. and NOBLE, E. P., eds. Biochemistry and pharmacology of ethanol. Vol. 1. New York; Plenum; 1979.

19. ISRAEL, Y., KALANT, H., ORREGO, H., KHANNA, J. M., PHILLIPS, M. J. and STEWART, D. J. Hypermetabolic state, oxygen availability, and alcohol-induced liver damage. Pp. 433–444. In: MAJCHROWICZ, E. and NOBLE, E. P., eds. Biochemistry and pharmacology of ethanol. Vol. 1. New York; Plenum; 1979.

20. CHIN, J. H. and GOLDSTEIN, D. B. Effects of low concentrations of ethanol on the fluidity of spin-labeled erythrocyte and brain membranes. Molec. Pharmacol. 13: 435–441, 1977.

21. CHIN, J. H. and GOLDSTEIN, D. B. Drug tolerance in biomembranes; a spin label study of the effects of ethanol. Science 196: 684–685, 1977.

22. HUNT, W. A. and MAJCHROWICZ, E. Alterations in the turnover of brain norepinephrine and dopamine in alcohol-dependent rats. J. Neurochem. **23:** 549–552, 1974.

23. DAVIS, V. E. and WALSH, M. J. Alcohol, amines and alkaloids; a possible biochemical basis for alcohol addiction. Science **167:** 1005–1006, 1970.

24. McCOMB, J. A. and GOLDSTEIN, D. B. Quantitative comparison of physical dependence on tertiary butanol and ethanol in mice; correlation with lipid solubility. J. Pharmacol. exp. Ther. **208:** 113–117, 1979.

25. KAKIHANA, R. and BUTTE, J. C. Ethanol and endoctrine function. Pp. 147–164. In: MAJCHROWICZ, E. and NOBLE, E. P., eds. Biochemistry and pharmacology of ethanol. Vol. 2. New York; Plenum; 1979.

3

Effects of Alcohol on Psychological Processes

Robert J. Pandina

The experimental evidence regarding the effects of small, medium and large amounts of alcohol on psychological functions of humans was summarized by E. M. Jellinek in *Alcohol, Science and Society* (1). The summary was based on a more extensive review (2) which distilled observations from about 200 research papers, the scientific literature on the topic at the time. In his summary Jellinek outlined four areas in which experimental psychology could, he believed, make substantive contributions to our understanding of the effects of alcohol on human behavior: the types of psychological functions affected, the physiological mechanisms mediating alcohol's effects, factors influencing the development of alcoholism or "compulsive" drinking, and differential alcohol sensitivity and tolerance development.

Jellinek concluded that only the first of his questions had been addressed while the others had not been explored. For Jellinek, some 35 years ago, there were more questions than solutions and the few attempts at solution were limited to documenting and cataloging the effects of alcohol on basic psychological processes. Alcohol affected these processes, he believed, by its actions on the central nervous system by mechanisms as yet undiscovered.

The focus of this chapter is to assess the progress made by experimental psychology in answering some of the questions Jellinek posed. I will also examine whether the questions are still relevant today; determine what, if any, new and important areas of inquiry have been opened; and examine possible blind spots in our research efforts.

Several reviews examining alcohol's effects on selected aspects of psychological functions are available (e.g., 3–13). The volume of primary research represented in the reviews is indeed staggering in comparison to the 200 or so reports available in 1945.

The writing of this manuscript was supported in part by grant No. AA 03509-03 from the National Institute on Alcohol Abuse and Alcoholism.

The dramatic increase in the experimental alcohol literature results, in part, from two factors. First, the rapid growth enjoyed by psychology which has seen its research interests spread to every aspects of the human condition; growth that has been characterized by research activity, theory revision and practical application. The second factor is the emergence of the subdiscipline of psychopharmacology which is an amalgamation of experimental psychology, neuropsychology, pharmacology and physiology. Psychopharmacology is concerned with changes induced in the central nervous system (a physiological–chemical entity) and parallel changes in what D. O. Hebb (14) termed the conceptual nervous system (an experiential–behavioral entity) when an organism ingests a psychoactive substance, that is, a "drug" such as alcohol, heroin, marijuana, diazepam (Valium), etc. Among the objectives of psychopharmacology are delineation of brain–behavior changes induced by a drug's actions, identification of mechanisms which subserve these changes, and analyses of the circumstances controlling the acquisition, maintenance, and extinction of drug-taking behavior. Given Jellinek's concerns about alcohol and drinking, one could surmise that he, in fact, was an early entrant into the field of psychopharmacology.

The growth of this subdiscipline has provided new approaches and techniques to apply to the study of alcohol. The news is not completely cheery since the growth has focused on a wide range of drugs, with alcohol more often than not being perceived as somewhat less attractive a target of research as compared to other drugs. Perhaps alcohol's lack of status results from the fact that, as a drug substance, its history with humankind is somewhat longer than many other substances; perhaps the reason is that many investigators, themselves alcohol users, found it difficult to study a beverage–drug with such positive utilitarian characteristics in their own lives and the lives of those around them. (After all, even experimental psychologists have their sacred cows!)

If activity is any measure of evaluation, then we have grounds to conclude that the scientific community has indeed responded to Jellinek's early chiding with a flurry of researches in every one of the four areas of his concern. Indeed, the beaches of science are strewn with alcohol research reports but many of the attempts have provided less than complete solutions to the questions; in fact, some have provided competing and even opposing solutions.

A Context for Evaluating Psychology's Contributions

The psychological functioning of the human organism can be parceled into several domains. These domains, the psychological functions they subserve and the phenomena they give rise to (e.g., learning, memory, decision making, problem solving, emotional expression, motivation, sleep) are directly related to the functioning of the central nervous system (CNS) and its peripheral apparati (e.g., the peripheral nervous system).

Considerable discussion could be devoted to detailing the relationship of the physical entity, the central nervous system, with the phenomenological (i.e., experiential) entity, the mind. For this chapter, it is sufficient to note that there is a direct, albeit complex, relationship between the central and the conceptual nervous systems, and that alcohol exerts its influences, in part, by altering the chemical functioning of the CNS (physical) with a resultant change in the psychological functioning of the CNS (experiential).

Psychological functioning in humankind may be divided into three mind-domains. The first is termed the *reflective* domain or the "thinking mind" and consists of those psychological processes involved in sensory detection, perceptual organization and cognitive integration. Through these processes the phenomena of the external world are organized and transformed into internal sets of complex knowing or "thought maps."

The second is termed the *affective* domain or the "feeling mind" and consists of those psychological processes which comprise arousal state, emotive appraisal and motivation (or drive) state. This system, using data filtered by the reflective system, signals to the organism that a change in external environmental circumstance has occurred, evaluates the change's potential significance for the organism's well being and primes the organism for action deemed appropriate to cope with the change as perceived and appraised. This system evaluates events and conditions in the environment as they pertain to the short- and long-term benefit of the organism and provides a set of feeling or evaluative maps of events occurring during the organism's life span.

The third can be conceived of as the *effector* domain or "the acting mind" and consists of those processes falling into three areas: reflexive action, simple motor sequencing and psychobehavioral programming. This system is perhaps most familiar; it is responsible for producing the overt actions or "behaviors" (including alcohol use) of the organism. This system, operating on the information provided by the reflective system and the evaluation and priming force of the affective system, allows the organism to operate in the environment. At the most complex level of this system (psychobehavioral programming) elaborate action maps representing complicated sequences of behaviors are maintained.

Both the reflective and affective systems possess as well sub-systems which provide information and evaluations of an organism's internal environment. These subsystems also provide the organism with feedback and evaluation of its own actions, that is, of the activities and status of the effector system. Data from such subsystems also determine an organism's actions.

These systems, functioning in synchrony (or, in the case of certain pathological states, perhaps out of synchrony) give rise to integrated behaviors which, taken together, are recognized as an organism's behavioral repertoire. Included in this somewhat arbitrary set is alcohol-using behavior.

An organism, however conceptually or physiologically construed, does not exist in a vacuum. Rather, all organisms live in the context of various environmental contingencies, some of which are relatively well defined (e.g., pay received for work performed) while others are less well defined (e.g., familial and other interpersonal dynamics, governmental policy). The context can influence behavior in an immediate, short term or phasic manner through the stimulus demands and contingent payoffs perceived. More indirectly, contextual variables can control behavior by inducing an expectancy set regarding the "demands–payoffs linkage" which can prime an organism for engaging in specific behavior patterns. Maps in each domain are malleable; both short- (phasic) and long-term (tonic) experiences may serve as modifiers.

One final point. Behavior develops across at least three forms of time: clock time, developmental (i.e., psychological) time, and maturational (i.e., physiological) time. Each form of time has its own special properties which are important to consider in any investigation, either of alcohol's effects or of the acquisition, maintenance and extinction of alcohol-using behavior. Thus the point in time (variously defined) in the life cycle of an organism at which one chooses to observe alcohol use or its effects can alter critically the conclusion made regarding alcohol's impact.

When defining alcohol use, differentiation should be made between alcohol, a drug with certain pharmacological properties, and drinking, a behavior pattern characterized by voluntary selection and ingestion of beverage alcohol.

Consideration of the effects of alcohol requires attention to certain parameters such as: dosage administered per unit of body weight, route of administration, rate of ingestion, absorption and metabolism rates, physiologic status of systemic systems (e.g., empty versus full stomach), differential tissue sensitivities (both in the periphery and in the CNS), frequency and duration of alcohol use (e.g., single vs multiple doses over time) (discussed by Roach in this volume).

The amount of alcohol in the blood (blood alcohol concentration, BAC) and the rate at which peak BACs are reached are important indicators of the amount of alcohol circulating in the CNS. In our discussion small amounts of alcohol are defined as those giving rise to BACs of 0.05% (i.e., grams of alcohol per 100 ml of blood) or less. To reach a BAC of 0.05%, a dose of 0.5 g of alcohol per kg of body weight would have to be consumed. (The amount of alcohol a 70 kg [155 lb] person would have to consume in a relatively short time to achieve this BAC would be about 3 oz of 86-proof whisky.) Medium amounts of alcohol are considered to be those giving rise to BACs between 0.05 and 0.10%; a dose of 1 g of alcohol per kg of body weight would yield a BAC of about 0.10%. Our average person would need to drink about 3 to 6 oz of 86-proof whisky to achieve BACs in the moderate range. Large amounts of alcohol are those giving rise to BACs of between 0.10 and 0.15% with the upper end of the range being achieved by an alcohol dose of about 1.5 g per kg of body weight (6 to 8 oz of whisky consumed in a relatively short time). Very large amounts are those giving rise to BACs in excess of 0.15%. These values are arbitrary but useful dose ranges to remember in relation to changes in the psychological processes purportedly controlled by CNS mechanisms.

It should also be remembered that alcohol use is a behavior and as such is subject to the same factors and conditions which control other forms of behavior (15, 16). Alcohol use is often considered a "special" behavior requiring special laws to explain its acquisition, maintenance and extinction, and related phenomena (e.g., alcohol tolerance, dependence, addiction). Thus, in a broad sense, the actions of alcohol-the-drug can be influenced by factors affecting drinking-the-behavior. In this regard, factors deserving attention include: a subject's prior experience and expectation regarding the effect on his or her behavior of a specific dose of alcohol, the over-all state of mind induced in a subject before, during and after drinking, the performance demands and environmental contingencies placed on the subject while under the influence of alcohol, and so on.

Alcohol's Effects on Selected Psychological Processes

Few general statements regarding the effects of alcohol on psychological processes in humans can be made with universal agreement. Most of the experimental literature is comprised of studies of "normal" healthy volunteers or subjects diagnosed by varying criteria as "alcoholic" who are tested in controlled laboratory situations. Few studies conducted in more naturalistic settings are available. Reference to results from animal studies is made only to support or extend results obtained with human subjects.

Reflective Functions

The greatest amount of information available pertaining to simple reflective functions has been obtained through work on the visual system. Research on other sensory modalities (audition, somesthesis, olfaction and gustation) has been limited. Laboratory research involving the visual, auditory and somesthetic systems has produced relatively consistent results: small and moderate doses of alcohol produce relatively small decrements in the ability to detect the presence of an appropriate stimulus. Large doses of alcohol appear to be required to disrupt significantly simple detection functions in these systems. The few reports on olfaction and gustation, however, indicate a somewhat greater disruption of simple detection functions at small and moderate dose levels.

As task complexity increases, somewhat smaller doses of alcohol appear to interfere with performance of simple reflective functions. Thus, moderate doses of alcohol have been shown to disrupt the ability to discriminate accurately between lights or sounds of different intensities and qualities; similarly, the ability to determine if two adjacent points on the skin have been stimulated simultaneously (i.e., "two-point" discrimination) is diminished. Moderate doses of alcohol also increase pain threshold although factors such as expectation of pain intensity can reduce alcohol's anesthetic effects.

Recent reports have reversed the earlier conclusions of Jellinek and McFarland (2) regarding restriction of the visual field (cf., tunnel vision). More recent studies have shown that peripheral vision is not impaired significantly unless very high doses of alcohol are consumed. The general principles suggested by Jellinek in his lecture regarding the effects of alcohol on the performance of complex tasks presumed to be indicative of higher-order reflective functions (e.g., information processing, memory, learning, problem solving) have stood the test of time rather well although some important challenges to his hypotheses have been mounted (e.g., 17). Jellinek surmised that the more complex the task, the greater the likelihood that smaller amounts of alcohol would cause performance impairment. He also suggested that, dose for dose, alcohol would be less decremental to performance of more familiar tasks and that skilled or practiced human subjects would be less likely to show decrements in performance.

The ability to respond appropriately and quickly to a simple sensory signal is, perhaps, the most fundamental of the complex reflective functions.[1] This ability is necessary in circumstances where a subject is instructed to respond to the presence of a signal (i.e., simple reaction-time tasks) or to respond quickly to one of several different signals presented at random (e.g., choice reaction-time tasks). In both cases the subject is asked to make a decision on the basis of information provided and indicate that decision by a simple response. Numerous studies (cf: 9) have confirmed and expanded Jellinek's suggestion that alcohol increases reaction time (i.e., slows reactivity) with moderate and large doses producing progressively longer reaction times. The more recent studies suggest that large doses of alcohol are usually required to produce relatively consistent and large performance decrements (i.e., greater than 19% of baseline). Doses of alcohol large enough to impede reaction time also increase the proportion of errors subjects make. Increases in false signals (responding in the absence of the cue) and in incorrect choices are among the most common errors. The nature and intensity of error appear to be related to the nature and complexity of the task and the response required. More demanding tasks and intricate responses appear to be more susceptible to debilitating alcohol effects at somewhat lower doses than non-demanding, simpler tasks.

Several more recent studies (e.g., 18–20) using complex information processing models have demonstrated that alcohol interferes with the ability to detect accurately a signal while simultaneously fixing attention on a standard stimulus or while occupied with an on-going task. Disruption is greatest with large doses or when the task requirements are more demanding. Alcohol appears to interfere with the ability to register, accept and process information under high demand instructions requiring rapid judgment or decision. Still unresolved is the question of differential effects of alcohol on various portions of the information-processing chain (e.g., registration vs central processing vs judgment).

While memory is probably a factor in studies of information processing, these studies alone do not directly test the effects of alcohol on this important process. In many respects memory is the

---

[1]Results of laboratory investigations of higher-order reflective functions are sometimes difficult to interpret because the test often depends on integrated sensory–motor performance. Given the documented impact of alcohol on tasks dependent on sensory–motor integration it is often difficult to tease out effects which are restricted to alcohol-related dysfunctions of the reflective processes. This caveat should be born in mind during this discussion.

cornerstone mechanism underlying the learning process—a process which is generally assumed to be disrupted by acute and chronic alcohol ingestion. Jellinek (1) devoted about one paragraph to learning (specifically to "memorizing") in his lecture; needless to say, the amount of literature available in this area has grown. Substantial effort has been devoted to understanding alcohol's effect on memory using visually and verbally presented material in various learning paradigms (21). Progress in this area has been impeded, in part, by a lack of consensus regarding the nature of memory itself.[2]

Moderate and large doses of alcohol interfere primarily with immediate and short-term memory. The greatest deficits appear to be in registration (imputing and encoding) and retention (consolidation and storage); retrieval mechanisms (accessing and recall) seem to be more resistant to alcohol's effects. Similar to the general findings in information processing, more complicated tasks and those dependent on self-constructed memorizing strategies appear to be more susceptible to alcohol. Alcohol's tendency to lower general arousal at medium and large doses may also contribute to memory deficits by lowering the significance of the information, or possibly the over-all significance of the learning episode.

Two special memory dysfunctions which may be related deserve attention. A *blackout* is defined as "amnesia, not associated with loss of consciousness, for all or part of the events that occurred during or immediately after a drinking session" (23, *p. 42*). Memory loss may be complete and encompass large blocks of time (hours or days) or fragmentary, encompassing relatively short intervals (e.g., portions of conversation, specific events in isolation). "En bloc" (complete) losses usually result in the inability to recall any events occurring during the interval while some portions of fragmentary losses may be regained if the individual is prompted regarding some portion of the lost events.

The blackout phenomenon has been reported to occur in alcoholics and nonalcoholics; the extent and frequency of occurrence, however, appear to be greater among alcoholics (24). The prerequisites and mechanisms of blackouts are not well understood.

---

[2]Although there are many competing views (22), traditional theories suggest that memory can be parceled into three stages: immediate, short- and long-term memory. Three interdependent phases (each involving several processes) are also postulated: registration (imputing and encoding of material via sensory mechanisms), retention (consolidation and storage of encoded material) and retrieval (accessing and recall of stored material). More contemporary views of memory emphasize the importance of context variables, complexity and significance of task, and attention and arousal states of learners.

Blackouts appear to result from achieving very high BACs in a relatively short time although other factors, such as a history of head injury, concurrent use of sedative-like drugs, may be important. Blackouts are generally considered short-term memory disruptions resulting from the inability to consolidate or transfer information from encoding to permanent storage (24). Alternatively, Mello (10) suggests that blackouts may result from alcohol's disruption of the registration phase—more specifically, the transfer of information from immediate to short-term memory. Both hypotheses predict that information obtained during a blackout (especially "en bloc" memory loss) will not find its way to permanent memory storage and therefore cannot be recalled.

Certain behaviors acquired or performed while under the influence of various drugs (e.g., alcohol, barbiturates, marijuana) may not be recalled, in part or whole, under nondrug conditions but may be retrieved if and when a drug state similar to that experienced during the initial learning episode is again induced (cf: 25). This phenomenon has been termed *state-dependent learning* (or alternatively, dissociative learning, state-dependent retrieval). This state differs from the blackout in that experiences are transferred to memory stores when the individual is intoxicated and are retrievable when the appropriate state of intoxication is reinstated. This phenomenon may be an important clue to an understanding of selective forgetting often reported in clinical observations of heavy drinkers and alcoholics. Evidence from animal experiments suggests that large doses of drugs are usually required to produce the phenomenon. The question of relationship, if any, between blackouts and state-dependent effect (including the possibility of a common mechanism) remains open.

Problem solving requires a relatively high level of cognitive functioning in that many abilities, e.g., signal detection, information processing, memory, are called on to function in an integrated fashion for a specific purpose. The choice of abilities and the level of integration depend on the nature of the task. As a result, it is difficult to characterize the effects alcohol has on problem solving in a simple fashion.

Problem solving involving simple arithmatic calculations appears to be affected adversely by moderate and large doses of alcohol. As task difficulty increases (e.g., addition of a constant to a given sum versus division of a sum by a constant) the effect of alcohol increases; the deficits are not necessarily large (decreases of 30% from baseline seem to be maximum) and are more prominent in measures of accuracy than of speed.

Problem solving involving manipulation of spatial relationships (e.g., Kohs' Block Design Test) appears to be easily disrupted by moderate and large doses of alcohol. However, tasks which require manipulations of verbal concepts in a manner considered to be dependent on somewhat more complex processes appear more resistant (9). The addition of stress to the testing situation appears to enhance alcohol's negative effects on problem solving performance. Thus, manipulation of arousal or emotive state of the subject may be an important factor to consider in relation to intellectual performance.

The few available reports regarding more complex problem solving present a somewhat different and controversial picture. Carpenter et al. (17) examined the effects of small, medium and large doses of alcohol and a placebo on the ability of social drinkers to solve unfamiliar complex problems involving the use of relatively sophisticated techniques. Small and medium doses had little, if any, deterimental effect on the number of problems solved; in fact, small doses had a modest facilitating effect. Only the large doses of alcohol produced marked decreases in the proportion of solved problems. Several measures of efficiency in problem solving (e.g., frequency of redundant responses, use of inappropriate manipulations) indicated that alcohol tended to decrease efficiency in linear fashion: larger doses tended to decrease efficiency to a greater extent, especially in terms of increasing general (and unnecessary) problem-solving activity. The findings were confirmed in a subsequent experiment, although other findings in the second study led to the conclusion that "defects in logical thinking seemed primarily due to loss of motivation, initiative or 'will power'" (26, *p. 381*). The latter observation suggests another potential factor involved in complex or "higher-order" problem solving.

Comparison of results from the last two studies, in contrast to those involving simple problem solving, highlights the difficulty of achieving a consensus of the effects of alcohol on higher-order cognitive functioning. The seemingly straightforward assumption that higher-order functions are more susceptible than simple sensory processes to the effects of alcohol may be misleading (3, 17). The difficulty may result from attempts to order problem-solving tasks linearly as a function of task complexity. The assumption of linearity may not be warranted since tasks may differ along some nonlinear (i.e., qualitative) dimension as yet undefined. This important and difficult area should be a prime candidate for future research. At present it seems too much a victim of early closure on the part of the research community.

The early interest in alcohol's impact on cognitive function has been refocused on the issue of long-term effects in alcoholics. A key issue is the degree to which prolonged ingestion of alcohol in large quantities leads to deficits in cognitive abilities in the sober state. Recently interest has focused on functional (i.e., "psychological") disabilities in contrast to those resulting from identifiable brain pathologies such as Wernicke–Korsakoff disease.

Rather than displaying deficits in global cognitive–intellectual capacities, as often assumed, alcoholics appear to show more specific dysfunctions (11, 27). For example, in tests of general intelligence, performance scores are somewhat lower than verbal scores even though full scale scores are within normal range. Specific deficits have been detected in perceptual synthesis tasks (e.g., Kohs' Block Design Test), spatial scanning and planning tasks (e.g., Trail Making Test of the Halstead–Reitan Battery), and spatial orientation (e.g., Rod and Frame Test).

A recent and important set of studies has focused on psychological deficits in social drinkers in the sober state (28, 29). Preliminary findings suggest that social drinkers may experience deficits in abstracting, adaptive abilities and concept formation and that, to some extent, the deficits appear similar to those observed in older individuals. Deficits also appear to be related to amount of alcohol consumed per drinking occasion. These investigations seem important and should be vigorously pursued. One important implication is the possibility that the deterioration observed in alcoholics is the extreme or end point of a continuum beginning at some critical (but within "normal" range) value of drinking intensity.

Effector Functions

Even relatively simple effector responses depend on and are synchronized with some form of sensory input. In this respect, information about effector abilities are derived from studies relying on tasks which require some form of effector–sensor interaction. This fact should be considered in evaluating alcohol's impact on effector functions. Simple functions include muscular strength and steadiness, finger-tapping speed and ocular-motor activity. Complex effector functions require elaborate sequencing of "voluntary" motor actions and greater amounts of effector–sensor interaction, such as tracking and controlling objects and automobile driving.

Alcohol in moderate doses appears to have minimal effect on muscular strength; even large doses produce small (though significant) deficits. Several studies have demonstrated that small doses may even increase strength, however, these reports tend to be in the minority. Muscular strength deficits have been reported among abstinent alcoholics (11); some recovery may occur as sobriety continues. The findings raise the possibility of long-term alcohol effects on muscular performance in the absence of dramatic acute effects. Finger-tapping speed (a motor performance impairment index of the Halstead–Reitan Test) is somewhat impaired after moderate and large doses of alcohol. Like muscle strength, this index has shown impairment in abstinent alcoholics (11).

One of the more persistent and uniform effects of alcohol is the induction of involuntary eye movements when the head is placed on its side (as it might be when an intoxicated individual lies down). The movements are characterized by slow upward drifts of the eyes followed by quick downward jerks. The phenomenon (known as positional alcohol nystagmus, PAN I) occurs after moderate to large doses of alcohol and can continue for up to 4 hr after peak BACs are reached. The effect appears to result from alcohol's disturbance of the "normal" coordination of the ocular-motor apparatus and the vestibular apparatus (sensory receptors here are responsible for processing information regarding bodily orientation relative to gravity). In humans, a second type of positional alcohol nystagmus (PAN II) has been identified in which an initial downward drift of the eyes is followed by upward compensatory jerks. This effect occurs after alcohol's direct action on ocular-motor and vestibular mechanisms has dissipated, often 8 to 10 hr after achieving BACs necessary for PAN I. The intoxicated condition sometimes referred to as "the spins" is probably related to PAN I.

Another reliable and sensitive index of alcohol's impact on effector functions is the impairment observed in the ability to stand steadily (feet parallel and shoulder width apart) with a minimum of body sway (the Romberg test). This behavior, though well rehearsed, results from active processes. Standing steadiness is usually tested with eyes opened and with eyes closed. When the eyes are closed there is usually greater body sway, even under nonalcohol conditions, as a consequence of depriving the subject of visual cues. Subjects tested while under the influence of alcohol display greater deficits in the eyes open than eyes closed condition presumably because of the difficulty in standing steadily with the eyes closed baseline (no alcohol) condition. Deficits can be observed at low and moderate doses; large doses give rise to very large deficits (30).

Indices of motor dexterity (e.g., moving rings from one peg to the next, placing pegs or pins in holes, sliding small objects through holes) have shown little (less than 10%) impairment at low and moderate BACs; significant impairment does occur at very high BACs. Other measures of effector coordination impaired at higher BACs include hand steadiness, finger to finger test (touching the two index fingers together with eyes closed), rail walking (walking heel to toe on narrow rails) and walking a straight line. Very large doses of alcohol seem to be required before marked impairment of gait (normal walking behavior) can be observed.

The ability to coordinate effector activity in response to sensory information has been the subject of numerous investigations. Many studies have centered on activities believed to be important components of automobile driving (for reviews see 3, 5, 31). Simple pursuit tasks in which subjects are required to maintain the position of a pointer relative to a moving target are disrupted by moderate doses of alcohol. Requiring the subject to use a steering wheel apparatus to keep a marker on a moving target (presumably a more difficult task) appears to produce larger deficits after moderate doses of alcohol. Still larger deficits can be induced if subjects are required to attend and respond to other discrete signals or instructions (e.g., pressing a button in response to a signal) while engaged in tracking tasks.

A number of laboratory studies which use tasks simulating driving situations have been conducted (e.g., 12). In one study (32) using a "traffic movie," impairment of "driving" was only found at high BACs; however, introduction of an additional attention task resulted in impairment at lower BACs.

Several investigators argue that such simulations do not adequately represent on-the-road driving (33). "Live" studies tend to find impairment of a wide range of components of driving behavior after small and moderate doses of alcohol. Disparity between laboratory and "live" studies raises a question regarding the applicability of laboratory results regarding components of a behavior (e.g., driving) to real world performance (34). One of the greatest difficulties in such extrapolations lays in obtaining an accurate description of a behavior and its component parts. Additionally, the context may influence performance. For example, faulty driving in a laboratory simulation does not carry the same real and imagined consequences as faulty driving on the road.

Studies of driving and airplane flying have also revealed the importance of skill level and practice in determining the disruptive effects of alcohol on complicated sensory–effector activities. Lovibond and Bird (35) compared the driving ability of highly skilled racing drivers with that of noncompetitive drivers. Several driving conditions were tested, including high speed cornering, negotiation of a slalom course, entering and parking in a garage. Over-all driving performance of noncompetition drivers was affected at relatively low BACs; competition drivers were not affected until relatively higher BACs were achieved. At those levels, competitive drivers' performances dropped below the nonalcohol performance of the noncompetitive group. A similar study[3] tested the effects of alcohol on the airplane flying ability of experienced nonprofessional and professional pilots. Nonprofessional pilots exhibited difficulty in tracking and committed significantly more procedural errors at lower doses of alcohol than did the professionals; low doses did result in serious errors even among the skilled professionals.

This brief review indicates that straight-forward determination of alcohol's impact on motor performance is not easy. There are both methodological problems and theoretical anomolies which make all but the most general of statements difficult to unequivocally support. For example, the fact that relatively simple and well-rehearsed activities (e.g., eye movements, standing steadily) are rather sensitive to alcohol has led some reviewers (9) to challenge the validity of the general principle that complex tasks are more likely to be adversely affected by alcohol than simpler tasks. Also challenged is the notion that practice while under the influence of alcohol does not always diminish deficits.

The validity of these challenges notwithstanding, one can observe that alcohol's effects on a given behavior cannot necessarily be predicted from observing behaviors thought to be from the same domain. Similarly, prediction of changes in "complex" behaviors can not necessarily be made by observing alcohol's impact on "simpler" behaviors purported to be components of the more complex behavior. Perhaps the one general statement that can be made at this time is that alcohol does not appear to facilitate effector behavior except under most unusual circumstances and then probably as a result of changes in affective (i.e., feeling) states.

One question which becomes of some importance remains to be raised and addressed by researchers. Given the effect of alcohol on the performance of such a wide range of effector functions in presumably "normal" or social drinkers, what, if any, are the cumulative or long-term effects of drinking on effector abilities?

Affective Processes

Since unrecorded time alcohol has been touted as a prime method

[3]BILLINGS, L. E., WICK, R. L., GERKE, R. J., et al. The effects of alcohol on pilot performance during instrument flight. (Federal Aviation Administration Rep. No. FAA-AM-72-4.) Washington, D.C.; 1972. Reported in Sharma, et al. (12).

of changing something called "feelings." The role of alcohol as an affect manipulator (i.e., an effective tool in controlling how we feel) is considered a central theorem for the acquisition and maintenance of drinking behavior. At the time of Jellinek's review few psychologically oriented studies were available which focused on what he termed "emotion, sentiments and moods" (1, p. 90). In fact, the now classic studies of Masserman had just begun (1, p. 85). Presently, the impact of alcohol ingestion on affective processes giving rise to arousal, emotive and motivational expressions and states accounts for a large slice of the alcohol research pie. Several recent reviews attest to the present scope and result of these efforts (e.g., 9, 10, 36–39).

The search for rules which govern alcohol's potential influence over affect is difficult because of the subjective and idiosyncratic nature of affective experience. By and large researchers and clinicians alike have relied on indirect estimates—such as verbal reports of subjects' experience, subject ratings of mood—which may not correspond to actual affective state changes. Reliance on such measures is not peculiar to the alcohol field. Few, if any, adequate methods are available for assessing affective states. Measurement is also made difficult as a result of the labile and transient quality of affective states; by their nature they are readily influenced by a wide range of internal and external events and circumstances which potentially hold significance for well being if not survival.

Alcohol's apparent biphasic effect on behavioral arousal (the "elation–depression" phenomenon) was perhaps one of the earliest observations of alcohol-induced changes in affective expression. Studies using a variety of techniques, most of which relied on self-report or observation of overt behavior, are relatively consistent in their conclusion (e.g., 9, 10): alcohol doses in the small to moderate range tend to produce signs of behavioral arousal including increased talkativeness, gregariousness, and other signs indicative of general excitation and heightened arousal. Larger doses appear to result in relaxation, decreased arousal and withdrawal while still larger doses result in drowsiness and sleep.

Subjective reports by experimental subjects, including responses to standard mood rating scales, indicate that small and moderate doses of alcohol are correlated with positive feeling states (e.g., happiness, euphoria, elation, relaxation, lowered anxiety) while larger doses appear to be connected to more negative feelings states (e.g., anxiety, depression, depressed detachment, irritability, general dysphoria). High doses are also often accompanied by physical symptoms of discomfort (e.g., nausea, "spins") especially in naïve and social drinkers. It is important to recognize that the full range of affective arousal changes can take place in the same subject as progression from sobriety to high levels of intoxication occurs (10).

The biphasic effect can be influenced by a variety of factors including the affective state at the time of intoxication (e.g., depressed vs happy), drinking history (e.g., social drinkers vs alcoholismic), rate of alcohol ingestion (e.g., slow vs rapid rise in BAC), circumstances (e.g., social vs laboratory setting), demand characteristics (e.g., subject required to perform difficult tasks vs drinking to "enjoy"), expectancy regarding alcohol's effects (e.g., anticipating "positive" vs "negative" affective change), type of instrument used to assess status (e.g., mood check vs experimenter observation vs "projective" technique), time after drinking when assessment is made (e.g., rising vs falling BAC). Drug action is superimposed on existing phasic (short-term) and tonic (long-term) behavior themes.

The exact mechanism or mechanisms through which alcohol exerts its biphasic behavioral effects are not known. Small doses of alcohol excite neural activity in certain locations in the CNS while large doses tend to depress CNS functioning (10). It has also been suggested that alcohol depresses certain inhibitory brain mechanisms located in the reticular activating system (a series of neural structures located in the brainstorm purported to be a central mechanism in the modulation of arousal) thereby releasing other areas from normal inhibitory control. It is hypothesized that release from inhibition (really, then, an indirect effect of a direct depressant action) and not direct stimulation is responsible for the apparent behavior-arousing effects of alcohol. The latter theory, while less parsimonious than the direct-stimulation hypothesis, has received much attention (34). It has the advantage of allowing alcohol to retain its status as a "depressant" (a position favored by Jellinek) and yet allows explanation of the behavioral arousal and emotive energizing which characterize alcohol's effects under a wide range of circumstances. Whether either position will stand the test of time is unknown.

Four specific phenomena related to arousal and emotive states deserve special discussion.

Sleep. The soporific (sleep-inducing) properties of alcohol have long been recognized, and alcohol has been recommended often as a mild sleeping preparation. However, even early scientific reports noted the disruptive effects of alcohol on sleep patterns (40). Alcohol in moderate and large doses decreases time to onset of sleep and results in restless activity and wakenings during the latter portions of the sleep episode.

Alcohol also appears to suppress rapid eye movement (REM) sleep, an activity associated with dreaming, especially during the first portion of the evening; a rebound or increase in such sleep periods during the latter portion of the sleep periods may occur. Disruption of REM sleep has been related to negative alternations in daytime (awake) moods and decreases in ability to perform routine activities. Some recovery of normal REM sleep patterns may occur with repeated use of alcohol; this result is somewhat equivocal and may occur only in moderate drinkers.

Sleep patterns of alcoholics also include REM suppression, frequent wakenings and restlessness. During alcohol withdrawal profound sleep disturbances (including insomnia, nightmares and hallucinations) may last for some time after the initial period of detoxification. Recovery of normal sleep patterns does occur; however, the time required is variable with disturbances lasting for upwards of 6 months in some patients. Alcohol withdrawal episodes frequently end with a period of profound deep sleep typically lasting for 12 hr or more.

Aggression. Alcohol has long been considered the "agent provocateur" in a wide range of aggressive acts (41). If recent estimates are accurate, alcohol use is implicated as a significant factor in nearly half of all violent crimes and highway fatalities. These observations are often cited as evidence of alcohol's role as a facilitator and possible inducer of violent and aggressive behavior (e.g., 10).

Conventional wisdom suggests that small amounts of alcohol decrease or inhibit aggressive behavior while larger amounts increase or disinhibit the behavior. However, the scientific literature presents a more complex picture, one in which characteristics of the individual and circumstances of the testing situation play an important albeit unclear role. For example, in studies[4] where relatively neutral instructions are given to subjects (e.g., 42) alcohol does not appear to enhance aggressive behavior. In group situations where conflict and competition are encouraged, small amounts of alcohol decreased and larger amounts heightened aggressive behavior (43–45). In a

[4]A popular paradigm for directly observing the impact of alcohol on human aggression involves observing the amount of noxious stimulation, often electric shock, subjects after various doses of alcohol are willing to deliver to others in various experimental situations. Typically, the experimenter controls the actual amount of "stimulation" delivered and manipulates various aspects of the situation. Situational manipulations are designed to increase or decrease the probability that aggressive behavior will occur.

study designed to separate pharmacological effects from those of expectancy or context, Lang et al. (46) found an increase in aggressive behavior only among those subjects who had been told they were to receive alcohol whether they received it or not, i.e., expectancy effects were strong. The results indicated that the assumption of a simple cause–effect relationship between drinking and aggression must still be considered speculative. A wide range of phasic and tonic factors (e.g., expectancies and contextual cues) may modulate aggressive expression in "live" situations.

Sexual Arousal. Experimental investigation of the effects of alcohol on sexual arousal has a recent history. Farkas and Rosen (48) and Briddell and Wilson (49), working with men college students, and Wilson and Lawson (50), working with women students, noted small increases in sexual arousal (assessed by physiological responsivity) with small amounts of alcohol and marked decrease in arousal with higher doses. In the latter two studies, attempts to influence arousal by manipulating subjects' expectancies about alcohol's effect on sexuality had no systematic impact; in both studies subjects also reported a belief that alcohol would lead to increased arousal even when presented with "contrary data" by the experimenters.

In a subsequent study Wilson and Lawson (51) demonstrated the importance of subjects' expectancy regarding sexual arousal on the effectiveness of alcohol to influence arousal. Subjects who expressed a prior belief that heterosexually-oriented erotic films would increase sexual arousal all experienced increased arousal when shown the films whether or not they actually received alcohol; surprisingly, subjects who believed that they had received alcohol before viewing the film showed significantly greater arousal irrespective of the actual amount received. The authors suggested that the study "demonstrates the importance of cognition–mediating processes in alcohol's influence on behavior" (51, *p. 594*).

Rubin and Henson (52) also found evidence of cognitive control over sexual arousal while under the influence of alcohol. In this study, subjects were asked to inhibit sexual arousal to erotic films and to become aroused voluntarily in the absence of any arousing film. Low and moderate doses of alcohol significantly interfered with sexual arousal to film stimuli, but not with the ability to inhibit or induce arousal in the absence of explicit films. Large doses of alcohol resulted in greatly decreased sexual arousal in response to films and decreased ability to become aroused in the absence of overt stimuli, but did not interfere with the ability to inhibit arousal to films. Interestingly, over half of the subjects receiving alcohol believed after the experiment that alcohol enhanced sexual responding even though their experience had been otherwise.

The trend of the evidence suggests that alcohol, particularly in large amounts, decreases sexual arousal. However, at low or moderate amounts of alcohol, sexual responsivity—objectively and subjectively assessed—depends on the context of the drinking episode and the expectancies of the drinkers.

Tension Reduction. It has been long held that one of the most significant effects of alcohol is its ability to reduce tension (39). Since the initial observations of Masserman (53) and Conger (54) of alcohol's effects in reducing fear and enhancing conflict resolution in animals, a variety of experimental paradigms have been used (e.g., escape, avoidance, conditioned suppression, audiogenic seizure induction, frustration induction). Cappell (6, 39) has reviewed the studies and concluded that the supporting evidence is thin. The hypothesis has not held up well in studies of alcoholics (55–57); it appears that increased dysphoria accompanies higher levels of intoxication and occurs in later stages of a drinking episode. Dysphoria may occur even in alcoholics who, prior to a drinking episode, anticipate improvement in their affective state.

Should the hypothesis be discarded? Not necessarily. The problems of studying "tension" experimentally are enormous, and an adequate model of tension may be more complex in construction and dynamic principle than the existing models. In addition, Nathan et al. (57) have suggested differential memory of different stages in drinking episodes with negative affect of later stages more likely to be blocked and earlier, positive affect stages retained. The positive remembrances result in renewed expectations of positive affect change achieved through drinking.

The tension reduction hypothesis survives despite shaky experimental support because it is a "good sense" explanation of why people drink. (Hardly enough evidence to guarantee a long and healthy tenure for any hypothesis.) We have no viable alternative answer to the question, "Why do people drink?" Greatly needed is investigation of affective change occurring during drinking in nonalcoholics.

What does the drinker find reinforcing in the drinking episode? What changes in affect, positive and negative, occur in each stage of a drinking episode? What satisfying events result in the establishment of drinking styles? Answers to these and related questions may lead to a better understanding of fundamental principles underlying acquisition and maintenance of the full range of drinking behaviors, including its pathologies.

Concluding Comments

There is little question that Jellinek's challenge for more research activity has been met. However, the attention did not come solely from his urgings or from any sustained and special interest in alcohol or drinking behavior. Indeed, much of the activity has resulted from the general growth and expansion of experimental and clinical psychology. What is the value of information obtained from so many diverse studies in providing a relevant psychological framework for understanding alcohol use in the general behavioral repertoire of organisms?

Psychological studies have contributed much to our understanding of which functions are affected by different amounts of beverage alcohol. The way in which functions are affected have also been described but with a lesser degree of certainty. In terms of mechanisms of action the picture is still less clear; in many instances understanding is largely at the level of description as opposed to explanation. Theories unifying the findings into cohesive themes are generally lacking.

This may be due more to the nature of the research results as they have unfolded than from any scarcity of psychologists proposing unifying principles. Herein may lie a subtle yet important contribution: the complexity of the results obtained to date dispels the notion that alcohol effects can be easily explained by simple unidimensional propositions (e.g., "alcohol is a depressant" or "alcohol acts like an anesthetic"). Alcohol's effects on psychological processes do not seem to be predictable solely from its pharmacological properties and actions, or from cognitive and affective properties of the organisms, or from drinking context or broad environmental factors. The dynamic interaction of these and other factors plays an important and perhaps critical role in the determination of alcohol's action.

Knowledge of the mechanisms and rules of the dynamic interaction would permit formation of general principles which could be used in predicting alcohol's effects. At present, our understanding is sufficient for only a narrow range of relatively simple equations.

Perhaps the most intriguing and complex equations revolve around the question, "Why do people drink alcohol?" Unfortunately, answers to this and related questions are absent. Given the multitude of reports recounting the deleterious effects of alcohol on a wide range of functions, one wonders what positive consequences could possibly be obtained which would lead to and sustain the level of drinking behavior observed in so many individuals in society today. This query is especially pertinent as long standing hypotheses regarding motivations for drinking, such as the tension reduction hypothesis, come under critical scrutiny. The void in this central problem area is disturbing and may possibly be due to the disregard by theory makers of the value of drinking behavior to the individual.

In any event, the question "Why do people drink alcohol?" deserves special attention. One positive trend in approach is emerging. Several groups of investigators are studying drinking behavior using a prospective approach to document the history of drinking behavior as it unfolds in its natural milieu. These studies, while complex and difficult to manage, hold the promise of identifying significant factors related to the development of drinking behavior, and its pathologies. In the years to come, these studies may provide important new solutions to the questions posed by Jellinek 35 years ago.

Looking over psychological studies of the action of alcohol, one can conclude that many pebbles of knowledge have been gathered and that, in fact, some have been fashioned into crude outlines of a structure. If the construction is to move toward completion, our current efforts may require a serious reassessment in order that techniques suitable to the task might be judiciously selected and parsimoniously applied.

REFERENCES

1. JELLINEK, E. M. Effects of small amounts of alcohol on psychological functions. Pp. 83–94. In: Alcohol, science and society; twenty-nine lectures with discussions as given at the Yale Summer School of Alcohol Studies. New Haven; Quarterly Journal of Studies on Alcohol; 1945.
2. JELLINEK, E. M. and McFARLAND, R. A. Analysis of psychological experiments on the effects of alcohol. Quart. J. Stud. Alc. 1: 272–371, 1940.
3. CARPENTER, J. A. Effects of alcohol on some psychological processes; a critical review with special reference to automobile driving skill. Quart. J. Stud. Alc. 23: 274–314, 1962.
4. CARPENTER, J. A. Effects of alcohol on psychological processes. Pp. 45–90. In: FOX, B. H. and FOX, J. H., eds. Alcohol and traffic safety. (U.S. Public Health Service, Publ. No. 1043.) Bethesda, Md.; 1963.
5. CARPENTER, J. A. Contributions from psychology to the study of drinking and driving. Quart. J. Stud. Alc., Suppl. No. 4, pp. 234–251, 1968.

6. CAPPELL, H. and HERMAN, C. P. Alcohol and tension reduction; a review. Quart. J. Stud. Alc. **33:** 33–64, 1972.

7. KALANT, H. Some recent physiological and biochemical investigations on alcohol and alcoholism; a review. Quart. J. Stud. Alc. **23:** 52–93, 1962.

8. LINNOILA, M. Effect of drugs and alcohol on psychomotor skills related to driving. Ann. clin. Res., Helsinki **6:** 7–18, 1974.

9. WALLGREN, H. and BARRY, H., 3d. Actions of alcohol. 2 vols. New York; Elsevier; 1970.

10. MELLO, N. K. Alcoholism and the behavioral pharmacology of alcohol; 1967–1977. Pp. 1619–1673. In: LIPTON, M. A., DiMASCIO, A. and KILLAM, K. F., eds. Psychopharmacology; a generation of progress. New York: Raven; 1978.

11. TARTER, R. E. Empirical investigations of psychological deficits. Pp. 359–393. In: TARTER, R. E. and SUGARMAN, A. A., eds. Alcoholism; interdisciplinary approaches to an enduring problem. Reading, Mass.; Addison–Wesley; 1976.

12. SHARMA, S., ZIEDMAN, K. and MOSKOWITZ, H. Alcohol effects on behavioral performance. Pp. 79–89. In: BLUM, K., BARD, D. L. and HAMILTON, M. G., eds. Alcohol and opiates; neurochemical and behavioral mechanisms. New York; Academic; 1977.

13. KISSIN, B. and BEGLEITER, H., eds. The biology of alcoholism. Vol. 2. Physiology and behavior. New York; Plenum; 1972.

14. HEBB, D. O. Organization of behavior. New York; Wiley; 1949.

15. SOLOMAN, R. L. An opponent-process theory of acquired motivation; the affective dynamics of addiction. In: MASER, J. D. and SELIGMAN, M., eds. Psychopathology; experimental models. San Francisco; Freeman; 1977.

16. SOLOMAN, R. L. The opponent-process theory of acquired motivation; the costs of pleasure and benefits of pain. Amer. Psychol. **35:** 691–712, 1980.

17. CARPENTER, J. A., MOORE, O. K., SNYDER, C. R. and LISANSKY, E. S. Alcohol and higher order problem solving. Quart. J. Stud. Alc. **22:** 183–222, 1961.

18. HUNTLEY, M. S., JR. Effects of alcohol and fixation-task difficulty on choice reaction time to extrafoveal stimulation. Quart. J. Stud. Alc. **33:** 89–103, 1972.

19. MOSKOWITZ, H. and SHARMA, S. Effects of alcohol on peripheral vision as a function of attention. Hum. Factors, N.Y. **16:** 174–180, 1974.

20. KING, H. E. Ethanol induced slowing of human reaction time and speed of voluntary movement. J. Psychol. **90:** 203–214, 1975.

21. BIRNBAUM, I. M. and PARKER, E. S., eds. Alcohol and human memory. Hilsdale, N.J.; Lawrence Erlbaum Associates; 1977.

22. THATCHER, R. W. and JOHN, E. R. Foundations of cognitive processes. New York; Wiley; 1977.

23. KELLER, M. and McCORMICK, M. A. A dictionary of words about alcohol. New Brunswick, N.J.; Rutgers Center of Alcohol Studies Publications Division; 1968.

24. GOODWIN, D. W. The alcoholic blackout and how to prevent it. Pp. 177–183. In: BIRNBAUM, I. M. and PARKER, E. S., eds. Alcohol and human memory. Hilsdale, N.J.; Lawrence Erlbaum Associates; 1977.

25. OVERTON, D. A. State-dependent learning produced by alcohol and its relevance to alcoholism. Pp. 193–217. In: KISSIN, B. and BEGLEITER, H., eds. The biology of alcoholism. Vol. 2. Physiology and behavior. New York; Plenum; 1972.

26. KASTL, A. J. Changes in ego functioning under alcohol. Quart. J. Stud. Alc. **30**: 371–383, 1969.

27. TARTER, R. E. Psychological deficit in chronic alcoholics; a review. Int. J. Addict. **10**: 327–368, 1975.

28. PARSONS, D. A., ed. Cognitive dysfunction in alcoholics and social drinkers: problems in assessment and remediation. Quart. J. Stud. Alc. **41**: 105–186, 1980.

29. PARKER, E. S. and NOBLE, E. P. Alcohol consumption and cognitive functioning in social drinkers. Quart. J. Stud. Alc. **38**: 1224–1232, 1977.

30. SCHNEIDER, E. W. Characteristics of alcohol ataxia. Ph.D. dissertation, Rutgers University; 1970.

31. Vermont symposium on alcohol, drugs and driving. Saf. Res., Chicago. **5**: 106–223, 1973.

32. MOSKOWITZ, H. The effects of alcohol performance in a driving simulator of alcoholics and social drinkers. (Institute of Transportation and Traffic Engineering, Report ENG.-7025.) Los Angeles; 1971.

33. HUNTLEY, M. S. Alcohol influences upon closed-course driving performance. J. Saf. Res., Chicago **5**: 149–164, 1973.

34. PERRINE, M. W. Alcohol influences on driving-related behavior; a critical review of laboratory studies of neurophysiological, neuromuscular and sensory activity. J. Saf. Res., Chicago **5**: 165–184, 1973.

35. LOVIBOND, S. H. and BIRD, K. Danger level—the Warwick Farm project. Pp. 299–305. In: KILOH, L. G. and BELL, D. S., eds. Proceedings of the 29th International Congress on Alcoholism and Drug Dependence. Melbourne; Butterworths; 1971.

36. BARRY, H., 3d. Motivational and cognitive effects of alcohol. J. Saf. Res., Chicago **5**: 200–221, 1973.

37. FREED, E. X. Alcohol and mood; an updated review. Int. J. Addic. **13** 173–200, 1978.

38. RUSSELL, J. A. and MEHRABIAN, A. The mediating role of emotions in alcohol use. J. Stud. Alc. **36**: 1508–1536, 1975.

39. CAPPELL, H. An evaluation of tension models of alcohol consumption. Pp. 177–209. In: GIBBINS, R. J., ISRAEL, Y., KALANT, H., POPHAM, R. E., SCHMIDT, W. and SMART, R. G., eds. Research advances in alcohol and drug problems. Vol. 2. New York; Wiley; 1975.

40. POKORNY, A. D. Sleep disturbances, alcohol and alcoholism; a review. Pp. 233–260. In: WILLIAMS, R. L. and KARACAN, I., eds. Sleep disor-

ders; diagnosis and treatment. New York; Wiley; 1978.

41. CARPENTER, J. A. and ARMENTI, N. P. Some effects of ethanol on human sexual and aggressive behavior. Pp. 509–543. In: KISSIN, B. and BEGLEITER, H., eds. The biology of alcoholism. Vol. 2. Physiology and behavior. New York; Plenum; 1972.

42. BENNETT, R. M., BUSS, A. H. and CARPENTER, J. A. Alcohol and human physical aggression. Quart. J. Stud. Alc. **30:** 870–876, 1969.

43. TAKALA, M., PIHKANEN, T. A. and MARKKANEN, T. The effects of distilled and brewed beverages; a physiological, neurological and psychological study. (Finnish Foundation for Alcohol Studies, Publ. No. 4.) Helsinki; 1957.

44. BOYATZIS, R. E. The effect of alcohol consumption on the aggressive behavior of men. Quart. J. Stud. Alc. **35:** 959–972, 1974.

45. TAYLOR, S. P. and GAMMON, C. B. The effects of type and dose of alcohol on human aggression. J. Personality social Psychol. **32:** 169–175, 1975.

46. LANG, A. R., GOECKNER, D. J., ADESSO, V. J. and MARLATT, G. A. The effects of alcohol in male social drinkers. J. abnorm. Psychol. **84:** 508–518, 1975.

47. NATHAN, P. E. and LISMAN, S. A. Behavioral and motivational patterns of chronic alcoholics. Pp. 479–522. In: TARTER, R. E. and SUGARMAN, A. A. eds. Alcoholism; interdisciplinary approaches to an enduring problem. Reading, Mass.; Addison–Wesley; 1976.

48. FARKAS, G. M. and ROSEN, R. C. Effect of alcohol on elicited male sexual response. J. Stud. Alc. **37:** 265–272, 1976.

49. BRIDDELL, D. W. and WILSON, G. T. Effects of alcohol and expectancy set on male sexual arousal. J. abnorm. Psychol. **85:** 225–234, 1976.

50. WILSON, G. T. and LAWSON, D. M. Effects on sexual arousal in women. J. abnorm. Psychol. **85:** 489–497, 1976.

51. WILSON, G. T. and LAWSON, D. M. Expectancies, alcohol and sexual arousal in male social drinkers. J. Abnorm. Psychol. **85:** 587–594, 1976.

52. RUBIN, H. B. and HENSON, D. E. Effects of alcohol on male sexual responding. Psychopharmacology, Berl. **47:** 127–134, 1976.

53. MASSERMAN, J. H., JACQUES, M. G. and NICHOLSON, M. R. Alcohol as a preventive of experimental neuroses. Quart. J. Stud. Alc. **6:** 281–299, 1945.

54. CONGER, J. J. The effects of alcohol on conflict behavior in the albino rat. Quart. J. Stud. Alc. **12:** 1–29, 1951.

55. MENDELSON, J. H., LaDOU, J. and SOLOMON, P. Experimentally induced chronic intoxication and withdrawal in alcoholics. Pt. 3. Psychiatric findings. Quart. J. Stud. Alc., Supp. No. 2, pp. 40–52, 1964.

56. NATHAN, P. E. and O'BRIEN, J. S. An experimental analysis of the behavior of alcoholics and non-alcoholics during prolonged experimental drinking; a necessary precursor of behavior therapy? Behav. Ther., N.Y. **2:** 455–476, 1971.

57. NATHAN, P. E., GOLDMAN, M. S., LISMAN, S. A. and TAYLOR, A. A. Alcohol and alcoholics; a behavioral approach. Trans. N.Y. Acad. Sci. **34:** 602–627, 1972.
58. MCNAMEE, H. B., MELLO, N. K. and MENDELSON, J. H. Experimental analysis of drinking patterns of alcoholics; concurrent psychiatric observations. Amer. J. Psychiat. **124:** 1063–1069, 1968.

4

In Other Cultures, They Also Drink

Dwight B. Heath

Whatever else it may be, ethanol is undoubtedly the most widely used psychoactive drug in the world, and it is probably the oldest. Although it has been known and used in most societies throughout the world, there is no universal use, meaning, or function for alcohol. Patterns of belief and behavior with respect to alcoholic beverages vary enormously from one population to another, with diametrically contrasting attitudes and norms often strongly held by human beings whose attitudes toward many other things are similar, whose languages may be closely related, and who live in similar environments at a comparable level of technological development. One does not have to know much about anthropology or sociology to recognize that, however important the biochemical, physiological, and pharmacokinetic functions of alcohol may be, sociocultural factors are also important if we are to understand the complex interrelations of alcohol and human behavior.

It is also noteworthy that beliefs and behaviors that have to do with drinking in different cultures are not merely of desultory interest to a few individuals who are curious about quaint and exotic customs, or to those academicians who feel compelled to analyze anything that comes to their attention. The beliefs and behaviors that relate to alcohol are often unlike those that relate to other beverages, especially in the frequency and the extent to which they are a focus of strong emotions. In almost every society, there are rules about who may drink, where and when drinking may take place, how much of what beverages may be drunk, and so forth. The rules may be different for different beverages, and they are almost always different for people of different social statuses. (Status is used here in the general sociological sense to refer to categories such as age, sex, and marital status, as well as to occupation, prestige, etc.) However detailed the prescriptive rules may be in terms of specifying what should be done, there is usually another set of proscriptive rules, telling what is forbidden, that are at least equally specific. We are familiar with these in our own society, for example: many people feel that children should not drink, although adults are often encouraged to do so; drunkenness may be tolerated among young men but deplored among women; drinking that is typical at a New Year's Eve party would be totally out of place at a religious ceremony.

It is important that these rules not merely exist, but that people often feel very strongly about them. Even those groups that do not use alcohol are not neutral about it; on the contrary, the few populations around the world that have done away with drinking have strong moral as well as legal sanctions against it.

Although it is often characterized as "the neutral spirit" in chemical terms, alcohol is subject to many and intense restrictions, presumably because of the rapidity and intensity of the effects on human behavior. The fact that these effects are so variable from one population to another lends special significance to a sociocultural approach which attempts to illuminate and account for both the uniformities that do occur cross-culturally and the variation that sometimes seems even more striking among the many cultures throughout the world.

Historical Perspectives

Nobody knows when or where our ancestors first recognized the association between the process of fermentation and the unusual feelings that came from drinking the product. In a sense, it would be futile to look for the "invention" of brewing or vintning at some specific time and place. Fermentation is a natural process that occurs commonly without any human intervention, and other creatures occasionally take advantage of it. We sometimes see birds who have become intoxicated from eating overripe fruits or berries, and other animals are reported to have become drunk from naturally fermented foods. In a similar manner, it is probable that the first homebrew was drunk by humankind either by accident or under the press of profound hunger or thirst—perhaps a batch of gruel that had "gone bad" or a juicy fruit that was on the verge of rotting. In this process, ethanol accumulates as a waste-product of the micro-organisms that have been acting on the natural sugars. Many varieties of yeast are free-floating in the air at any time and work readily on sap, juice, or other sources of carbohydrates. In a sense, the elaborate arts and industries of brewing and wine-making can be viewed simply as refinements of this natural process.

Wherever it first occurred, beer making had apparently become an important activity by the time that the first villages of farmers appeared both in the Near East and in Mesoamerica about 7–10,000 years ago.

Even in the relatively simple and uncomplicated lives of ancient peoples, ambivalence about drinking and drunkenness was apparent. We know little about the details of drinking patterns or attitudes until the emergence of writing, but one of the earliest papyri with Egyptian hieroglyphs is a letter from a father to his son, warning that too much beer can be dangerous, in terms of his saying something indiscreet or in terms of wasting his money. The first written system of laws that we know, Hammurabi's Code decreed for Babylonia around 4000 years ago, devotes considerable attention to alcoholic beverages. Not only was there concern for consumer protection, in detailed discussion of units of measure and adulteration, but there was also concern for controls in such prosaic terms as restricted hours for wine-shops, specifications of who may work there, and so forth. The importance of wine in classical Greece (1) and Rome (2) is evident in the fact that the names of the gods Dionysus and Bacchus survive even today whenever we refer to dionysian revels, bacchanals, or use a related term to refer to a context of heavy drinking (3).

The Bible contains many references to wine, most of which are favorable rather than condemning (4). Islam's prohibition on drinking (5) is more recent than the Koran. A wide variety of alcoholic beverages played important roles in the pharmacopoeia of ancient India (6), and the supposed temperance of contemporary Chinese differs markedly from the widespread and highly esteemed drunkenness that figures importantly in most of the novels and histories written in that country during centuries past.

Differences in customs among various populations were noted early, and accounts written by explorers and other travelers often include vivid descriptions of drinking, drunkenness, and drunken comportment. Some observations were obviously colored by ethnocentrism and religious prejudice, especially those of missionaries who were undoubtedly often shocked by what they saw, and who also aimed to emphasize the problems that they faced in their efforts at conversion. Others are matter-of-fact in tone and seem plausible even centuries later, as is the case with Tacitus among the Germanic tribes almost two thousand years ago.

For that matter, we need not look far afield for evidence that drinking, like other aspects of culture, has changed markedly within the much shorter span that is often referred to as "modern history." The "gin plague" that Hogarth so vividly portrayed in eighteenth-century London (7) seems to have had a counterpart in colonial America where cheap locally produced rum was used as currency, especially in wages for the poor who moved to the cities in search of work.[1] "Hard-drinking" as a characteristic of the frontier is not a creation of Hollywood; it is amply evidenced in a wide range of documentary sources, many of which are from unbiased observers (8). The depth of popular feeling about alcoholic beverages is also illustrated in this country's so-called "noble experiment" with Prohibition. It is obvious that a variety of religious, ethnic, and other sociopolitical considerations were involved in the rapid spread of the Temperance Movement (9) but it is remarkable nonetheless that the only Constitutional amendment that ever addressed work-a-day activities of the American populace also became the only Constitutional amendment to be repealed. More recently, we have seen changes in patterns of consumption that are widely recognized, for example, a marked increase in wines and vodka at the expense of whisky. At the same time, both official and popular expressions of alarm over the apparent changes in the drinking population—especially among women and young people—show that concern often runs ahead of scientific knowledge.

Ethnic Variation within the United States

As the fallacy of the "melting pot" theory of cultural assimilation has become more widely recognized, social scientists have been conducting more detailed and insightful studies among several of the populations that remain diverse and vital within the United States. A few examples of the range of variation in drinking patterns may be worth examining briefly to underscore the importance of sociocultural factors in relation to alcohol use and related problems.

Even a layperson who knows little about alcohol and less about anthropology is likely to have some beliefs and "knowledge" (whether accurate or not) about social and cultural variation in drinking patterns. For example, it is widely believed that people of Irish–Catholic background tend to drink more than, say, those of German–Jewish heritage. The stereotype is often extended to include the idea that the Irish–Catholic population has an exceptionally high rate of alcohol-related problems, and the German–Jewish population an exceptionally low rate, both in comparison with the population at large.

In fact, sociological and epidemiological data from the middle of this century strongly supported these views. It is ironic that pop-

---

[1]PINSON, A. The New England rum era; drinking styles and social change in Newport, R.I., 1720–1770. Brown University Department of Anthropology Working Papers on Alcohol and Human Behavior, No. 8, 1980.

ular stereotypes, which are so uncritically adhered to by many people, are often inappropriately rejected by others on an equally uncritical basis. If we accept the word "stereotype" as applying to favorable as well as unfavorable characterizations, and if we recognize it as referring to a modal or typical (but not universal) pattern, then there is often a sound scientific basis for stereotypic statements. The problem arises when only negative characteristics are applied to a population, or when it is presumed that every member of a population fits the stereotype. In short, it is not bigotry to note that, although it is adult men who drink among Irish–Americans, that ethnic category is significantly over-represented among problem drinkers (10). By the same token, scientific evidence supports the popular view that, although both sexes and all ages of American Jews drink, that category is significantly under-represented among problem drinkers (11).

The different meanings of drinking and drunkenness can perhaps be appreciated more easily among such populations who participate in "mainstream American culture" and with whom many of us have close and sustained contact than among primitive tribes who may seem remote if not irrelevant to the everyday concerns of most of us. Among Irish youth, whisky drinking is disapproved until late adolescence, when it becomes an important adjunct to sociability, and drunkenness is considered an acceptable, almost inevitable, outcome of habitual carousing. It is expected that one will "drown his sorrows," or equally to "celebrate his fortunes" with heavy drinking in the company of male companions away from his family, and such behavior often creates tensions when he comes home. By contrast, Jewish children are introduced to wine and distilled spirits at an early age, taking small quantities as they participate in religious rituals around the family table. Drunkenness is generally scorned by Jews, and is even labelled an un-Jewish state of being. A succinct summary and interpretation of these contrasts is available (12).

During recent years, there is considerable controversy over the question of Jewish sobriety in this country, and it may be that both our stereotype and our sociological evidence are inaccurate (13). There is increasing anecdotal and impressionistic evidence of drinking problems among American Jews, and a few progressive synagogues have shown concern by establishing groups of Alcoholics Anonymous (14). Some observers speculate that, during the decades since the major systematic social research was conducted on the subject, Jews have become increasingly "acculturated" in the sense of adopting more of the patterns of beliefs and behavior of the "mainstream" (presumably middle-class Protestant) population, and that different drinking patterns and associated problems are part of

that general process. Others suggest that diminishing religious orthodoxy and widespread loosening of family bonds are such that the
ritual connotations of alcohol have diminished, while its secular uses
in stimulating sociability—e.g., helping one to relax, as an adjunct
to business transactions—have increased in ways that make drinking
problems more likely. It is also possible that Jews with drinking
problems are not a new phenomenon at all, but that more are willing
to admit to problems in a context of diminishing social stigma. The
kind of research that might shed light on the relative importance of
those alternate interpretations is only now being conducted (15), so
it would be premature to point to any one as "the reason." For that
matter, it may well be that all are important, with various ones more
important in affecting the behaviors and decisions of different individuals. The crucial point is that these explanations are basically
sociocultural in nature, having to do with attitudes and values that
are learned.

Recent research on Americans of Armenian ancestry[2] provides
an illuminating comparison with Jewish patterns. Like the Jews,
Armenians of both sexes begin to drink small amounts of alcohol at
an early age and in the company of their families. Unlike the Jewish
tradition, however, no specifically religious or other ritual meanings
are associated with such drinking. Instead, alcohol is introduced
simply as a good thing to be taken in moderation—often with the
explicit as well as implicit warning that, like many good things, it
can be dangerous when used too much. Alcohol-related problems
were virtually absent in the large population studied.

Studies among various other immigrant populations including
Italians (16), Hispanics from several Caribbean countries,[3] Poles,[4]
and others illustrate not only that attitudes and values toward drinking vary, but also that they change in various ways during successive
generations in this country. Native American populations show a
wide range of variation, from almost uniform abstinence to exceptionally high rates of drinking and drinking problems (17). Even
more dramatic illustrations of the degree of difference that can be

[2]FREUND, P. J. Armenian-American drinking patterns; ethnicity, family and religion. Brown University Department of Anthropology Working Papers on Alcohol
and Human Behavior, No. 5, 1979.

[3]GORDON, A. J. Cultural and organizational factors in the delivery of alcohol
treatment services to Hispanos. Brown University Department of Anthropology
Working Papers on Alcohol and Human Behavior, No. 7, 1979.

[4]FREUND, P. J. Polish-American drinking; an historical study in attitude change.
Brown University Department of Anthropology Working Papers on Alcohol and
Human Behavior, No. 9, 1980.

found in drinking and its consequences can be garnered from cultures that may have less immediate relevance to current social and political issues in the United States but are no less important as examples of the range of human experience with alcohol.

Drinking Patterns around the World

Differences in the nature of drinking and drinking problems, even among the most cosmopolitan populations of the world, came sharply into focus in the 1950s when a committee tried to formulate a definition of alcoholism that would be acceptable and relevant among members of the newly formed United Nations. Their first effort reflected their recognition of variation; they labeled as alcoholism "any form of drinking which in its extent goes beyond the traditional and customary 'dietary' use, or the ordinary compliance with the social drinking customs of the whole community concerned" (18).

Unfortunately, although their simplistic attempt at cultural relativity must be appreciated, that usage did not go very far, even in comparing problems among modern industrial societies that share common religious, historical, and other traditions. Even without shifting the focus to societies that evolved unaffected by the mainstream of Western civilization, it is easy to point out that major differences in the ways that people drink are linked not only to traditional use or social drinking customs, but also to different views about the nature of alcohol and the nature of alcoholism. In France, for example, it is commonplace to encounter senior citizens who are fearful of drinking water, and who do not consider wine alcoholic. Many among them are physiologically addicted to alcohol even though they have never been intoxicated. This kind of addiction may come to light only when an individual is hospitalized and, for the first time in years, does not have access to wine throughout the day. A worker may be accustomed to having wine for breakfast and taking successive wine-breaks throughout the morning. A liter of wine with lunch and more wine-breaks during the afternoon can maintain a blood alcohol concentration that is addicting even if not intoxicating. More wine with dinner and perhaps brandy afterward mean that the liver has little rest throughout the night, and the body, accustomed to alcohol at a fairly constant level of between 0.04 and 0.10%, would evince symptoms of withdrawal when deprived of alcohol, even in a person who had never drunk fast enough to reach the level of concentration that results in slurred speech, gross failure in motor coordination, and so forth.

A major concern of anthropologists is the description and

understanding of patterns of belief and behavior among all members of the human community. This often involves paying attention to remote bands or tribes of a few hundred people in the same ways that one would to a complex and modern nation-state. In a sense, comparisons among peoples are the nearest that we can come to conducting major long-term experiments in the social sciences. If we view the world as a vast "natural laboratory," the many and diverse cultures that have developed throughout time and space represent a wide range of "natural experiments" in human adaptation, and it is possible to identify specific variables that distinguish one from another. Obviously, we cannot control that encompassing environment to the degree that investigators in the natural sciences control the experimental environments in which they work, nor can we manipulate variables in the ways that they often can. Nevertheless, it is important in considering the great variation in beliefs and behaviors to recognize that they are not just a patchwork array of quaint and curious customs, but that each is integrated within a context of social, ideological, symbolic, and other behavioral actions and meanings (19).

It is usually easy for an anthropologist to catch the attention of an audience by offering them a smorgasboard of exotic, and sometimes titillating, ethnographic tidbits chosen from different cultures throughout the world. Although that may be enjoyable, it is not a very effective way to hold an audience or to teach people about the meanings and functions of alcohol. For those purposes, I will discuss a few conceptual and theoretical points, and try to relate them to scientific and practical implications for better understanding the interrelations of alcohol and human behavior.

The same World Health Organization committee issued a new definition of alcoholism just a few months later, partly in recognition of the French pattern (*alcoolisation*), partly in recognition of the Finnish representatives' concern about public safety as the major issue among drinking problems (especially when violent fights result from binge drinking during a brief season of affluence among rural workers who spend their winters in sober isolation from each other), and partly in recognition of a wide range of other kinds of problems. The focus was shifted and somewhat sharpened, to emphasize manifest problems or trouble as the distinguishing feature: "Alcoholics are those excessive drinkers whose dependence upon alcohol has attained such a degree that it shows a noticeable mental disturbance or an interference with their bodily or mental health, their interpersonal relations, and their smooth social and economic functioning; or who show the prodromal signs of such development" (20). The emphasis on "interference," problems, or trouble is congenial for those of us who are sometimes uncomfortable with the controversy that has developed around the disease concept of alcoholism, and it is a useful measure in evaluating drinking and its outcome in any culture, regardless of how the people themselves view alcohol as a substance.

It is widely recognized that ethanol is at the same time a food, a poison, and a drug. It should come as no surprise that various peoples around the world give different emphasis to these characteristics, so that the meanings and uses of it vary enormously, and the consequences are similarly varied (21).

Although we now recognize the dangers of "empty calories" in distilled spirits, heavy drinking can be nutritious and healthful in some societies. Those in our society who substitute the high caloric value of distilled spirits for other foods often suffer malnutrition or undernutrition. By contrast, Indians in the Andean highlands of Bolivia or Peru derive an abundance of vitamin C from the maize beer that they drink, although it is lacking in the rest of their diet. Recent detailed analyses demonstrate that their chicha, like the maize beer made by native peoples throughout much of eastern and southern Africa, is also exceptionally rich in protein. People who deplore what they view as a "waste" of grain in those preparations overlook the fact that such fermented beverages literally draw protein out of the air, in the form of yeasts and their by-products. Much the same advantage can be cited for a wide variety of other kinds of homebrews, including palm wine in western Africa and parts of Oceania, manioc beer in tropical South America, rice wine in much of Asia, and so forth. In many of the societies where these beverages are commonplace, they constitute a major portion of the diet for people of both sexes and all ages (22), and intoxication tends to be either rare or an occasional deliberate, approved, and socially contextualized event.

It is also commonplace in such societies that the fermented drink is highly esteemed as a commodity, and often imbued with additional symbolic values. Drinking is an important aspect of hospitality in those parts of the world, and drinking together often demonstrates that an argument has been settled or an agreement has been reached. Among peasant farmers who take turns helping each other with periodic tasks that require massed labor such as land-clearing and harvesting, the host is normally expected to provide beer for his unpaid neighbors, as well as to help each of them in turn. (This is similar to, in American history, the barn-raising, husking bee, and other cooperative efforts.)

The toxic quality of alcohol is not so widely recognized, and, in fact, without distillation or at least mass production, it is difficult to imagine that one could be poisoned by alcoholic beverages. It may be a blessing in disguise that nonliterate peoples around the world rely heavily on fermented drinks but have rarely perfected distillation. Although it is fallacious to ignore the alcoholic content of beer and wines, especially in an industrial society where production and distribution allow access to amounts that would be impossible in a context of domestic production, the prevalence of drinking problems is in approximate proportion to the consumption of distilled beverages. (My own work among the Camba of Bolivia [23] is an important exception, but only in the sense that they have no problems with the abundant rum they drink, and not because they have problems with beers or wines.)

The importance of alcohol as a drug is recognized by many peoples throughout the world. As a simple relaxant in small quantities, it is widely used and appreciated. There are some accounts of its being used, again in limited quantities, to prepare warriors for battle, as well as to refresh workers as they labor through a long day. In the early years of the fur-trade in northeastern North America, it appears as if the Iroquois used brandy as they had previously used self-imposed partial starvation, as a means of assuring that a young man achieve the hallucinatory "vision" that would be the basis of his personal link with supernatural powers throughout the rest of his life (24). The Tarahumara Indians of northern Mexico are said to be so modest that both men and women need to drink before they can engage in sexual intercourse, even with a long-term spouse. There are, of course, many groups in which drunkenness is sought as a means of "forgetting one's troubles," or "feeling strong," or "being a man." But the effects of drinking and drunkenness are so varied that the pharmacology of ethanol plays only a minor role. A large-scale historical and ethnographic review of the literature led a psychiatrist and his colleague to "conclude that drunken comportment is an essentially *learned* affair" (25, *p. 88*).

Another significant aspect of alcohol that tends sometimes to be overlooked in our own culture is the extent to which it is used as a symbol. It is for this reason that beliefs as well as behaviors, and meanings as well as uses, have been stressed throughout this chapter. Before citing a few examples from primitive cultures, however, we need to remember that, however secular and "scientific" our society may be, we often attach symbols that have no relation to the substance. One need not engage in a theological disquisition about the philosophy of transsubstantiation to appreciate that the wine used in the Christian ritual of communion has a very different significance from that used by college students on a picnic, and both of those differ markedly from that used by a "wino" huddled in a cold doorway.

There is no doubt that alcoholic beverages are often thought to be gifts of the gods. The ancient Aztec religion was explicit about this (26), and their contemporary Mexican descendants often refer to pulque, their agave-sap homebrew, as "the milk of the Virgin" (27). In a large number of religions, alcoholic beverages are favored offerings to the gods, and it is not uncommon for drunkenness to be viewed as a religious experience (28).

Just as alcohol sometimes mediates between human beings and supernaturals, it often has special meanings in relations between people. In some societies, protocol requires that a single container be used for drinking and that everyone present drink only in strict sequence according to social status or age, whereas in other communities it is important that all who are present drink simultaneously. Among various Pueblo Indian groups in the southwestern United States, alcohol is virtually tabooed for fear that it might unleash forces that would disturb the harmonious relations that are of transcendental importance in the relations among human beings and also between human beings and the rest of the world. By contrast, some other nearby tribes approve of drinking to the point of drunkenness because of the almost spiritual exhiliration that it affords. Although no society has formulated the rule explicitly, there is abundant evidence that drunkenness often provides a period of "time out" (25) in which many of the normal rules of social propriety are suspended, and people can "blow off steam" in ways that would otherwise not be allowed.

A number of other aspects of the interrelation of alcohol with human behavior deserve to be studied in world-wide perspective, but this is not the context for such an effort. Much of what is usually called "human nature" is not shared by more than a small group of human beings, and the several cultures that are still vital throughout the world reflect almost as many "human natures" as there are societies. It should be no surprise that conflict is by no means a common adjunct to drinking, or that aggressive behavior—whether verbal or sexual or punitive—is similarly unknown among drinkers in many areas. Maudlin sentimentality is rare as an outcome of drunkenness, just as is the view that a drunk is humorous. In other publications, I have reviewed this range of variation in some depth (29–31).

Cross-Cultural Studies

It may be more useful to summarize the general propositions that have emerged from large-scale surveys of alcohol-related behavior, and to show how the myriad details collected in many anthropological studies have yielded theoretical insights that are of interest to people in many other disciplines as well. In fact, it is ironic that anthropology has probably contributed more specific and explicit answers to the fundamental question, "Why do people drink?" than has any other field of study. We do not pretend to have found the golden key that will unlock the knotty problems that remain, but partial answers are probably better than none, even if those partial answers may subsequently be revised or qualified by further study. In a sense, that is the essence of the scientific method.

Probably the best known and most widely quoted statement on the subject is Donald Horton's: "the primary function of alcoholic beverages in all societies is the reduction of anxiety" (32, *p. 223*). That conclusion grew out of a detailed correlation of specific details on alcohol use and on various indices of anxiety, among the 56 societies for which there was adequate information in the early 1940s. Subsequent explorations of world-wide associations between drinking and other aspects of culture are less well known, although they reflect major improvements in research methodology as well as significant increases in both the quantity and quality of data available.

One such study used the same sample as Horton, but focused on elements of social organization (33) rather than anxiety. The degree of stability and formal structure of organizations, especially among corporate groups of kinsmen, was found to be directly correlated with the importance of sobriety in societies.

A similar approach applied to 139 societies, with meticulous attention to sampling, fullness of data, and other methodological issues that had been criticized in the earlier studies, lay special emphasis on the relationship between patterns of child-rearing and the expectations of normal adult behavior (34). Among many interesting conclusions, the one that drew most attention was the direct association between heavy drinking and drunkenness and discontinuity between childhood socialization in a dependent mode and demands for independence on the part of adults.

More recently a psychologist used thematic analysis of folktales from societies throughout the world in support of his thesis that "Men drink primarily to feel stronger" (35, *p. 334*), showing how different expressions of power are associated with rates of drinking.

Although specialists hotly debate the strengths and limitations of this kind of research, and we recognize that correlations do not necessarily imply causal relationships, it is noteworthy and appropriate that the nearest approximations to general statements about reasons for the widespread and enduring importance of alcoholic beverages are firmly based on evidence that comes from primitive societies as well as from our own.

Issues and Prospects

The original edition of *Alcohol, Science and Society* is probably the first context in which a broad cross-section of the alcohol constituency, including those interested in treatment, education, and prevention as well as in research, were given any systematic insight into the importance and implications of the fact that in other cultures they also drink. The present context seems appropriate for a brief outline of some of the key questions that remain and of some current work that holds promise for the future.

It is gratifying that the contribution of sociocultural comparison to alcohol studies has been so widely accepted. Recognition that different populations drink differently and have different attitudes toward drinking, drunkenness, and drunken comportment is probably general among those who deal with drinking and drinkers, in whatever capacity. At the same time, there are other immediate and practical lessons that can be learned and that may affect both attitudes and actions of those who are concerned about alcohol use and related problems.

One of the most basic among such lessons is so obvious as to seem almost banal when it is spelled out: the contemporary vernacular of "different strokes for different folks" contains an idea that is not only profoundly simple but at the same time simply profound. Among the few things that we have learned with any certainty in the course of studying alcohol is that a wide range of facilities, approaches, and treatments should be available so that choice among alternatives is possible. This is true not only among individuals but also, in a larger sense, among populations whose values and attitudes often differ markedly.

Although the fact may be considered patently obvious, it should be recognized that some who would agree at the verbal level do not behave in a way that fits with such a view. For example, it is fruitless to expect a man to admit that he is helpless in the face of alcohol if he has been reared from childhood to consider it unmanly and weak to act in any way other than assertively. Similarly, the kind of outspoken confrontation and frank group discussion that is so effective for some would be intolerable for others who decline to talk about anything that their relatives might consider shameful or embarrassing. Such basic attitudinal differences occur among large minority groups in cities and towns throughout the United States, and not merely among isolated tribes in remote areas.

Other clear implications can be found in the virtual universality with which drinking problems appear to be increasing among various populations around the world. It is evident that this trend cannot be linked to any single political or economic system.

There is realistic hope that more and better reporting and analysis will soon be available with respect to the ethnography of alcohol. Descriptions of drinking patterns in other cultures have been sketchy and imprecise, with even the most detailed reports tending to emphasize modal or "typical" patterns, giving us little idea of the range of variation that occurs within a population. Such variation may be unimportant in a few societies, but in others there may be significant differences on the basis of age, sex, social status, occupation, or some other criterion. It is also probable that in some other societies, as in our own, there are significant differences among individuals quite apart from any such categorical differences. Similarly, detailed data on quantity and frequency of intake are extremely rare, as are nutritional and chemical analyses of the beverages themselves. Other gaps in the data become apparent as anthropological studies of drinking are compared with other kinds of studies, and as ethnographic case-studies are used as bases to test hypotheses cross-culturally.

Until recently, all of the accounts of drinking in primitive societies were by-products of research that had very different aspects of life as their focal concerns. It is only within the past couple of years that a few anthropologists have set out with the express purpose of systematically investigating alcohol and its place in the lives of various peoples. In this light, one might view the corpus of ethnographic reporting as remarkably thorough and accurate—given the incidental nature of the data-gathering—even if it is not so comprehensive or precise as we might wish (36).

As alcohol and its interaction with human behavior become more widely recognized as appropriate topics for research by social scientists, and as more investigators are trained in ways that alert them to the transdisciplinary approaches and problems that are important in alcohol studies, we can expect more and better reporting to provide an increasingly relevant and useful body of cross-cultural information.

Even on the basis of our present limited information, however, there are a number of significant generalizations that deserve mention when we take a global overview of the meanings, uses, and effects of alcoholic beverages.

Probably the outstanding conclusion is that alcoholism—even in the very general sense of problems associated with drinking—is rare in most of the societies throughout the world, even among populations where drinking is an everyday occurrence and occasional drunkenness is common. This may relate to the fact that, in most societies, drinking is essentially a social act. As such, drinking is usually embedded in a context of values, attitudes, and other norms. To a significant extent, the effects of drinking are shaped by those values, attitudes, and norms, as well as by the social setting in which the drinking takes place. It is apparent that consistency in norms, however strict they may be, tends to result in better mental health and social integration than occurs with ambivalence or other inconsistency. Like drinking behavior, drunkenness is also affected by values, attitudes, and norms, to such an extent that drunken comportment is more shaped by learning than it is by the alcohol content of the beverage consumed. Despite the occasional occurrence of aggression and other troubles in social relations, the value of alcohol as promoting integration and facilitating social solidarity is emphasized by most peoples.

We know from experience that piecemeal borrowing of traits from one culture and trying to fit them into another can be difficult. But we also know from experience that the values, attitudes, and norms that have been associated with drinking in our own culture have resulted in grave and increasing problems for individuals and for society at large. For those who are concerned about prevention, it may be time for us to try to profit from the experience of other cultures.

REFERENCES

1. MCKINLAY, A. P. The "indulgent" Dionysius. Trans. Amer. Philolog. Ass. **70:** 51–61, 1939.
2. JELLINEK, E. M. (YAWNEY, C. D. and POPHAM, R. E., eds.) Drinkers and alcoholics in ancient Rome. J. Stud. Alc. **37:** 1718–1741, 1976.
3. ROLLESTON, J. D. Alcoholism in classical antiquity. Brit. J. Inebr. **24:** 101–120, 1927.
4. DANIELOU, J. Les repas de la Bible et leur signification. Paris; La Maison Dieu; 1949.
5. DROWER, E. S. Water into wine; a study of ritual idiom in the Middle East. London; Murray; 1956.

6. RAVI VARMA, L. A. Alcoholism in Ayurveda. Quart. J. Stud. Alc. **11:** 484–491, 1950.

7. COFFEY, T. G. Beer Street, Gin Lane; some views of 18th century drinking. Quart. J. Stud. Alc. **27:** 669–692, 1966.

8. RORABAUGH, W. J. The alcoholic republic; an American tradition. New York; Oxford University Press; 1979.

9. GUSFIELD, J. R. Symbolic crusade; status politics and the American Temperance Movement. Urbana; University of Illinois Press; 1963.

10. STIVERS, R. A hair of the dog; Irish drinking and American stereotype. University Park; Pennsylvania State University Press; 1976.

11. SNYDER, C. R. Alcohol and the Jews; a cultural study of drinking and sobriety. (Rutgers Center of Alcohol Studies, Monogr. No. 1.) Glencoe, Ill.; Free Press; 1958.

12. BALES, R. F. Cultural differences in rates of alcoholism. Quart. J. Stud. Alc. **6:** 480–499, 1946.

13. BLAINE, A., ed. Alcoholism in the Jewish community. New York; Commission on Synagogue Relations; 1980.

14. ZIMBERG, S. Sociopsychiatric perspectives on Jewish alcohol abuse; implications for the prevention of alcoholism. Amer. J. Drug Alc. Abuse, N.Y. **4:** 571–579, 1977.

15. BLUME, S., DROPKIN, D. and SOKOLOW, L. The Jewish alcoholic; a descriptive study. Alc. Health Res. World 4 (No. 4): 21–26, 1980.

16. BLANE, H. T. Acculturation and drinking in an Italian-American community. J. Stud. Alc. **38:** 1324–1346, 1977.

17. LELAND, J. Firewater myths; North American Indian drinking and alcohol addiction. (Rutgers Center of Alcohol Studies, Monogr. No. 11.) New Brunswick, N.J.; 1976.

18. WORLD HEALTH ORGANIZATION. EXPERT COMMITTEE ON MENTAL HEALTH. Report on the first session of the Alcoholism Subcommittee. (WHO Technical Report Series, No. 42.) Geneva; 1952.

19. HEATH, D. B. A critical review of the sociocultural model of alcohol use. Pp. 1–18. In: HARFORD, T. C., PARKER, D. A. and LIGHT, L., eds. Normative approaches to the prevention of alcohol abuse and alcoholism. (NIAAA Research Monograph No. 3; DHEW Pub. No. ADM-79-847.) Washington, D.C.; U.S. Govt Print. Off.; 1980.

20. WORLD HEALTH ORGANIZATION. EXPERT COMMITTEE ON MENTAL HEALTH. Second Report of the Alcoholism Subcommittee. (WHO Technical Report Series, No. 48.) Geneva; 1952.

21. MARSHALL, M., ed. Beliefs, behaviors and alcoholic beverages; a cross-cultural survey. Ann Arbor; University of Michigan Press; 1979.

22. GASTINEAU, C. F., DARBY, W. J. and TURNER, T. B. eds. Fermented food beverages in nutrition. New York; Academic Press; 1979.

23. HEATH, D. B. Drinking patterns of the Bolivian Camba. Quart. J. Stud. Alc. **19:** 491–508, 1958.

24. CARPENTER, E. S. Alcohol in the Iroquois dream quest. Amer. J. Psychiatry **116:** 148–151, 1959.

25. MacAndrew, C. and Edgerton, R. B. Drunken comportment; a social explanation. Chicago; Aldine; 1969.

26. Gonçalves de Lima, O. El maguey y el pulque en los códices mexicanos. Mexico City; Fondo de Cultura Económica; [ca. 1956].

27. Madsen, W. and Madsen, C. The cultural structure of Mexican drinking behavior. Quart. J. Stud. Alc. **30**: 701–718, 1969.

28. Washburne, C. Primitive drinking; a study of the uses and functions of alcohol in preliterate societies. New York; College and University Press; 1961.

29. Heath, D. B. A critical review of ethnographic studies of alcohol use. Pp. 1–92. In: Gibbons, R. J., Israel, Y., Kalant, H., Popham, R. E., Schmidt, W. and Smart, R. E., eds. Research advances in alcohol and drug problems. Vol 2. New York; Wiley; 1975.

30. Heath, D. B. Anthropological perspectives on alcohol; an historical review. Pp. 42–101. In: Everett, M. W., Waddell, J. O. and Heath, D. B., eds. Cross-cultural approaches to the study of alcohol; an interdisciplinary perspective. The Hague; Mouton; 1976.

31. Heath, D B. Anthropological perspectives on the social biology of alcohol; an introduction to the literature. Pp. 37–76. In: Kissin, B. and Begleiter, H., eds. The biology of alcoholism. Vol 4. Social aspects of alcoholism. New York; Plenum; 1976.

32. Horton, D. J. The functions of alcohol in primitive societies; a cross-cultural study. Quart. J. Stud. Alc. **4**: 199–320, 1943.

33. Field, P. B. A new cross-cultural study of drunkenness. Pp. 48–74. In: Pittman, D. J. and Snyder, C. R., eds. Society, culture, and drinking patterns. New York; Wiley; 1962.

34. Bacon, M. K., Barry, H., 3d. and Child, I. L. A cross-cultural study of drinking. Quart. J. Stud. Alc., Suppl. No. 3, 1965.

35. McClelland, D. C., Davis, W. N., Kalin, R. and Wanner, E. The drinking man. New York; Free Press; 1972.

36. Heath, D. B. and Cooper, A. M. Alcohol use and world cultures; a comprehensive bibliography of anthropological sources. (Addiction Research Foundation Bibliographic Series No. 15.) Toronto; 1981.

5

Drinking among American Indians

Edwin M. Lemert

American Indians perhaps more than any other population are celebrated in fact and fiction for their misadventures with alcohol. For the most part Indians living north of Mexico were aboriginally without liquors and hence had no cultural experience nor preparation for its effects when introduced to alcohol by European explorers, soldiers, trappers and traders. These effects, if we are to accept historical accounts by missionaries, early travelers and traders, were socially disruptive, disastrous to their welfare or both, following from the penchant of the Indians to drink to excess whenever liquor was made available, and to engage in violence when intoxicated. The idea became fixed that American Indians were given to intemperance and uncontrolled drinking. In time a "firewater myth" grew up conveying the idea that American Indians were constitutionally or even genetically prone to react differently to alcohol ingestion, in contrast to non-Indians.

More recently evidence has been cited to challenge the historical portrayal of orgiastic drunkenness among American Indians and to insist that the picture has been overdrawn. Thus MacAndrew and Edgerton (1), after winnowing historical accounts, took exception to the notion that "Indians can't hold their liquor" and showed that not all American tribes drank and that among those that did there were instances in which the people drank peaceably. Another review of the literature pertaining to the firewater myth also reveals considerable variation in the reports of Indian drinking behavior when it is separated into specific items or in response to particular questions (2). One of the better known exceptions to the firewater myth, native Americans who largely resisted the allure of alcohol by instituting methods for its control, were the Pueblo peoples of the Southwest.

It should be noted that MacAndrew and Edgerton did not actually challenge the assertions that American Indians drink excessively, i.e., to a state of drunkenness. They were more concerned in dispelling the idea that drunkenness is a form of disinhibition produced by alcohol itself: they argued that Indians do not necessarily turn to aggression or debauchery as the result of intoxication.

Actually the idea was well documented some years ago by Ruth Bunzel (3) in her classic study of the role of alcohol in two cultures of Central American Indians. Among one, the Chamula, aguardiente was drunk primarily for a sense of warmth, conviviality and to acquire a sense of irresponsibility. Aggression while intoxicated was minimal and no sexual promiscuity occurred. This stood in sharp contrast to the violence and sexual deviance associated with intoxication among the Chichicastenange with whom they were compared.

After weighing the discussions of these issues it must be said that whether American Indians are given to violence or venery when in their cups puts the questions about their drinking rather narrowly. This is particularly true when attention is shifted to some of the larger facts distinguishing their drinking from that of other groups. Thus, by statistical reckoning, Indians have an inordinately high crime rate, much above that for our population as a whole. Their death rate from liver disease greatly exceeds the national rate. Deaths from accidents and Indian homicide rate are also well above those for the nation. How much of the deviance or "social pathology" is attributable to the modes of alcohol consumption is a question not readily answered. However, the beliefs of many health officials and also of Indian leaders themselves that excessive drinking is among their most serious problems are not easily dismissed.

Trying to relate these over-all or aggregate representations, mostly statistical, to the actual patterns of drinking of modern-day Indians, keeping in mind the need to give the provenience of the data, is no easy task. In reviewing the relevant literature it is important to heed a number of cautions. One is that the data, while by no means fragmentary, nevertheless are scarcely comprehensive in the sense of covering the diversity and differentiation of American Indian peoples, ranging as they do from Iroquois structural steel workers in the East, to Navaho sheep ranchers in the Southwest, to Kwakiutl salmon fishermen on the Northwest Coast. Few studies if any give complete information on differential participation in drinking, not only by age, sex and social status, but even more simply by those who drink and those who do not, or by those who drink to intoxication and those who do not. Few investigations systematically compare the drinking behavior of Indians with that of Whites having similar occupational status, e.g., industrial workers, government employees, herders, loggers and fishermen. Care also must be taken in generalizing from drinking studies without reference to their historical context: drinking among Indians as among other groups undergoes change, particularly under present-day conditions of high mobility and increasing detribalization.

General Theoretical Considerations

The broadest approach to the study of Indian drinking will not exclude the possibility that genetic, constitutional or physiological differences may dispose Indians to some special vulnerability to the effects of alcohol. Even though attempts to discover a connection between racial traits and behavior have not been availing, there may be some such association between race and the susceptibility to disease. Hence, for those who see alcoholism as a disease, racial differences may be significant information.

Beyond such biological considerations are social and cultural factors commonly thought to influence Indian drinking. A number of studies by anthropologists give strong weight to cultural factors in their interpretations, so much so that possible physiological effects of differences in alcohol consumption are downplayed or ignored. Thus the reality of drinking tends to be limited to that which is defined by culture or cultural symbols (4).

This view omits at least one important consideration, namely, the fact that part of the differences found in patterns of drinking are differences in the amounts of alcohol consumed in given periods of time and the frequency with which these periods recur. The latter have direct significance when the research interest lies in the consequences of intoxication, such as accidents or liver impairment. The cultural approach in its extreme form also fails to speak to individual differences in indulgence in alcoholic beverages and it sheds almost no light on the growth of alcoholism.

As will be shown, there are a number of novel and inventive theories devised to explain Indian drinking patterns but most have the disadvantage of being based on studies of single cultures. As yet no theoretical model has appeared which incorporates the factors which researchers in the field have singled out as significant aspects of alcohol consumption in primitive settings, including the nature of alcohol's social and physiological effects, the pattern of drinking, its function and meaning, its social context, and the processes by which the pattern changes or is selectively followed by individuals.

The Patterns of Indian Drinking

Impressions of drinking by Indians are strongly influenced by its visibility, particularly among Indians of the western United States. Even so, it is still fair to say that American Indian drinking is distinctive and that it deviates substantially from accepted norms of drinking in the dominant White society. Much of the literature points to a pattern of "drinking to get drunk," passing out, sharing of bottles and peer pressure to drink. The Indian drinking group tends to be made up of men of similar ages. Consumption, often of cheap wine, is rapid "blitz" type drinking, continuing until the supply is exhausted or the drinkers pass out. Or, it may go on to become an extended binge. Drinking in and about automobiles or while driving is not uncommon among Indians of the Southwest. In the Northwest region drinking aboard fishing boats or in small town or village bars along the coast is common practice, as well as drinking on reservations.

On-reservation drinking by Indians occurs in connection with rituals, ceremonials, pow wows, and native dances in areas where these survive. Rodeos, loggers' competitions and soccer matches also become occasions for considerable drinking by Indians as well as by Whites. Small drinking parties in houses on reservations are also frequent. It is difficult to say if or how many Indians drink alone, but to the extent that it does happen it is atypical.

Indian drinking in towns and urban areas is done in bars, parks, back alleys, vacant lots, fields and public places. Skid Row-type bottle gangs are formed by Indians as well as by White drinkers. There is a perceptible pattern of exclusiveness in group drinking by Indians and a preference for drinking on their own "territory": congregating in certain bars, in fields or behind structures such as trading posts. Some bars, such as these found in Gallup, New Mexico, have a number of Indian regulars, "bar flies," who after heavy indulgence pass out and sleep in the vicinity. Such behavior, of course, contributes strongly to the stereotype of the "drunken Indian."

Participation in Drinking

Generally speaking, a lower proportion of the Indian population is currently drinking than is true of the national population. Some data for given tribes show as little as 42% of Indians drink compared to somewhere around 70% of the nation's adult population. At the same time, however, Indians have a much higher proportion of heavy drinkers than do other groups in our society. Indians also rank much higher than the rest of the population in the proportion who are abstainers: many, perhaps half, are not life-long nondrinkers but are former heavy drinkers who have given up alcohol. This may explain the lower rates of drinkers found in the Indian population as a whole.

A goodly proportion of Indian women also have a history of heavy drinking but have come to be abstainers, a proportion that may be higher than for men.

Typically, the Indian heavy drinker is a young male, unemployed, without job skills and without career plans. As he grows older in many instances he reaches a point at which he decides to abstain; apparently this is achieved without overwhelming difficulty or with less difficulty than is true for White problem drinkers.

As Indians become more acculturated, particularly those who are better educated and have steady employment in white-collar jobs, such as working for the government, they tend to take on the "White man's way of drinking." They drink quietly and moderately in bars and restaurants or they drink cans of beer beside their television sets in the evening.

Concomitants of Indian Drinking

One fairly well established fact is a very high arrest rate for Indians in comparison with other groups in American society, estimated to be 12 times that of the general population (5). The bulk of the arrests are for alcohol-related offenses: driving under the influence, drunkenness, disorderly conduct, and violations of the liquor laws. While Indians considerably exceed comparable rates for these offenses among the White population, the most outstanding difference is in their rates of arrests for public drunkenness. Indians also have a very high recidivism rate or "revolving door" pattern of arrests for drunkenness. Such arrests are higher than expected in less urbanized areas but they reach extremely high levels in urban centers. Thus in rural Humboldt County, California, 3 times as many Indians were apprehended for drunkenness in 1974 than expected but the rate was 26 times higher than expected for Indians in Sacramento County (6).

The arrest rate of Indians for serious "index" crimes runs about three times that of Whites and their rate of commitment to prisons is six times that of Whites, equalling that of Blacks (7, p. 141). Undoubtedly a high proportion of more serious crimes by Indians are alcohol-related but the nature of the connection between drinking and the commission of crime is unclear. But neither is it clear in the cases of alcohol-connected crimes committed by Whites and others. Some observers, such as Indian agents, have stated that practically all serious crimes by Indians are carried out while under the influence of alcohol but this is scarcely a sufficient explanation of why they commit them. In research among Southwest tribes no necessary relationship could be found between rates of homicide, suicide and levels of drinking; rates for the former have remained relatively constant over time despite changing rates of alcohol consumption (8, *p. 98 ff.*).

Intoxication may serve as a catalyst, precipitant or preparatory rationalization for homicide, suicide or engaging in violence, and possibly property crimes; however, the potential for deviance must already exist for alcohol to activate it and for violent crimes there must be victims available. The most that can be said is that drunken aggression is more likely to occur in some Indian cultures than in others. Likewise it does not happen randomly; as Kluckhohn (9, *p. 94 ff.*) noted years ago, drunken fighting occurs just where you would expect it to in Navaho social structure.

Excessive drinking contributes considerably more to health problems of Indians than in the general population. Age-adjusted death rates for cirrhosis of the liver among Indians in 1974 were 66.8 per 100,000, compared to only 12.9 in the White population, a ratio of nearly five to one. Deaths from accidents, particularly automobile accidents, were over 3 times more frequent among Indians than among Whites (10, *p. 33*).

Some reports have singled out the apparent association of Indian drinking with a variety of social problems: child abuse, infant mortality, domestic strife, loss of jobs, financial difficulties and psychiatric disorders. Yet there is no good way to assess the extent to which drinking may be a contributing factor, primarily because Indians have so many problems that it is a difficult or even impossible task to sort out their various interrelationships.

It should be noted that while Indians taken as a whole surpass Whites in their rates of alcohol-related problems, there is nevertheless great variation in such rates between regions and between tribes in the same region. Eastern Indians, such as the Cherokee and the Creeks, have much lower rates of alcohol problems than those in the west. But even in western Indian areas there are some tribes which have markedly lower frequencies of alcohol pathologies than those surrounding them. Death rates from cirrhosis of the liver are much higher in Hopi Indians than in Navaho, despite the fact that public drinking is much more common among the latter (8, *p. 100*). Access to a supply of liquor may very well affect such differences in death from liver disease, suggested by the fact that the rate tends to increase with the proximity of Indian tribes to towns.

Alcoholism

We have noted that there are relatively more heavy drinkers in the

American Indian population than is true for the nation as a whole. A Los Angeles sample survey (11) has shown that 50% of Indians representing 80 tribes were classifiable as moderate to heavy drinkers. Even more revealing was the finding that half of the men drinkers engaged in fights while intoxicated, experienced blackouts, and had trouble with the law. In a comparative study made in Colorado, Indians were found to consume 7 times as much alcohol as Anglos, to get drunk 7 times as often, and to have 6 times as many problems associated with their drinking (2, *p. 184*).

Although evidence is plentiful that American Indians more often than others have serious drinking problems, a question remains as to whether they suffer from alcoholism as it is commonly recognized and diagnosed in the White population: in the sense of addiction, insuperable craving, or inability to abstain from its use, or "gamma" alcoholism (13). Disagreement on the question hinges on what criteria are deemed critical in defining alcoholism. When behavioral criteria alone are employed, Indians admitted to hospitals and treatment clinics resemble White drinkers diagnosed as alcoholics. However, efforts to test and equate the sociopsychological features of Indian drinking pathologies with that of Whites have not been successful (14). The problem drinking of Indians appears to have a different sociocultural meaning from that of Whites. The apparent absence of guilt and complex attitudes towards their drinking, the lack of community rejection of the heavily drinking Indian, plus in many cases the ability to stop drinking without great difficulty (sometimes with the aid of religion) favors the idea that Indian alcoholism is qualitatively different from that of non-Indians (8).

Explanations of Indian Drinking

Racial

Some new credence has been given to the stereotyped idea that Indians are more vulnerable than others to the effects of alcohol following the discovery of the cutaneous flush among oriental peoples. Research has shown that Japanese, Taiwanese and Koreans have a higher incidence of flushing after consuming alcohol than do Caucasians (15). The flushing also has been found among Eskimos and American Indians, along with evidence that there may be differences in the rates of alcohol metabolism between American aborigines and Caucasians (16). Attempts to discover the biochemical correlates of the flushing phenomena have not yielded much conclusive data and flushing is not yet understood, although orientals in experiments have attested to some special psychological reactions to alcohol: dizziness, sleeplessness, anxiety, weakness, and warmth, which may be termed, if somewhat pedantically, diseuphoric. They are perhaps better summarized by a Chinese–Hawaiian woman abstainer who once told me that liquor made her "get red in the face, self-conscious and uncomfortable."

Flushing is not an insuperable obstacle to drinking; for example, a newly promoted Japanese industrial relations worker in Hawaii "practiced" for six months to learn to drink whisky at luncheons and conferences he was expected to attend. A further negative note is that Japanese do not seem to be uncomfortable when they become intoxicated drinking hot saki at teahouse parties.

In summary, current knowledge does not give much support to the hypothesis that there is some kind of racially induced vulnerability to the effects of alcohol. Defining and isolating racially homogeneous populations for study is difficult; American Indians have a large admixture of Caucasian and Negro genes in their gene pools, so much so that study samples cannot be assumed to be genetically similar. Furthermore, racially mixed Indians appear to share the same drinking behavior and problems as those who meet the criteria of "racial purity," and in actuality may suffer more problems. A final point hard to reconcile with a racial theory of vulnerability is that while American Indians and Chinese and Japanese are all mongoloid in race, the former have high rates of alcohol problems, while the latter do not.

Anxiety Reduction

The notion that alcohol is a kind of psychological reagent which dissolves inner restraints and "disinhibits" so that human beings more readily indulge in socially unacceptable behavior is of ancient vintage. A more modern and somewhat different version of this belief is that alcohol intoxication functions to reduce anxieties about threatening aspects of human existence. An early classic attempt to demonstrate a relationship between degrees and kinds of anxiety felt by primitive peoples and the extent of inebriation among them was made by Horton (17). Drawing on data from the Human Relations Area Files he looked for measures of association between anxieties about hunger, war and acculturation and the occurrence of drunkenness in a large number of primitive societies. More specifically he raised a secondary question of anxiety about the consequences of drunkenness, namely aggression and sexual promiscuity, assuming that the incidence of drunkenness would reflect some kind of counterbalance between the anxieties it assuaged and penalties for aggression and sexuality it released. No valid conclusions could be reached despite the theoretical inventiveness of the formulation.

Although Horton's theory was not borne out, anxiety about aggression may well be a motive for drinking to intoxication among native Americans. A number of writers, beginning with Hallowell (18, 19), have noted that the American Indians they studied were socialized by indirect methods to be patient, forbearing, polite and nonaggressive in interpersonal relations with their own tribal people; yet by cultural training they could be utterly savage and ruthless towards outgroup enemies. One result was that hostility towards their own tribe members induced anxiety which, when dissolved by intoxication, led to intragroup aggression.

While anxiety may help to explain inebriety among Indians in a large number of situations it is by no means a sufficient explanation by itself and in some cases it may not be involved at all. These explanations of intoxication tend to be psychological and reductionistic and, among other shortcomings, they fail to locate the sources of anxiety in social organization. A more comprehensive, structural explanation of heavy drinking was first proposed by Cheinisse (20) who advanced a Durkheimian view that alcoholism was a result of a lack of cohesion and solidarity in the religious organization of society. Much later Field (21) concluded from a cross-cultural study that loosely structured societies in which families and individuals were relatively isolated had more inebriety than more structured societies with strong corporate kin groups. A final study in this vein was the systematic effort carried out by Jessor et al. (12) to implement or test Merton's (22) conception of anomie and social structure in relation to deviant drinking. The study included a sample of Colorado Indians whose drinking and deviance was compared with that of Spanish-Americans and Caucasians. The basis of the study was field theory in which differential pressures and controls related to social status along with lack of normative consensus produced differences in deviant drinking, in this case highest for the Indians.

Generally, the anomie theory of deviance has been criticized for a failure to develop specific referents for anomie and the confusion of a structural condition with a psychological state. Also the theory is global, in that it makes a leap from an abstracted social structure to individual behavior without designating intervening processes that convert anomie into a drinking response. Snyder (23), after a careful review of the subject, concluded that while anomie may have some explanatory value it ignores the intervening importance of anxiety as a facilitating factor in drinking. Finally, as Heath (24) notes, there are societies such as the Navaho where heavy drinking tends to be associated with power, prestige and socioeconomic security.

Social Deprivation

The anomie theory of Indian drinking is closely akin to the social deprivation explanation of alcohol use by American Indians: that contacts and interaction of Indians with Whites and resultant social change have caused native institutions to attenuate and disappear, thus making it difficult for Indians to continue to express and fulfill their needs by traditional means. Alcohol and intoxication consequently become a means of compensating for the loss of cultural forms and meanings. Dozier (25) was one of the main proponents of this theory and as part of his argument he cited the Pueblo Indians as a relatively söber peoples. Ostensibly because they are not socially deprived, their institutions remain intact.

Drinking by Indians, according to the social deprivation theory, helps to promote social interaction in the absence of native culture. Hammer (26) and Curley (27) propose similar theories but show more specifically how intoxication lessens anxiety such that Indian men temporarily or in fantasy can reassert their lost social roles as hunters and warriors by aggressive acts or dominance over females.

Culturological

Cultural anthropologists are not likely to be very receptive to Durkheimian or narrow structural explanations of alcohol use and misuse by American Indians. They are more likely to favor such explanations as cultural symbols, values, stereotypes, traditions and socially transmitted patterns of behavior as influences shaping drinking of native peoples. Devereaux (28) for example, explained the absence of aggressive drunkenness among the Mohave Indians as being due to the fact that they "respect their own courage and do not admire the drunken bully." Mohaves pass out as a defense against unwanted aggression. Heath (29) showed how drunkenness was an integrated feature of Camba culture, expressing the high value the Camba gave to social interaction. Simmons (30) likewise gave importance to cultural values which drinking expressed in Peruvian mestizo culture. Finally, Lemert (31) demonstrated how ritualistic drinking and drunkenness became integrated features of the cultures of the Northwest Coast Indians, in many ways analogues of their feast and potlatch patterns.

One of the more provocative culturological theories applicable to Indian drinking is that of MacAndrew and Edgerton (1), who propose that inebriety tends to be patterned and is expressed within limits set by the culture and acquired by socialization. According to their theory, deviant behavior associated with intoxication must be understood as a product of culturally defining the situation as "time out" in which ordinary rules of conduct are suspended and deviance is licensed. In short, collective drunkenness is a kind of primitive corroboree. At the same time, the lack of sanctions against drunken depredations is part of the time-out explanation. The authors also imply that societies burdened with numerous rules and status restrictions are more likely than others to have a place for time-out alcohol indulgence. Hence the theory has some implicit structural and social control aspects which qualify its central idea that the definition of the situation accounts for drunkenness among native peoples.

The time-out hypothesis has more ramifications than are apparent on first reading. Among other things its authors contend that alcohol ingestion does no more than produce "sensorimotor incompetence" which is easily monitored and corrected or compensated. At the same time they seem to be saying something else when they state that drinking is a "warning signal to others that drunkenness and untoward behavior may follow."

Whether the existing data from cultural descriptions of Indian drinking are sufficient to conclude that alcohol ingestion affects the sensorimotor but not the higher cortical processes in human beings is debatable. Moreover, by calling drinking a warning signal MacAndrew and Edgerton may be referring to some of its more universalistic attributes, echoing Cavan's (32) statement that drinking is inherently precarious behavior. Elsewhere (33) I have expanded on this idea by proposing that alcohol intoxication alters the nature of human (symbolic) interaction. An important result of this is that the drinker interacts in a special psychosocial environment which attenuates or distorts the feedback process. I believe that this conception may help explain why it is more difficult to learn to drink in socially acceptable ways than it is to acquire other skills, and why, contrary to the MacAndrew and Edgerton thesis, it often is difficult for the individual to monitor his drinking.

I find it difficult to apply the time-out theory to explain the drinking patterns of contemporary American Indians. For one thing it is doubtful whether there are very many tribes or reservations with homogeneous tribal populations having similar culturally acquired conceptions of "drunkenness within limits." Moreover, the associated idea that American Indians do not hold each other accountable for aggressive acts committed when intoxicated is, for me, overly simplistic.

What seems more likely is that most American Indian groups do not have the means for enforcing accountability for drunken depredations by their own people and they are unlikely to resort to agencies of social control in the White community. It also is possible that this passivity is reinforced by the sober personality character- istics of many American Indians. However, this overlooks a rough kind of accountability that may operate when Indians are drunk: hostilities generated in the course of drinking may carry over from one drunken interlude to another.

I found that the Salish Indians with whose drinking I became most familiar had long memories of wrongs done them by others drunk or sober: an eye lost in a drunken brawl, a baby knocked out of its crib by a drunken father and burned on contact with the stove, a boy blinded by denatured alcohol given him by older men, a mother believed to have been pushed overboard from a boat at night by drunken companions and lost in tidal rapids. It was when they were themselves intoxicated that the Salish were most likely to con- front those who had aggressed against them and to seek their revenge.

Cultural Stereotypes and Indian Drinking

Some writers have dwelled on the instrumental function of the cul- tural stereotype of the "drunken Indian." One such theory combines elements of culturological theory with that of anomie and social dep- rivation associated with minority status (34). From this view getting drunk becomes one of several ways Indians can validate their "Indianness" or achieve success on their own terms in the face of an alien culture and a rejecting dominant society. One way is to become a respected person according to the ideals commonly accepted in the Indian community, by demonstrating dignity, respon- sibility, resourcefulness, respect for others and reciprocal generosity. Another alternative is to acquire and employ Indian expertise: knowledge of lore and dances, tracking and guiding for hunters and fishermen. Finally, Indians may assert their identities as leaders serving the Indian community and dealing with problems of living in a White world.

Indians unable to pursue these alternatives may elect to express themselves in stereotypical fashion by heavy drinking and acting the part of the "drunken Indian." This assumes that the Indian does not define his drunkenness as "bad" in the same way as Whites do, but

merely as a means to an end. Since Indians show no personal disapproval of drunkenness in others it becomes an acceptable vehicle
for protest against White domination, something further encouraged
by the tendency for law enforcement agencies to treat Indian drunkenness leniently.

This theory is a laudable move towards explaining some of
the differences of Indian participation in drinking but it overdraws
the facts to try to subsume the variety of Indian drinking patterns
under a single category of social protest. Drinking by Indians on
reservations and at parties in homes, sometimes in defiance of their
own morality and rules, is difficult to see as a symbolic protest against
White society. And, of course, to the extent that drinking still is
patterned or integrated with other activities in Indian culture an
explanation in terms of protest is not applicable.

Conclusion

The substantial evidence that drinking generally by American Indians is differentiated in form, function and meaning from other groups
in our population seems to favor adoption of a sociocultural model
for purposes of its explanation. But it is presumptuous to write as if
such a model has as yet been devised. Many if not most sociocultural
explanations of Indian drinking tend to be ad hoc conclusions about
drinking in particular Indian cultures. Taken together the variety of
elements they single out can become the basis of such a model, but
its construction remains to be accomplished.

Granting that social structure, patterned behavior, traditions
and cultural symbols help to shape the drinking of Indians, nevertheless certain of its aspects require attention to subjective processes
to be fully understood. These concern learning to drink, the evaluation of drinking experiences, individual differences with respect to
participation in drinking, increasing, controlling, or decreasing the
intake of alcohol, plus resolutions and action to abstain.

There is, of course, a considerable psychological literature
dealing with learning and the cognitive aspects of drinking behavior.
Unfortunately, not much of this has cross-fertilized the thinking and
research of observers of drinking among Indians and other native
peoples. The heavy commitment of psychologists to experimental
methodologies may account for the disinterest of social scientists in
their theories but this is scarcely an acceptable reason for failure of
the latter to make a place for cognitive processes in explanations of
Indian drinking. Some beginnings at remedying this omission have
been made in what is called variously "cognitive anthropology,"

"linguistic anthropology" or "ethnoscience" (35), but accomplishments in this new area at best are only rudimentary.

A final observation is that the usefulness of explanatory models developed by social scientists to enlighten native drinking will rest on the extent to which attention is given to the nature of alcohol and its physiology. MacAndrew and Edgerton (1) give some oblique recognition of this issue and recently Heath (29) has broken ranks with the more extreme culturologists by insisting that in explanations of drinking, "We cannot afford to ignore biology" i.e., "how the body reacts to alcohol." Madsen (36) speaks even more forcefully to the necessity to set aside the false mind–body dichotomy which balkanizes the thinking and research of alcohologists.

REFERENCES

1. MACANDREW, C. and EDGERTON, R. B. Drunken comportment; a social explanation. Chicago; Aldine; 1969.
2. LELAND, J. Firewater myths; North American Indian drinking and alcohol addiction. (Rutgers Center of Alcohol Studies, Monogr. No. 11.) New Brunswick, N.J.; 1976.
3. BUNZEL, R. The role of alcohol in two Central American cultures. Psychiatry **3**: 361–387, 1940.
4. MANDELBAUM, D. G. Alcohol and culture. Curr. Anthrop. **6**: 281–293, 1965.
5. STEWART, O. Questions regarding American Indian crime. Hum. Organiz. **23**: 61–66, 1964.
6. VALLO, D. Intertribal Council of California Rural Alcoholism Project. Sacramento, Calif.; 1974.
7. CRESSEY, D. Criminology. 10th ed. Philadelphia; Lippincott; 1978.
8. LEVY, J. L. and KUNITZ, S. J. Indian drinking; Navaho practices and Anglo-American theories. New York; Wiley; 1974.
9. KLUCKHOHN, C. Navaho witchcraft. Boston; Beacon Press; 1944.
10. U.S. INDIAN HEALTH SERVICE. OFFICE OF PROGRAM STATISTICS. Indian health trends and services—1974 ed. (DHEW Publ. No. 74-12009.) Washington, D.C.; U.S. Govt Print. Off.; 1974.
11. BURNS, M., DAILY, J. M. and MOSKOWITZ, H. Drinking practices and problems of urban American Indians in Los Angeles. Part I. Study description and findings. Preliminary report. Santa Monica, Calif.; Planning Analysis and Research Institute; 1974.
12. JESSOR, R., GRAVES, T. D., HANSON, R. C. and JESSOR, S. Society, personality and deviant behavior; a study of a tri-ethnic community. New York; Holt, Rinehart & Winston; 1968.
13. JELLINEK, E. M. The disease concept of alcoholism. Highland Park, N.J.; Hillhouse Press; 1960.

14. WESTERMEYER, J. Chippewa and majority alcoholism in the Twin Cities; a comparison. J. nerv. ment. Dis. **155:** 322–327, 1972.

15. WOLFF, P. H. Vasomotor sensitivity to alcohol in diverse Mongoloid populations. Amer. J. hum. Genet. **25:** 193–199, 1973.

16. FENNA, D., MIX, L., SCHAEFER, O. and GILBERT, J. A. L. Ethanol metabolism in various racial groups. Canad. med. Ass. J. **105:** 472–475, 1971.

17. HORTON, D. The functions of alcohol in primitive societies; a cross-cultural study. Quart. J. Stud. Alc. **4:** 199–320, 1943.

18. HALLOWELL, A. Z. Culture and experience. Philadelphia; University of Pennsylvania Press; 1955.

19. WHITTAKER, J. Alcohol and the Standing Rock Sioux. II. Psychodynamic and cultural factors in drinking. Quart. J. Stud. Alc. **24:** 80–90, 1963.

20. CHEINISSE, L. La race juive, jout-elle d'une immunité a l'égard de l'alcoolisme? Sem. méd. **28:** 613–615, 1908.

21. FIELD, P. B. A new cross-cultural study of drunkenness. Pp. 48–74. In: PITTMAN, D. J. and SNYDER, C. R., eds. Society, culture, and drinking patterns. New York; Wiley; 1962.

22. MERTON, R. K. Social theory and social structure. Rev. ed. New York; Free Press; 1957.

23. SNYDER, C. R. Inebriety, alcoholism, and anomie. Pp. 189–212. In: CLINARD, M. B., ed. Anomie and deviant behavior. Glencoe; Free Press; 1964.

24. HEATH, D. B. A critical review of the sociocultural model of alcohol use. Pp. 1–18. In: HARFORD, T. C., PARKER, D. A. and LIGHT, L., eds. Normative approaches to the prevention of alcohol abuse and alcoholism. (NIAAA Research Monogr. No. 3; DHEW Publ. No. ADM-79-847.) Washington, D.C.; U.S. Govt Print. Off.; 1980.

25. DOZIER, E. P. Problem drinking among American Indians; the role of sociocultural deprivation. Quart. J. Stud. Alc. **27:** 72–87, 1966.

26. HAMMER, J. Acculturation stress and the functions of alcohol among the Forest Potawatomi. Quart. J. Stud. Alc. **26:** 285–302, 1965.

27. CURLEY, R. T. Drinking patterns of Mescalero Apache. Quart. J. Stud. Alc. **28:** 116–131, 1967.

28. DEVEREUX, G. The function of alcohol in Mohave society. Quart. J. Stud. Alc. **9:** 207–251, 1948.

29. HEATH, D. B. Drinking patterns of the Bolivian Camba. Quart. J. Stud. Alc. **19:** 491–508, 1958.

30. SIMMONS, O. G. Drinking patterns and interpersonal performance in a Peruvian mestizo community. Quart. J. Stud. Alc. **20:** 103–111, 1959.

31. LEMERT, E. M. The use of alcohol in three Salish Indian tribes. Quart. J. Stud. Alc. **19:** 90–107, 1958.

32. CAVAN, S. Liquor license; an ethnography of bar behavior. Chicago; Aldine; 1966.

33. LEMERT, E. M. Paradigm lost or paradigm regained. J. Stud. Alc., Suppl. No. 8, pp. 34–45, 1979.

34. LURIE, N. O. The world's oldest on-going protest demonstration: North American Indian drinking patterns. Pp. 127–145. In: MARSHALL, M., ed. Beliefs, behaviors and alcoholic beverages; a cross-cultural survey. Ann Arbor; University of Michigan Press; 1979.

35. TOPPER, M. D. The cultural approach, verbal plans, and alcohol research. Pp. 379–402. In: EVERETT, M. W., WADDELL, J. O. and HEATH, D. B., eds. Cross-cultural approaches to the study of alcohol; an interdisciplinary perspective. Paris; Mouton; 1976.

36. MADSEN, W. Body, mind and booze. Pp. 217–225. In: EVERETT, M. W., WADDELL, J. O. and HEATH, D. B., eds. Cross-cultural approaches to the study of alcohol; an interdisciplinary perspective. Paris; Mouton; 1976.

6

Epidemiology: Alcohol Use in American Society

Don Cahalan

Alcohol, Science and Society did not have a chapter on the epidemiology of alcohol use, because scientifically-conducted quantitative studies of drinking behavior in America simply did not exist prior to that time. But the volume contributed materially to establishing the need for objective research on the drinking behavior of the general public, so that during the ensuing 35 years there have been a fair number of studies underwritten by the federal government. My objective here is first to mention some special problems in studying drinking behavior, and then to summarize some of the key findings and implications from a series of national surveys my colleagues and I have been conducting during the last 20 years.

Webster's Dictionary (Seventh New Collegiate version) defines epidemiology as "A science that deals with the incidence, distribution, and control of disease in a population." The word epidemiology as applied to alcohol reflects some difficulties in the ways in which our culture responds to the phenomenon of alcohol, because the term carries the surplus meaning that we must be talking about something that is "epidemic" in its spread and which therefore implies a real and present danger requiring much greater public attention and governmental funding for its prevention or control. Thus, bureaucrats like the term!

The emotional connotations of "the epidemiology of alcoholism" were driven home to me a few years ago when a small group of eminent researchers and federal program administrators met as an ad hoc group in Washington, under the auspices of Rutgers University, to try to define research and prevention objectives and strategies for their implementation. In no time at all, a couple of noted medical epidemiologists and a couple of sociologists who were specialists in alcohol studies were at swords' points, with the epidemiologists complaining that the sociologists must be in the pay of the alcoholic beverage industry because they failed to agree with the "obvious

conclusion from the facts" that alcohol should be prohibited altogether as a dangerous drug, and the sociologists' perceiving the
epidemiologists as being "rigid Carrie Nation prohibitionists." Fortunately cooler heads (especially Bob Straus, who was elected peace-
maker-chairman) papered over the ill-feeling and misunderstand-
ings; but these kinds of unnecessary conflicts continue to plague the
alcohol research field even today.

Our Social Research Group and its predecessors in Berkeley
and at George Washington University have been conducting interdisciplinary studies of drinking behavior for 20 years of a type that
might be termed "epidemiologic" because we use so many of the
tools of the epidemiologists. These include probability sampling of
surveys and longitudinal studies of the incidence and prevalence of
various types of drinking problems, and relating the findings to available statistics of social indicators of alcohol use to provide a picture
of American drinking practices. The studies began because of the
conviction of a small handful of researchers in the California State
Department of Public Health and in the National Institute of Mental
Health that objective studies of drinking behavior among the general
population were essential to dispel the misleading perspectives which
prevailed in a field populated primarily with concepts stemming
from clinical studies of generally "advanced-stage" alcoholics. I will
present a summary of a few of the findings which I think are the
most relevant for readers whose careers in alcohol research and evaluation are in the practicalities of planning and administering social
welfare programs in which alcohol is a ubiquitous ingredient.

Summary of U.S. Drinking Behavior Surveys

The series of national surveys on which I am reporting began in the
California Department of Public Health in 1960 under Wendell Lipscomb, Ira H. Cisin and Genevieve Knupfer, continued through
national surveys conducted by the George Washington University
under Professor Cisin in the mid-1960s and since by the Social
Research Group at the University of California in Berkeley. In all,
we conducted a dozen community and six national surveys. I will
focus my discussion primarily on the findings from our national surveys, particularly those reported in our three books (1–3), with
additional data comparing drinking behavior in 1967 and in 1979.[1]

[1]CLARK, W. B. and MIDANIK, L. Alcohol use and alcohol problems among U.S.
adults; results of the 1979 survey. Draft report to the National Institute on Alcohol
Abuse and Alcoholism. University of California Social Research Group, 1980.

Our survey program had three general phases. During the first few years, there was an emphasis on description of drinking practices and attitudes about drinking (1, 4, 5). Next there were a number of studies concentrating on specific problems related to drinking (2, 3, 6, 7). Current work concentrates on analysis of the latest (1979) survey and of longitudinal data on drinking behavior and problems within samples of individuals interviewed at two or more times over a span of years (1967–1969 vs 1973). All of the surveys have been strictly controlled (scientifically randomized) probability samplings of the general population aged 21 or older.

Drinking Behavior and Attitudes Toward Drinking

Historically, tabulations of U.S. beverage statistics (based on tax-paid withdrawals, not survey data on individuals) by Mark Keller and his Rutgers colleagues over the years show that there have been some profound changes in apparent per capita consumption over a very long-term perspective, but also lengthy periods of little change. Data from 1850 to 1976 show that consumption apparently increased sharply just after the turn of the 20th century and remained high until just before Prohibition began in 1920. Between Repeal (1933) and the beginning of World War II, however, total apparent per capita consumption rebounded to pre-1900 levels, where they remained for nearly 20 years. About 1960, total apparent per capita consumption rose significantly, showing a 30% gain between 1961 and 1971, but there has been virtually no change in total per capita sales (in terms of absolute alcohol) since 1971 (8, *p. 7*).

About 85% of the consumption of alcohol around 1850 was in the form of distilled spirits, whereas by the turn of the century about 50% was in the form of beer. The shift has been attributed not to the Temperance Movement so much as to heavy immigration of people from traditionally beer-drinking countries (9).

Related to such historical changes in the cultural backgrounds of our citizenry, our national surveys, as well as the work of other U.S. researchers (9–13),[2] consistently reflect clashes in values between U.S. groups (and within individuals) concerning drinking. The generally uninformed state of knowledge and the ambivalent attitudes about drinking found in our national surveys fit well with our national heritage, especially the classic conflict between the Protestant Ethic values and the hard-drinking behavior of a large number of our cit-

[2]LOUIS HARRIS AND ASSOCIATES, INC. American attitudes toward alcohol and alcoholics. Report prepared for the National Institute on Alcohol Abuse and Alcoholism, 1971.

izenry, from the time of the frontiersman to the contemporary suburban cocktail circuit. The prevalence of guilt feelings about drinking is reflected in our 1969 national survey which found that while 68% of the respondents reported that they drank at least once a year, 35% said that "nothing good" could be said about drinking, three-fourths (including a majority of the heavier drinking men) said that they thought drinking "does more harm than good," and three-fourths rated alcoholism as a serious health problem (1). Another instance of ambivalence about drinking was noted in our 1974 California survey, when vast majorities reported that they would like to see tougher treatment of drunken drivers and very expensive automobile liability insurance for heavy drinkers, but only small to medium-sized minorities would favor measures to reduce consumption of alcohol (such as closing bars earlier, increasing taxes, or rationing of alcoholic beverages) (14, *p.* 388).

In our first national survey of 1964–1965, we found the highest proportions of heavier drinkers to be among those about age 40, of lower social status, living in larger cities and in the Middle Atlantic, New England, and Pacific areas, and of Irish, British or Latin–American extraction (1). The groups with the highest proportions of drinkers, however, were not the groups with the highest proportions of heavier drinkers. For example, while Jews and Episcopalians had the lowest proportions of abstainers, they also had extremely low ratios of heavy or problem drinkers.

Abstainers were more likely to be older, of below-average income, from the South or rural areas, of native-born parentage, and from conservative Protestant denominations. Alcohol appeared to be a greater threat to those of lower economic and social status, possibly because these groups are more vulnerable and economically insecure than those of higher status.

In examining the social correlates of drinking we found that younger people and those of higher socioeconomic status reported that their parents drank frequently and approved of drinking, and that parents' permissiveness about drinking was generally correlated with the person's drinking. Heavy drinking among women was almost always associated with heavy drinking on the part of their husbands.

The respondents in this 1964 survey reported that, in general, alcoholic beverages were served more often when they met with people with whom they worked than when they were with their close friends. Drinking occurred most often with friends (including friends from work), next most often with family members, and least often alone. Even heavy drinking was "social drinking" in the sense that the solitary heavy drinker was relatively rare.

Recent Trends in Drinking Behavior

Two surveys, conducted a dozen years apart, provide an opportunity to note any marked changes in drinking practices in recent years. The first national survey which asked questions on both drinking practices and drinking problems was a randomized follow-up conducted in 1967 of about half ($N = 1359$) of a national probability sample interviewed in 1964–1965 about their drinking habits and attitudes (1, 2). A separate national survey of 1772 persons aged 18 and older covering most of the same items on drinking behavior and problems was repeated in 1979, permitting comparisons over a 12-year span.

The findings in the tables below are summarized from the preliminary 1980 report of Clark and Midanik[1] to the National Institute on Alcohol Abuse and Alcoholism.

Quantity and Frequency of Drinking

In both surveys, detailed questions were asked concerning the quantity and frequency of drinking wine, beer, and distilled spirits during the past year, and the following categories were constructed:

Abstainers: those who do not. drink alcoholic beverages or who drink them less than once a year.

Light Drinkers: those who may have three or four drinks on occasion of one beverage type—wine, beer or distilled spirits—but who never have more than one or two drinks per occasion of the other two beverage types. Light drinkers report that they drink on less than ten occasions per month.

Moderate Drinkers includes three distinct types: (1) those who never have more than three or four drinks per occasion of but one beverage (as is the case for light drinkers) but who do have more than ten drinking occasions per month; (2) those who may drink five

TABLE 1.—*Types of Drinkers among the Men and Women Respondents in the 1967 and 1979 National Probability Surveys, in Per Cent*[a]

	Men		Women		Total	
	1967 (N=751)	1979 (755)	1967 (607)	1979 (1003)	1967 (1358)	1979 (1758)
Abstainer	20	25	36	40	29	33
Light	23	20	39	31	32	25
Moderate	44	44	24	27	32	35
Heavy	13	12	2	3	7	7

[a]Percentages are weighted and may not total 100 due to rounding; slight variations occur because of nonresponse, etc. Respondents were aged 21–59 in the 1967 survey, 18+ in the 1979 survey.

or more drinks per occasion of but one beverage type, never having more than three or four of another regardless of the number of drinking occasions per month; (3) those who drink five or more drinks of at least two beverage types, but who report fewer than ten drinking occasions per month.

Heavier Drinkers: those who sometimes drink five or more drinks per occasion of at least two beverage types *and* who report that they drink on at least ten occasions per month.

Table 1 shows the proportions of the various types of drinkers, by sex, from the two surveys. In the 1979 survey, 25% of the men and 40% of the women said that they had not had any alcoholic drinks within the past year. The proportions compare closely with those from a number of surveys conducted between 1946 and 1975, as summarized by Hyman et al. (15).

The differences between the levels of self-reported drinking in the two surveys were very small indeed, with slightly more abstainers and moderate drinkers and fewer light drinkers in 1979 as compared to 12 years earlier.

Definitions of Drinking Problems

Because both national surveys had similar problem measures, responses may be compared on nine scales (Table 2). A positive score on any scale indicates respondents reported that some item on that scale happened to them within the last year.

TABLE 2.—*Prevalence of Drinking Problems Reported in the Past 12 Months, among Men and Women, 1967 and 1979 National Probability Surveys, in Per Cent*[a]

	Men		Women		Total	
	1967 (N=751)	1979 (762)	1967 (608)	1979 (1010)	1967 (1359)	1979 (1772)
Health	6	4	5	2	5	3
Belligerence	5	8	3	4	4	6
Friends	2	3	b	1	1	2
Symptomatic drinking	11	20	5	9	8	14
Job	3	7	2	2	2	4
Law, police, accidents	1	2	b	1	b	1
Binge drinking	1	1	b	b	1	1
Psychological dependence	49	26	29	17	37	21
Spouse[c]	1	2	0	b	1	1

[a]See footnote *a*, Table 1.
[b]Less than 0.5%.
[c]Problems in the last 2½ years in the 1967 survey, 3 years in the 1979 survey.

Health Problems. Respondents who reported that drinking had been harmful to health or that a physician had advised him or her to cut down received a positive score on the health scale. While the differences are not statistically significant, fewer men and women reported any health problem connected with drinking in 1979 as compared with 1967.

Belligerence Associated with Drinking. Getting into a heated argument after drinking, as an indicator of belligerence, is slightly higher among respondents in the 1979 than in the 1967 survey (6 and 4%, respectively). The difference is more apparent among men than among women.

Problems with Friends. Having one's friends suggest that the respondent cut down on drinking was experienced by very few respondents in both surveys. As noted by Cahalan and Cisin (16), problems with friends may be reduced by some heavy drinkers because they tend to associate primarily with people who condone their drinking behavior.

Symptomatic Drinking. Items within this scale refer to signs of physical dependence and loss of control suggestive of Jellinek's gamma alcoholism: drinking to relieve a hangover, having difficulty in stopping drinking, having blackouts or lapses of memory, skipping meals while drinking, tossing down drinks quickly to achieve a quicker effect, sneaking drinks and drinking before a party to make sure one has enough alcohol. In the total samples, reports of symptomatic drinking increased from 8 to 14% in 1979. Among men, the increase was dramatic, with reports of symptoms increasing from 11 to 20%. Less of an increase was found among the women in the 1979 survey. Whether the changes are real or due to sampling variation cannot be answered at this point.

Job Problems. Five items describe problems associated with the job which directly relate to alcohol use—having lost or nearly lost a job because of drinking, having co-workers suggest that one cut down on drinking, feeling that drinking may have hurt chances for promotion, raises or better jobs, staying away from work or going to work late because of a hangover, and getting "high" or "tight" while on the job. The prevalence of job problems in both the 1967 and 1979 surveys is low: 2 and 4%, respectively. The slight increase in the latter is due primarily to the men respondents (3 to 7%).

Problems with Law, Police and Accidents. This scale is composed of four questions, which include arrests for drunken driving, arrests for intoxication, being involved in an accident in which the respondent was hurt, and being in an accident in which someone else was hurt or property was damaged. Less than 1% of the respondents in either survey reported problems in this area.

Binge Drinking. Related to Jellinek's concept of loss of control, this scale measures the prevalence of being intoxicated for several days at a time at least once in the last 12 months. Very few respondents in either survey can be characterized as binge drinkers.

Psychological Dependence. Drinking to alter a mood (to help nerves, to forget, to forget worries, to cheer up, to reduce tension and nervousness) comprises the psychological dependence scale. Thirty-seven per cent of the total sample in 1967 reported having had at least one of these experiences in the last 12 months. Fewer respondents (21%) reported any positive response to this scale in 1979. The decrease is evident in both the men and women respondents; however, the men reported more use of alcohol to change a particular mood.

Problems with Spouse. This scale refers to having had a spouse leave or threaten to leave the respondent during the last 2½ years in the 1967 survey or during the last 3 years in the 1979 survey. Very few people (1%) interviewed in either survey reported such an occurrence.

A Combined Typology of Alcohol Use and Alcohol Problems

Several summary tables below show alcohol use and some less severe alcohol problems by various sociodemographic characteristics of the survey population. The emphasis is on variations in drinking and problems among different segments of the population and not on the rates of more severe, and much more rare, problems due to drinking.

The measure of amount of drinking (Quantity–Frequency scale) used in this section was constructed from the combined number of reported drinks per month of beer, wine and distilled spirits. Beginning with the largest amount of each beverage the respondent could recall drinking on an occasion during the past 12 months, we asked the frequency of drinking that large a quantity (and all lesser quantities down to the minimum of 1 or 2 drinks per occasion). This scale has the advantage of being readily understood and of placing all cases on a continuum of consumption ranging from zero to over 600 drinks per month. The greatest disadvantage of the measure is that it combines people who drink the same number of drinks per month but who may drink in greatly differing ways: thus the drinker who has 2 drinks per night is categorized with the drinker who has 14 drinks on one night. For some purposes this collapsing of disparate patterns is to be avoided (6).

Different cutting points were used for different purposes with this scale. Initially, however, we divided the data on alcohol use over the past 12 months into three categories: (1) Those who did not drink alcoholic beverages during that time; (2) Those who reported up to 60 drinks per month; and, (3) Those who reported 60 or more drinks per month.

We also constructed measures of problems related to drinking. The scales should not be taken as direct indicators of rates of "alcoholism" because fairly low cutting points in the total range on each score were chosen here to illustrate the variation across groups.

The Social Consequences Scale includes reports that, in the past 12 months, one or more of the following events occurred due to drinking: (1) A spouse, a friend, or a relative of the respondent either threatened to break off the relationship, or actually did so, because of the respondent's drinking, or friends advised the respondent to cut down on drinking; (2) Police questioned or warned the respondent about drinking, or the respondent was arrested for drunkenness or drunken driving; (3) Drinking contributed to the respondent being involved in an accident (automobile or other) in which someone was hurt or property was damaged; (4) People at work indicated that the respondent should cut down on drinking, or the respondent felt that drinking had cost him or her a chance at a raise, a promotion, or a better job, or the respondent had lost a job, or nearly lost one, because of drinking.

The Alcohol Dependence Scale refers to a report of two or more of the following during the past year: skipping meals when drinking; sneaking drinks; morning drinking; drinking before a party to "get enough"; blackouts; gulping drinks; hands shaking after drinking. This scale is not identical to the symptoms scale discussed in the previous section, thus rates based on the two are not comparable even though some of the same items are used in both scales.

The Loss of Control Scale refers to reports during the past 12 months of one or more of the following: fear that one was alcoholic; attempts to cut down or quit drinking, but being unable to do so; sometimes kept on drinking after promising not to; found it difficult when drinking to stop before becoming completely intoxicated.

For convenience, positive responses to either or both the Loss of Control Scale and the Dependence Scale are combined in the accompanying tables.

Table 3 presents the results on amount of drinking and alcohol problems by sex and age groups. First note that abstention is not uncommon in the U.S. population; again, 33% of the total, 25% of the men and 40% of the women, did not drink in the 12 months preceding the interview (and recall that the proportions have not changed greatly in recent years). The body of the table notes what has been found in numerous surveys: the proportion of abstainers is greater among older than younger age groups, and is relatively higher among women than men.

As expected, young men include relatively more heavier drinkers among their number than do older men. Among women the same pattern may be seen, but in an attenuated form. Note that the rates of social consequences and loss of control and dependence follow the same pattern: men more than women and the young more than the old report these problems. As noted by Clark and Midanek,[2] Blane (17) contains a discussion of this often observed pattern. The popular idea that most kinds of alcohol problems are most common among the middle aged is not supported by surveys of the general population, although it is supported by surveys of clinic populations.

The right hand column of Table 3 shows the proportion of drinkers in each age and sex category who reported some degree of loss of control or dependence during the past 12 months. (This increases the percentages in all categories in proportion to the number of abstainers in the particular group.) Women drinkers aged 18 to 20 show the most marked difference—nearly a quarter report dependence or trouble controlling drinking by these measures, a proportion which is exceeded by only three of the age groups among men, and by none of the other age groups among women.

Tables 4 and 5 contain data on drinking and on problems within categories of family income and respondents' education. Cahalan (2), Cahalan and Room (3) and earlier work by Knupfer and Room (18), Mulford (19) and Bailey et al. (20) have shown in several populations that income and education are related to amounts of drinking and to problems associated with drinking; however, the patterns found vary somewhat, perhaps due to differences in the measures of social class variables. Generally, high rates of abstention from alcohol are found more often in the lower income and education groups. Problems due to drinking are not quite parallel, since somewhat higher rates of problems are found than would be expected given the proportions of heavy drinkers. Some of this may be seen

TABLE 3.—*Drinking Patterns and Drinking Problems, by Age and Sex, 1979 National Probability Survey, in Per Cent*[a]

Age	N	Abstainers	1-60 Drinks/Month	60+ Drinks/Month	Social Consequences	Loss of Control or Dependence	Loss of Control or Dependence (% Drinkers)
Men							
18–20	37	5	79	17	15	35	37
21–25	82	10	54	36	13	25	28
26–30	87	20	50	29	10	25	32
31–40	154	25	55	19	8	16	21
41–50	107	27	52	21	2	8	11
51–60	130	32	51	17	3	5	8
61–70	91	38	53	8	5	6	11
70+	72	41	45	13	4	2	4
Total Men	762	25	54	21	7	15	20
Women							
18–20	52	31	64	5	5	16	24
21–25	130	15	78	6	6	13	16
26–30	125	30	65	5	3	7	10
31–40	208	27	65	9	5	8	12
41–50	137	43	46	10	4	5	9
51–60	143	50	46	4	1	4	8
61–70	102	61	38	1	0	0	0
70+	103	61	39	0	0	0	0
Total Women	1010	40	54	5	3	6	10
Totals	1772	33	54	13	5	10	15

[a]Percentages are weighted and may not total 100 due to rounding, nonresponse, etc. From Clark and Midanik.[1]

TABLE 4.—*Drinking Patterns and Drinking Problems by Family Income and Sex, 1979 National Probability Survey, in Per Cent*[a]

Family Income[b]	N	Abstainers	1–60 Drinks/Month	60+ Drinks/Month	Social Consequences	Loss of Control or Dependence	Loss of Control or Dependence (% Drinkers)
Men							
Under $4000	32	37	44	19	17	21	34
$ 4000–5999	37	39	44	17	8	13	21
6000–7999	46	26	49	25	12	18	24
8000–9999	63	43	36	22	11	15	26
10,000–14,999	130	31	51	18	8	11	16
15,000–19,999	117	21	53	25	9	22	28
20,000–24,999	79	16	63	21	1	19	23
25,000–29,000	53	26	61	13	2	8	11
30,000–34,999	55	15	51	34	5	6	8
35,000–39,999	23	16	64	19	3	8	10
40,000+	60	5	76	19	5	14	15
Women							
Under $4000	97	68	32	c	1	7	22
$ 4000–5999	63	49	42	9	0	6	12
6000–7999	70	45	49	6	7	10	19
8000–9999	88	37	62	1	c	3	4
10,000–14,999	154	43	51	6	3	4	7
15,000–19,999	124	36	58	6	3	7	11
20,000–24,999	106	34	58	8	3	3	5
25,000–29,000	70	28	67	6	4	3	11
30,000–34,999	31	26	56	18	0	8	11
35,000–39,999	19	5	91	4	9	13	13
40,000+	52	13	83	4	5	16	18

[a]See footnote *a*, Table 3.   [b]Before taxes.   [c]Less than 0.5%.

TABLE 5.—*Drinking Patterns and Drinking Problems by Education and Sex, 1979 National Probability Survey, in Per Cent*[a]

Education	N	Abstainers	1-60 Drinks/Month	60+ Drinks/Month	Social Consequences	Loss of Control or Dependence	Loss of Control or Dependence (% Drinkers)
Men							
Less than 7 years	49	50	38	12	5	7	15
Some junior high (7–9 years)	99	34	50	16	8	11	17
Some high school (10–11 years)	94	25	50	25	13	17	22
High school graduate	229	26	51	23	8	15	21
Some college	130	18	63	19	6	21	25
College graduate	70	14	63	22	1	11	13
Some graduate work	24	14	66	20	0	13	16
Graduate degree	53	12	68	20	0	8	10
Women							
Less than 7 years	57	88	9	4	8	8	(5)[b]
Some junior high (7–9 years)	128	56	42	3	1	1	2
Some high school (10–11 years)	136	50	40	10	4	10	19
High school graduate	347	36	59	5	3	6	9
Some college	191	22	72	6	3	10	13
College graduate	82	21	76	3	0	3	4
Some graduate work	24	11	78	11	0	7	16
Graduate degree	26	20	68	12	0	0	9

[a]See footnote *a*, Table 3.
[b]Too few cases to calculate percentage; *N* is shown in parentheses.

in Tables 4 and 5 but, because of the small number of cases in some categories, the findings are tentative. The proportion of heavier drinking increases in concert with education among the men; it is less discernable among the women. The rates of social consequences are lower among those with more education, but the difference is not significant, nor is the difference in loss of control or dependence between the lower and higher educational categories. Higher income is related to amount of drinking only in regard to the number of abstainers. No other relationship between heavier drinking (column 3) and income is apparent in either sex in these data.

Table 6 shows drinking patterns and problems among the nine census areas of the country. Substantial regional differences in drinking patterns have been a common finding in past surveys. Cahalan and Room (3) contains a detailed exposition of these differences, which are related to the traditional "dryness" and "wetness" of each area. There are several explanations for the observed differences: regions differ in their relative proportions of ethnic groups which differ in drinking patterns. Religious affiliation is also differentially distributed, with more abstaining groups represented in southern and eastern regions of the country. Differences in the "dryness" of an area are at least potentially an influence on those who are non-drinkers mainly because the laws and regulations controlling the purchase and use of alcoholic beverages, as well as in the prevailing behavior within an area, may restrict the number of opportunities to use alcoholic beverages.

The data in Table 6 illustrate the relative "wetness" of the northern, eastern, southern and western states. But note also that rates of problems due to drinking do not parallel the proportions of nondrinkers. As found in other national surveys, the rates of problems among those who do drink is quite high in the southern areas, which we commonly associate with "dryness." An unexpected finding is the higher proportion of drinkers (of both sexes) with problems living in the mountain regions. In our earlier study (3) the mountain region was somewhat in the middle range; however, both surveys have small (and thus less reliable) samples for the region.

Drinking and Demographics

Drinking patterns grouped by *religious affiliation* (Table 7) did not always show, in 1979, the patterns found in earlier studies. For instance, past research leads us to expect that there will be few abstainers, and even fewer drinkers with problems, among Jews

TABLE 6.—*Drinking Patterns and Drinking Problems by Geographic Region and Sex, 1979 National Probability Survey, in Per Cent*[a]

Geographic Region	N	Abstainers	1–60 Drinks/Month	60+ Drinks/Month	Social Consequences	Loss of Control or Dependence	Loss of Control or Dependence (% Drinkers)
Men							
New England	43	8	66	27	6	10	11
Middle Atlantic	138	22	57	21	6	8	10
East North Central	133	24	54	22	6	14	18
West North Central	64	32	47	21	6	21	30
South Atlantic	128	36	52	12	6	7	11
East South Central	49	49	33	18	7	17	40
West South Central	57	22	55	23	3	26	34
Mountain	41	35	40	25	16	30	47
Pacific	109	10	66	24	10	21	23
Women							
New England	58	25	67	8	0	8	11
Middle Atlantic	174	29	66	5	3	7	9
East North Central	165	35	56	8	2	7	11
West North Central	75	44	53	3	1	4	8
South Atlantic	181	59	40	b	1	2	4
East South Central	75	76	21	3	2	3	11
West South Central	95	47	45	8	5	8	16
Mountain	40	41	46	13	11	18	30
Pacific	147	20	74	6	5	7	9

[a]See footnote *a*, Table 3.
[b]Less than 0.5%.

TABLE 7.—*Drinking Patterns and Drinking Problems by Religious Affiliation and Sex, 1979 National Probability Survey, in Per Cent*[a]

Religious Affiliation	N	Abstainers	1–60 Drinks/Month	60+ Drinks/Month	Social Consequences	Loss of Control or Dependence	Loss of Control or Dependence (% Drinkers)
Men							
Catholic	187	9	61	31	8	16	18
Liberal Protestant	139	24	53	20	6	14	19
Protestant Fundamentalist	325	36	50	14	7	14	22
Jewish	20	39	55	6	1	1	2
None	49	18	60	22	2	15	18
Women							
Catholic	256	23	69	8	5	7	10
Liberal Protestant	199	32	63	5	3	7	10
Protestant Fundamentalist	478	57	39	4	2	5	11
Jewish	24	28	68	4	0	12	17
None	32	21	70	9	8	15	20

[a]See footnote *a*, Table 3.

(1, 3).[3] While the 1979 survey indicates moderate to low rates of problems, however, the proportion of Jewish abstainers (39% of the men, 28% of the women) is higher than expected.[4] In 1979, as expected, the proportion of abstainers is quite high among the Fundamentalist Protestant groups, with Catholics, Liberal Protestant groups and those with no religious affiliation containing relatively fewer numbers of abstainers and relatively higher proportions of heavy drinkers. Among the Fundamentalist Protestants who do drink, however, the proportion of drinkers with problems is quite high; as high or higher, in fact, than have been found among traditionally heavy drinking groups (3). The pattern emerges in Table 7, but is not as clear as one might wish.

Ethnic identification was obtained by asking respondents: "What country did your ancestors come from?" In Table 8 this crude measure does produce some of the results we have been led to expect from past studies. The Irish include a larger-than-chance proportion of drinkers with problems. (Past work by Cahalan and Room [3] suggests that further dividing the Irish into Protestants and Catholics will show the Catholics to have the higher problem rate of the two.)

In the present survey, those who identify with Italy have low rates of abstinence and low rates of drinking problems as well. Those who identify with Great Britain and with Germany tend to have rates of both drinking and drinking problems which are intermediate between the groups mentioned above.

Changes in Drinking-Problem Status over Several Years

Longitudinal studies are crucial to understanding cause–effect relationships because only with studies of individuals at two or more points in time is it possible to determine which events and states of mind preceded which later events. Accordingly, a primary objective of this series of surveys by the Social Research Group has been to assess the correlates of change in drinking practices and problems over time.

A probability sample of San Francisco men aged 21–59 initially interviewed in 1967 and reinterviewed in 1972 first made it possible for us to measure detailed changes over a period of several years (7). The traditional expectations based on popular beliefs of

[3]CLARK, W. B. Contextual and situational variables in drinking behavior. University of California Social Research Group staff paper, May 1977.

[4]But the finding may be due to sampling error.

TABLE 8.—*Drinking Patterns and Drinking Problems, by Country of Origin and Sex, 1979 National Probability Survey, in Per Cent*[a]

Country of Origin[b]	N	Abstainers	1–60 Drinks/Month	60+ Drinks/Month	Social Consequences	Loss of Control or Dependence	Loss of Control or Dependence (% Drinkers)
Men							
United States	154	30	52	18	9	17	24
England, Scotland, Wales	113	31	52	17	10	14	20
Ireland	62	25	36	38	7	20	26
Italy	50	6	66	28	0	3	3
Eastern Europe	53	16	71	13	2	9	11
Germany	113	24	57	19	4	14	18
Women							
United States	220	44	52	4	3	7	12
England, Scotland, Wales	143	44	51	5	2	5	9
Ireland	112	28	67	5	7	8	12
Italy	56	36	64	0	0	1	2
Eastern Europe	72	30	63	7	3	10	14
Germany	148	31	61	8	2	9	14

[a]See footnote *a*, Table 3.
[b]The country which most ancestors came from. Only those countries selected by 15 or more respondents of each sex are reported.

Jellinek's models (21) is that alcoholics will accumulate increasing numbers of problems over time, with the "early symptoms" of progressive alcoholism not being replaced by later symptoms, but being merely added to problems already existing.[5] The San Francisco findings, however, were at variance with the conventional expectations, because there was not found to be much "snowballing" of problems with the passage of time. While the proportion of respondents who no longer had the same problem at Time 2 (reinterview) was quite high, ranging from 50 to 96%, people with *any* specific drinking problem at Time 1 tended to have *some* type of problem at Time 2, although not necessarily the same problem. In addition, those with specific problems at Time 1 tended to have high rates of heavy intake or binge drinking at Time 2.

We conclude from this San Francisco study that continuity of specific problems over time is rather low, but that the probability of future involvement in some alcohol problems—but not necessarily the same ones—is increased if alcohol-related problems develop. Thus the "progressive disease" concept of alcoholism is open to question. The fact that those who have drinking problems at one time tend to have drinking problems of varying kinds at a subsequent time, however, may imply that environmental factors play a considerable part in determining the type of problems that may occur.

A national probability sample measuring change over 4 years was also conducted as part of this same series. In this survey, 725 men aged 21–59 were first interviewed in 1969 and again 4 years later. The findings bore out the findings in the San Francisco survey of a high turnover or change in drinking problems: about as many respondents shifted into, or out of, a drinking-problem status as stayed in a drinking-problem status (22).

Our analysis of the longitudinal data is still in progress, but we do have sufficient evidence to conclude that the "symptom" or "prodromal" status of specific drinking problems as predictors of later and more serious problems has been considerably overrated, since the data suggest that an "early warning" prediction based on such "symptoms" will yield a substantial number of false positives among respondents who will never get into serious trouble. We believe that the high rates of turnover in drinking problems will be found to be associated with specific environmental circumstances and life events; and we are now conducting an intensive analysis of

[5]ROOM, R. Assumptions and implications of disease concepts of alcoholism. Presented at the 29th International Congress on Alcoholism and Drug Dependence, Sydney, Australia, February, 1970.

the impact of life events or environmental circumstances on changes in drinking problems.

Considerations for Future Research

Much useful research has been done during the years since the first edition of *Alcohol, Science and Society*, much of which has been summarized between the covers of this new edition. I hope the next 35 years will see rapid increases in collaboration among behavioral scientists, and between behavioral and biomedical scientists, in analyzing the environmental aspects of alcohol use, including more work along the following lines:

1. Cross-Cultural Studies. A paper written by my colleague and successor, Robin Room, for the World Health Organization's report, *Alcohol-Related Disabilities* (22) notes the similarities and differences in drinking patterns between the U.S. and other countries, by sex, social class, urbanization, and geographic region. He makes a very strong case for more longitudinal studies of drinking practices to assess the etiology of drinking problems on an international basis.

2. Special Studies of Youth Aged 12–17. Thus far, although Blane and Hewitt (23) have summarized a lot of piecemeal surveys of youthful drinking, there has not been one single across-the-board national probability sampling of youthful drinking habits. The formation of drinking and nondrinking habits at this age is obviously crucial to drinking behavior later in life; but there has been no broad-gauge study of youthful drinking habits, even though there have been many national surveys with much more sensitive questions on the use of other drugs.

3. Research of the Positive or Pleasureable Aspects of Alcohol Use. Many behavioral science research projects begin with the intention of objectively describing all aspects of drinking behavior; but before long the emphasis gravitates almost exclusively to the negative or hurtful effects of alcohol. If there is any truth to the adage that we need to "know the enemy" to deal with him, we need to know a great deal more about how alcohol promotes sociability and relaxation, and how it makes the economic wheels go round, if the society is going to be able to intervene to direct some of alcohol's seductiveness into more constructive channels.

4. Examination of the Alcohol-Related Health and Social Welfare System. These studies could be designed with more of an eye to understanding the interaction of social and economic ecologies involved in the treatment of alcoholism and the employment of social services to deal with the consequences of excessive use of alcohol, than merely to attempting to assess which types of services seem to be efficient or inefficient. Parkinson's Law implies that health and welfare services that can be utilized will be over-utilized; but there does seem to be something more than Parkinson's Law operating in such phenomena as the "silting up" of detoxification centers with alcoholic recidivists, as described in the excellent preliminary research by Dr. Alex Richman at Mount Sinai Hospital in New York.[6]

5. *Studies of The Dynamics of Organizations and Institutions in the Alcohol Field.* The rapid growth and casualties in this controversial field have been many over recent years and a systematic study should help us do a better job in the future. In a recent report (24) I summarize what seem to be some of the main factors operating on organizations in the alcohol field these days; and Carolyn Wiener (25) has dealt with many similar concepts about alcohol-related organizational behavior. These preliminary efforts need to be supplemented by intensive, in-depth studies of a wide range of alcohol-related organizations. Ideally, in view of the controversial nature of such a study of politically powerful groups, the research should be funded by independent foundations rather than by either the government or the alcoholic beverage industry. Such studies should be attractive to foundations that are interested in facilitating the growth of organizational behavior research in general, which has been a sadly under-developed field.

REFERENCES

1.   CAHALAN, D., CISIN, I. H. and CROSSLEY, H. M. American drinking practices; a national study of drinking behavior and attitudes. (Rutgers Center of Alcohol Studies, Monogr. No. 6.) New Brunswick, N.J.; 1969.

2.   CAHALAN, D. Problem drinkers; a national survey. San Francisco; Jossey-Bass; 1970.

3.   CAHALAN, D. and ROOM, R. Problem drinking among American men. (Rutgers Center of Alcohol Studies, Monogr. No. 7.) New Brunswick, N.J.; 1974.

4.   CISIN, I.H. Community studies of drinking behavior. Ann. N.Y. Acad. Sci. **107**: 607–612, 1963.

5.   KNUPFER, G., FINK, R., CLARK, W. and GOFFMAN, A. Factors related to amount of drinking in an urban community. (California Drinking

[6]RICHMAN, A. "Recidivism" in substance abuse programs. Presented at the Symposium on Recidivism in Substance Abuse Programs at the American Psychiatric Association meeting, Toronto, May 1977.

Practices Study, Rep. No. 6.) Berkeley; California Department of Public Health; 1963.

6. KNUPFER, G. Epidemiologic studies and control programs in alcoholism. V. The epidemiology of problem drinking. Amer. J. publ. Hlth **57:** 973–986, 1967.

7. CLARK, W. B. and CAHALAN, D. Changes in problem drinking over a four-year span. Addict. Behav., Oxford **1:** 251–259, 1976.

8. U.S. NATIONAL INSTITUTE ON ALCOHOL ABUSE AND ALCOHOLISM. Alcohol and health; third special report to the U.S. Congress from the Secretary of Health, Education and Welfare, June 1978; technical support document. Ernest P. Noble, editor. (DHEW Publ. No. ADM 79-832.) Washington, D.C.; U.S. Govt Print. Off.; 1979.

9. U.S. NATIONAL INSTITUTE ON ALCOHOL ABUSE AND ALCOHOLISM. First special report to the U.S. Congress on alcohol and health from the Secretary of Health, Education and Welfare. (DHEW Publ. No. HSM 72-9099.) Washington, D.C.: U.S. Govt Print. Off.; 1971.

10. GUSFIELD, J. R. Status conflicts and the changing ideologies of the American temperance movement. Pp. 101–120. In: PITTMAN, D. J. and SNYDER, C. R., eds. Society, culture, and drinking patterns. New York; Wiley; 1962.

11. BACON, S. D. Alcohol and complex society. Pp. 78–93. In: PITTMAN, D. J. and SNYDER, C. R., eds. Society, culture, and drinking patterns. New York; Wiley; 1962.

12. HABERMAN, P. W. and SHEINBERG, J. Public attitudes toward alcoholism as an illness. Amer. J. publ. Hlth **59:** 1209–1216, 1969.

13. MULFORD, H. A. and MILLER, D. E. Public definitions of the alcoholic. Quart. J. Stud. Alc. **22:** 312–320, 1961.

14. ROOM, R. and SHEFFIELD, S., eds. The prevention of alcohol problems; report of a conference. Sacramento; California Health and Welfare Agency, Office of Alcoholism; 1974.

15. HYMAN, M. M., ZIMMERMANN, M. A., GURIOLI, C. and HELRICH, A. Drinkers, drinking and alcohol-related mortality and hospitalizations; a statistical compendium. New Brunswick, N.J.; Center of Alcohol Studies, Rutgers University; 1980.

16. CAHALAN, D. and CISIN, I. H. Drinking behavior and drinking problems in the United States. Pp. 77–115. In: KISSIN, B. and BEGLEITER, H., eds. The biology of alcoholism. Vol. 4. Social aspects of alcoholism. New York; Plenum; 1976.

17. BLANE, H. T. Middle-aged alcoholics and young drinkers. Pp. 5–38. In: BLANE, H. T. and CHAFETZ, M. E., eds. Youth, alcohol, and social policy. New York; Plenum; 1979.

18. KNUPFER, G. and ROOM, R. Age, sex, and social class as factors in amount of drinking in a metropolitan community. Social Probl. **12:** 224–240, 1964.

19. MULFORD, H. A. Drinking and deviant drinking, U.S.A., 1963. Quart. J. Stud. Alc. **25:** 634–650, 1964.

20. BAILEY, M. B., HABERMAN, P. W. and ALKSNE, H. The epidemiology of alcoholism in an urban residential area. Quart. J. Stud. Alc. **26:** 19–40, 1965.

21. JELLINEK, E. M. The disease concept of alcoholism. Highland Park, N.J.; Hillhouse Press; 1960.

22. ROOM, R. Measurement and distribution of drinking patterns and problems in general populations. Pp. 61–87. In: EDWARDS, G., GROSS, M. M., KELLER, M., MOSER, J. and ROOM, R., eds. Alcohol-related disabilities. (World Health Organization, Offset Publ. No. 32.) Geneva; 1977.

23. BLANE, H. T. and HEWITT, L. E. Alcohol and youth—an analysis of the literature; 1960–1975. (NITS No. PB-268-698.) Springfield, Va.; National Technical Information Service; 1977.

24. CAHALAN, D. Why does the alcoholism field act like a ship of fools? Brit. J. Addict. **74:** 235–238, 1979.

25. WIENER, C. The politics of alcoholism; building an arena around a social problem. Edison, N.J.; Transaction Books; 1980.

7

On Defining Alcoholism:
With Comment on Some Other
Relevant Words

Mark Keller

"A word to the wise is sufficient," the old proverb assured. For the acutest of the alchohol-related problems we are not yet so wise that one word can suffice. The words for It—we shall consider what It is—proposed by a group of "Investigators" convened by the World Health Organization, are alcohol-dependence syndrome (1).

(They mean alcoholism.)

The It we talk about, the It about which we are concerned and active, the It at the center of alcohol-related problems, has been known as alcoholism for about a century. And when, in remembered time, we said alcoholism, we thought we knew what we were talking about.

Not that there has ever been a clear-cut universally accepted definition. In the first formal attempt at a lexicon of *The Alcohol Language* (2), later in the first *Dictionary of Words about Alcohol* (3), many pages were devoted to the definition of alcoholism and cognate terms. Neither the approved or preferred definitions in these books, nor other precisely elaborated and logically rationalized definitions (e.g., 4, *p. 7;* 5, *p. 51;* 6; 7), nor the definitions in standard and medical dictionaries, satisfied everybody. Many many writers felt a need to formulate their own definitions.[1]

Nevertheless the great majority of the definitions, both formal and informal, evidence a consensus about It. In spite of differences in detail and emphasis, most of the definitions recognize that we are talking about the condition of people who cannot help repetitively drinking amounts of alcohol, usually enough to cause intoxication,

---

[1]See, for example, the wealth of citations in the first (3) and in the revised edition (8) of the *Dictionary of Words about Alcohol*, or almost any book and numerous articles, popular and nonpopular, about alcoholism.

which harm them. Many of the definitions incorporate—what is actually a separate issue—the further idea that such a condition is a disease, and some also incorporate the irrelevant issue of progression of symptoms.

In spite of the fact that most of the definitions were composed not by lexicographers but by amateurs striving to explain what they mean, most of them succeeded in expressing the same essential central idea.

An interesting distinction is made by some of the most sophisticated researchers. They would accept the foregoing conception of alcoholism but distinguish alcohol addiction. For addiction they require the occurrence of a withdrawal syndrome when the alcoholic temporally reduces his blood alcohol level (e.g., 9, 10).

I thought I worded it well 20 years ago when I proposed this as a "medical" definition: "Alcoholism is a psychogenic dependence or a physiological addiction to ethanol, manifested by the inability of the alcoholic consistently to control either the start of drinking or its termination once started" (11).[2]

I think the Cooperative Commission on Alcoholism too put it well when they defined it this way: "[Alcoholism is] a condition in which the individual has lost control over his intake in the sense that he is consistently unable to refrain from drinking or to stop drinking before getting intoxicated" (6).[3]

This is worth repeating: We all had a pretty good idea of what we were talking about when we said alcoholism, or when we used the derived noun alcoholic, meaning a person who has alcoholism. What, then, has happened to produce such a "confusion of tongues" that the new edition of the *Dictionary of Words About Alcohol* devotes more than a dozen pages to definitions of alcoholism, and that the W.H.O. Investigators proposed to abandon the term in favor of alcohol-dependence syndrome?

And what happened to alcohol addiction?

It is useful to recall the development of the relevant terminology over historic time.

Ancient or classical writers used words that are universally translated as drunkenness. The people who practiced it or had it were drunkards. These terms appear in the Bible[4] and in Seneca

[2]But today I would omit the psychogenic-physiological dualism (12).

[3]I was a member of the Commission but did not see this text before it was published. I would have transposed "consistently" and "unable."

[4]"Woe to the crown of pride, to the drunkards of Ephraim . . ." Isaiah 28:1. "Thou shalt be filled with drunkenness and sorrow, with the cup of astonishment and desolation . . ." Ezekiel 23:31.

(13). In early English literature Chaucer (14th century) used dronk-elewe, in referring to Seneca's conception, with a definite meaning of addiction to alcohol as a mental illness (14). By the 19th century "inebriety" was the favored descriptor; those who manifested it were not merely inebriated—they were inebriates. A Swedish public healthist, Dr. Magnus Huss, in 1849 coined alcoholism (15) as descriptor of a special condition of inebriates—what was later to come under Bowman and Jellinek's "chronic alcoholism" umbrella (16, 17). But the world had needed the right word for It, and alcoholism was it. For alcohol was at the center of the disease, and the suffix *ism*, in medical terminology, designates an established or chronic morbidity (vide automatism, glycoptyalism, hypothyroidism). So, almost immediately, the world chose to ignore Doctor Huss' specifications and adopted alcoholism as the ideal name for It.

The Itness of alcoholism was perhaps first enunciated in 1866 by a French physician, M. Gabriel, in his doctoral dissertation, *Essaie sur l'alcoolisme*, which treats the condition unmistakably as an addiction, a disease, and public-health problem (18). Gabriel was soon followed in this usage in many languages: Danish has alkoholisme; Dutch, alcoholisme; English, alcoholism; Finnish, alkoholismi; German, alkoholismus; Italian, alcolismo; Norwegian, alkoholisme; Polish, alkoholizm; Portuguese, alcoolismo; Russian, alkogolism; Serbo-Croatian, alkoholizam; Slovene, alkoholizem; Spanish, alcoholismo; Swedish, alkoholism.

Thus the medical world gained a linguistically logical name for It, and alcoholism was admitted to the foremost systematic nomenclatures of disease (19–23). In parallel, alcoholic became a noun designating a person afflicted with the disease, with like nosological variants in other languages, as French alcoolique, German alkoholiker, Italian alcoolista, Swedish alkoholist.[5]

In all the systematic nomenclatures the meaning of alcoholism has been effectively clear: a disease marked by (sometimes the indication is "caused by") repetitive gross alcohol intake. Where the system includes explanatory detail, as in the *Diagnostic and Statistical Manual* of the American Psychiatric Association (22), the text clarifies that the intake must be ample and repetitive in some unusual degree.

In their massive review of the psychiatric–biological literature

---

[5]How much better had English gone with Italian and Swedish and adopted alcoholist for the noun instead of alcoholic. Then we would not see such a linguistic lapse as "She is alcoholic," instead of she is *an* alcoholic, for no one would say "She is alcoholist." But probably it is too late to urge the change. Would that great good fellowship change its name to Alcoholists Anonymous?

on the effects of alcohol on the human, Bowman and Jellinek in 1941 distilled a substantial consensus about It (16, 17). They found nevertheless some confusion around the terminology of It and of the often associated physical and mental disorders. In a semantic tour-de-force they opted for the "chronic alcoholism" of Doctor Huss as the term for the associated diseases; It they designated by alcohol addiction. For the latter choice they had the best of reasons: All the expert descriptions converged on addiction, a drug addiction whether explicitly or by implication, and the drug was alcohol.[6]

A history of languages will show the development of popular fancies for new or modified words and phrases, and that languages evolve according to popular preferences. No authorities can impose linguistic usages.[7] The logical dicta of Bowman and Jellinek, with editorial support from the *Journal of Studies on Alcohol*, did not "take." People, including professionals, preferred alcoholism over alcohol addiction. Often they felt a need to ensure that their meaning would be understood, that they meant alcoholism, by prefixing the tautological "chronic." Professionals also felt a need to say chronic, because even doctors of medicine sometimes illiterately said acute alcoholism when they meant alcohol intoxication. Prefixing chronic seemed to avert misunderstanding, to ensure that they meant alcoholism, not mere alcohol intoxication.

The foremost scientific periodical specialized for alcohol-related knowledge attempted to resolve the problem anew. Tauto-logical "chronic alcoholism" was banned from its pages (except in quotation); It could be called either alcoholism or alcohol addiction. Ambiguous "acute alcoholism" was likewise banned; drunkenness, alcohol intoxication, or alcohol poisoning could replace it. Yet the same authors who wrote (or were converted to) alcoholism in the *Journal of Studies on Alcohol* reverted to chronic alcoholism in a next article in another journal.[8]

For the condition that was not certainly alcoholism but had some resemblance to it in the way alcohol was being used, a con-venient traditional word was available: inebriety, which could include alcoholism; and the people were, then, inebriates. Inebriety was old fashioned, however, dusty from the Latinizing 19th century, des-

---

[6]They did not say alcoholic addiction. I dissuaded them from that by warning that I would then insist, for consistency, on druggic addiction. With the same tactic I dissuaded them from alcoholic intoxication, pointing out that some people get intox-icated by water.

[7]A few sacrosanct words may be exceptions.

[8]Or was it that other editors reconverted them?

tined to be replaced by problem drinking; and the people would be problem drinkers.[9]

Alcoholism is too attractive a word to have been left to the scientists, scholars and professionals. Even without the "alcoholism movement" of the 1930s–1940s the popular press was featuring it in preference to old-fashioned drunkenness or inebriety. The word became common. Doctor Huss' Latin-Swedish neologism passed into everyday international speech and thereby experienced the fate of many ordinary words: It was made more and more useful. Its meaning was popularly enriched—and thereby technically impoverished. It ceased to be solely a diagnostic term for a condition identified by specifiable symptoms. And popular usage is not without influence on professionals.[10]

The meaning of alcoholism became soft, vague, nontechnical. The term came to be applied to prealcoholismic behaviors, to getting drunk sometimes by people who would never make it into alcoholism, to manifest troubles with drinking which have always been common in young people during the stage between adolescence and settling down, to heavy drinking, excessive drinking, deviant drinking and unpopular drinking. In short, to any drunking, sometimes even to youthful drinking (often called alcohol use). Robert Straus wittily observed (24) that the number of alcoholics in the United States nearly doubled in a single day, when a Federal agency casually redefined alcoholism.

No wonder that the group of W.H.O. Investigators concluded that the word had lost its usefulness as a term for It. They proposed that henceforth It should be called alcohol-dependence syndrome. We should remember that the latter term means what alcoholism used to mean before it was popularized. But E. M. Jellinek had anticipated them.

The W.H.O. group might have adopted alcohol addiction. But in the W.H.O. addiction had become a dirty word. Dependence was the tenderhearted (or soft-minded) substitute. Hence, not alcohol addiction but alcohol dependence. But alcohol dependence, like alcohol addiction, sounds like a real disease—at least like a psychological disease. So far-going a commitment could not be made in a group that included vocally skillful social scientists who look askance at "medicalization" and "clinicization." Hence, alcohol-dependence

[9]In time the word-fatteners, inspired by intoxification, would expand it to problematic drinkers.

[10]Why else do sociologists study "pot," psychiatrists call themselves "shrinks," physicians say they prescribe "downers," nurses administer "shots"?

syndrome. But the revised edition of the *Dictionary of Words about Alcohol* (8) treats alcohol addiction, alcohol dependence, alcohol-dependence syndrome, and alcoholism, as synonymns.

Jellinek's ingenious invention of the alcoholisms (25, 26) was an attempt to restore order out of the chaos in the meanings of alcoholism. Referring to the fact that some people "who never become addicted" were being labeled as alcoholics, Jellinek wrote (25): "In order to do justice to these . . . differences, we have termed as alcoholism *any use of alcoholic beverages that causes any damage to the individual or society or both.*" Five lines later, on the same page, he added: "such a vague definition is useless."[11] This, then, was his rationale for proposing to substitute the several species of alcoholism labeled with Greek letters.

By now the W.H.O. group has noted that there really are no alcoholisms. The Jellinekian species are "culturally, environmentally or personally patterned manifestations of the fundamental alcohol-dependance syndrome" (1). But the alcoholisms were fathered by a Great Man; his idea will not easily be dismissed. The alcoholisms, especially gamma and delta, achieved a popularity rivaling that of alcoholism itself. They are destined for long survival. It is well to remember, therefore, that both gamma and delta behaviors are manifested by alcoholics in the Americas, that the type is usually caused by circumstances, and that some alcoholics change, over time, from one to the other. We should remember too that in alpha alcoholism, which he did not consider a disease, Jellinek was describing what he himself had earlier delineated as a preparatory stage of heavy drinking—the prealcoholic or prodromal phase in his original phaseology (27)—preceding the loss of control that marks off the onset of the disease proper. This is an important clarifying distinction. At least some of those who would redefine alcoholism as not a disease are confused by the fact that prealcoholismic manifestations which do not constitute the disease are called alcoholism—be it with a Greek-letter prefix.

Along with the popularization and semantic destabilization of alcoholism came a new attack on the conception of alcoholism as a disease. This topic, this event, is directly relevant to the meaning of alcoholism.

It is necessary to note that this is a new attack. The denial of the status of disease to alcoholism is not new. Drunkenness was since ancient times regarded as a folly, a vice, a sin, a willful wrong-

---

[11]Nevertheless, people who have apparently taken a course in speed reading have seriously quoted the italicized phrase as Jellinek's definition of alcoholism!

doing. Civil and religious authorities did not recognize a distinction between willful drunking and the uncontrollable resort to alcoholic drink (11) of those who are "enslaved to the habit" (13). When in the 19th century humane American physicians[12] advocated treatment rather than punishment for those inebriates who manifested what Dr. Benjamin Rush had called a disease already in the 18th century (28) the Reverend Mr. Todd thunderously asserted before an assembly of the Connecticut clergy that drunkenness is "a vice, not a disease" (29). But the recent attack on the concept of alcoholism as a disease is not a renewal of that conflict.

Social scientists are chiefly the new revisionists of the historical disease conception.[13] From impressive results of national sample surveys they describe a continuum of alcohol-intake quantities and of related troubles (30, 31), but note that they cannot differentiate between alcoholics and other drinkers; they can only place drinkers into various quantitative categories over a range of alcohol intakes. From the standpoint of their methods and analyses they are right. Alcoholism is a medical diagnosis. It is not surprising if social scientists cannot frame diagnostic categories on the basis of population-survey data. But they should then refrain from antidiagnostics. Their findings permit them to say modestly that they are able to identify categories of drinkers. They should then speak only of drinkers and drinking, withholding judgments about alcoholism. Yet if they would say that drinking, or heavy drinking, or drinking-with-troubles, or heavy-drinking-with-troubles, or escape drinking, or heavy-escape drinking, or heavy-escape-drinking-with-troubles, is not alcoholism, I could agree. But their data do not allow any conclusions about the nature of alcoholism.

Of course social scientists as citizens have the right to argue that they as well as doctors of medicine have the right to determine what shall be designated as disease. If they would argue from the standpoint of the definition of disease that no amount or form of alcohol intake should be defined as disease, they would be on the sound ground of expressing a belief, as sound as the Reverend Mr. Todd in expressing the same belief. To argue from their data, however, that there is only a range of drinkings and that therefore the only difference between the most moderate and the most intemper-

[12]For example, Dr. Albert Day, Dr. T. D. Crothers, Dr. Lucy M. Hall, in early volumes of the *Quarterly Journal of Inebriety* in the fourth quarter of the 19th century.

[13]Some of them, however, ignore history and attack the disease concept as something newly invented by the "alcoholism movement" in the 1940s and 1950s.

ate is in categorical degree, excluding the possibility of a pathological process, is to go beyond what their data allow.[14]

The basis for thinking that alcoholism is a disease has been
detailed elsewhere (32). Here it is enough to emphasize that the
conception of disease should not be applied to mere heavy drinking,
or mere misbehaving with alcohol, or mere getting into trouble on
account of drinking, or mere getting drunk x times. The conception
of alcoholism as disease applies only to those who manifest the
symptoms of addiction. It is a diagnostic problem.

Not all social scientists wish to withdraw alcoholism from the
category of diseases and from the clutch of the health professionals.
A sophisticated sociologist, perhaps with the longest sociological
experience in the study of alcohol-related phenomena, and a contributor to the original *Alcohol, Science and Society*—Selden Bacon—
has recently written (33): "Addiction to alcohol is defined by a slowly
developing pattern of altered alcohol intake, related to increasing
social difficulties and continuing emotional disturbances," and highlighted by a steadily growing dissocialization. Bacon emphasizes the
function of the social group, the effectors of social control, in allowing and even socializing the deviant drinking, in failing to avert the
"loss of control over drinking." But he recognizes too that social
factors alone do not explain why some people enter and complete
the process of becoming addicted while most drinkers, including
many who go some way into the process, stop short of addiction.
And he recognizes the importance of the pharamacological action of
alcohol. This sociological viewpoint is not alienated from that of the
editors and authors of a book titled the *Biological Basis of Alcoholism* (34).

Given the conception of addiction as the essential characteristic of alcoholism, it is worth reemphasizing that there is no distinction between alcohol addiction, alcohol dependence, alcohol-
dependence syndrome, and alcoholism. In a nonauthoritarian world
with a nonauthoritarian language, everyone may apply whichever
term seems most palatable.

Obviously those who believe that the mark of addiction is a
withdrawal syndrome will wish to reserve alcohol addiction for those

---

[14]I would wonder whether they would equally argue that the number of red cells
in the blood is only a matter of degree, ranging from less than 2 million to over 5
million per cubic centimeter, so anemia is not a disease; or that the amount of time
people are willing to spend outdoors is only a matter of degree, ranging from a
preference of 100% to none at all, so agoraphobia is not a disease; or that the appetite
for food is only a matter of degree, ranging from voracity to disgust, so anorexia
nervosa is not a disease.

patients who exhibit the withdrawal symptoms. Others make an interesting distinction between psychic (psychological) dependence (or alcoholism) and physical (physiological) addiction (with a withdrawal syndrome). I believe that this is an artifical distinction, that addiction is addiction is addiction (12). Those who speak of psychological addiction are merely saying, by distinguishing it from physical addiction, that the alcoholic has not yet shown a withdrawal syndrome; or else they would be saying that some ghostly aspect of the organism is addicted, not the organism itself. Those who speak of physical addiction are merely saying that the alcoholic has shown a withdrawal syndrome. But it is unsafe to base a diagnosis on an inconsistent symptom which, as Mendelson and his colleagues have shown, can sometimes appear and on a later occasion fail to appear in the same patients (35). Until a biological test is established, the diagnosis of alcoholism must rest on the most pertinent behavioral symptoms, especially the reliable inference of loss of control—i.e., the disablement from consistently controlling whether to drink and when to stop if drinking is started. This principle renders the withdrawal syndrome immaterial in the diagnosis of alcoholism, and the distinction between psychic and physical addiction nugatory.

How should alcoholism be defined now?

A straightforward definition from the revised *Dictionary of Words about Alcohol* (8) seems the most practical for general use:

"Repetitive intake of alcoholic beverage to a degree that harms the drinker in health or socially or economically, with indication of inability consistently to control the occasion or amount of drinking."[15]

This definition only describes what alcoholism is from the perspective of an observer of the alcoholic's behavior. The full definition of a disease should state not only its manifestations but its cause as well. Bypassing the superficial notion that alcohol is the cause of alcoholism, it is necessary to address the question why it is that some people drink enough to become addicted but not the vast majority of drinkers. At this time it must be recognized that the etiology of alcoholism is unknown. In the nomenclature of the American Medical Association (21) alcoholism is classified among the diseases of the psychobiological unit of unknown site and etiology. There is a plausible etiological hypothesis: Learning effects changes in the central nervous system. Some drinkers learn to resort to alcohol automatically, reflexly, and incontinently, in response to

---

[15]Besides this most general formulation the *Dictionary* (8) provides many additional definitions of alcoholism and its synonyms, some derived from specified disciplines and professions, pertinent to various interests and purposes, and in total occupying more space than this chapter.

internal or environmental cues (36–39). The change in the central nervous system that then imposes the self-harming behavior is defined as a pathology. Alcoholism therefore must be classified as a disease (40) and the terms alcoholism, alcohol addiction and alcohol dependence should be reserved for the behavior that reflects this pathological alteration.

Alcoholism, and its satellite terms, alcohol addiction and alcohol-dependence syndrome, have been discussed at length because they are prime words in the field of alcohol studies and the realm of alcohol-related actions. The misuses of alcoholism, even the misuse of the meaning of alcoholism which is reflected in such a phrase as "alcoholism is not a disease," justifies the proposal of the W.H.O. group, previously noted, to abandon it in favor of some new term, such as alcohol-dependence syndrome, with a specified meaning. Experience with usage in language suggests, however, that this logical proposal will not avail. Alcoholism will likely survive. What is needed is to persuade users of the word to reserve it for its appropriate meaning: alcohol addiction or alcohol dependence.

Those who believe that conditions manifested only in behavioral symptoms are not diseases would make their position clear by saying "alcohol addiction, or alcohol dependence, is not a disease." For, indeed, alcoholism as now often misused, to include a range of problem-related drinking by people who are not addicted, is not a disease.

Because so much ambiguity, confusion and discord have developed around the word alcoholism it has been unavoidable to devote an excessive amount of attention to it—yet by no means with exhaustion of the subject. It is, finally, worth noting that except in Dr. Howard W. Haggard's Foreword the word alcoholism is rare in *Alcohol, Science and Society* as published in 1945 (41).[16]

The more common word then and therein was inebriety. And no wonder. The subject discussed, for the most part, then as now, was more than the disease alcohol addiction. The subject was all the problem-causing behaviors related to alcoholic beverages. The old-fashioned word inebriety served well for the purpose. Yet before long inebriety was replaced by problem drinking. This latter term, vague catch-all that it is, has proved to be universally useful precisely because it could encompass all the problem-related behaviors and effects of the use of alcoholic beverages. The W.H.O. group did not like problem drinking and wished it begone (1), but like alco-

---

[16]In contrast, in the updating preface to the 6th printing, in 1954 (42), alcoholism occurs on every page, often many times.

holism it is too useful to be abandoned either by the public or by professionals—especially by professionals. If it should decline, however, it will only be when replaced by a far less desirable term—the opprobrious, vindictive, pejorative term "alcohol abuse."

"Alcohol abuse" did not appear in *Alcohol, Science and Society*. In fact, in the American and British literature before 1970 "alcohol abuse" hardly occurred except occasionally in an English-language article from the Scandinavian countries. In 1970 a political document, the Hughes Act, which created a National Institute on Alcohol Abuse and Alcoholism, practically forced "alcohol abuse" on America.

The creation of the National Institute, with more money to spend than the alcohol-concerned people had dreamt of, literally popularized "alcohol abuse," with catastrophic effect on scientific discourse. In the next decade it became common to see, in journals that claim to be scientific or scholarly, reports of studies of "alcohol abuse" without any pretense at definition.[17] In fact, "alcohol abuse" is being used with the meanings of alcohol addiction, alcohol dependence, alcohol intoxication, alcohol misuse, alcoholism, deviant drinking, drunkenness, excessive drinking, heavy drinking, problem drinking, or any alcohol intake that evokes disapproval. For a writer who knows what he is talking about or who intends not to be ambiguous or obscure, "alcohol abuse" is an avoidance term. The nondiagnostic term problem drinking is still available. It has a formal meaning: Heavy, deviant or implicative drinking that causes private or public harm—that is seen to cause problems for the drinker or for others (8). When it is impossible to be more specific it is reasonable to use this portmanteau term, which may include alcoholism.

Francis Bacon warned of the intellectual peril in selecting the words that express our thoughts: The words tend to modify the thought (43). If we call a man a chronic alcoholic, how is that different from calling him an alcoholic? (Can an alcoholic not be chronic?) The difference is that chronic adds a smirch; chronic is repulsive. It denigrates: He is "a chronic." If we call a man an "abusive drinker," is it any different from calling him a problem drinker? Problem drinker may suggest that he too may have his prob-

---

[17]To avoid insulting authors and editors I cite a few examples from the recent literature by title only: *(1)* "Visual averaged evoked responses and platelet monoamine oxidase activity as an aid to identify a risk group for alcoholic [sic] abuse." *(2)* "Recoverability of psychological functioning following alcohol abuse." *(3)* "Amenability to treatment in an adolescent inpatient alcohol abuse program." *(4)* "Life expectancy after liver injury through alcohol abuse." *(5)* "Measuring the costs of alcohol abuse." *(6)* "Alcohol abuse in the U.S. and the U.S.S.R."

lem. "Abusive drinker" and "alcohol abuser," twin offspring of "alcohol abuse," are inherently nasty. They blame. "O ye wise, be cautious with your words" (44).

The Introduction to the original *Dictionary of Words about Alcohol* (3) asked, "But what is alcohol that a dictionary should be composed for its special interest"? The answer is in a word missing in that edition (but present in the second revised edition): Alcohology. There is an alcohology. There is a field, bigger than just a discipline, of studies on alcohol, and a language specialized to that field. Alcohologists were not yet recognized as such in *Alcohol, Science and Society*, though the word and the proposal to recognize a special discipline of alcohology were already published before 1903 (45). Now it is no longer possible to retreat from this recognition (46, 47). At least alcohologists, those who are substantially occupied with study of alcohol-related phenomena, owe it to their posture as scientists and scholars, to the standards of scientific communication, to word their thoughts meaningfully and precisely.

There is a difference between saying "alcohol abuse" or problem drinking or something more exact. As for alcoholism, we do know what It is.

REFERENCES

1. EDWARDS, G., GROSS, M. M., KELLER, M., MOSER, J. and ROOM, R., eds. Alcohol-related disabilities. (WHO Offset Publ. No. 32.) Geneva; World Health Organization; 1977.
2. KELLER, M. and SEELEY, J. R. The alcohol language; with a selected vocabulary. (Brookside Monogr. No. 2.) Toronto; University of Toronto Press; 1958.
3. KELLER, M. and MCCORMICK, M. A dictionary of words about alcohol. New Brunswick, N.J.; Rutgers Center of Alcohol Studies Publications Division; 1968.
4. HEWITT, D. W. Alcoholism; a treatment guide for general practitioners. Philadelphia; Lea & Febiger; 1957.
5. PITTMAN, D. J. Interdisciplinary considerations in alcoholism research. Pp. 48–58. In: PITTMAN, D. J., ed. Alcoholism, an interdisciplinary approach. Springfield, Ill.; Thomas; 1959.
6. COOPERATIVE COMMISSION ON THE STUDY OF ALCOHOLISM. Alcohol problems; a report to the nation. Prepared by T. F. A. Plaut. New York; Oxford University Press; 1967.
7. AMERICAN MEDICAL ASSOCIATION. Manual on alcoholism. Rev. ed. Chicago; 1973.
8. KELLER, M., MCCORMICK, M. and EFRON, V. A dictionary of words about alcohol. 2d ed. New Brunswick, N.J.; Rutgers Center of Alcohol Studies Publications Division; 1982.

9. MENDELSON, J. H. Biological concomitants of alcoholism. New Engl. J. Med. **283**: 24–32, 71–81, 1970.

10. GOLDSTEIN, D. B. Physical dependence on alcohol in mice. Fed. Proc. **34**: 1953–1961, 1975.

11. KELLER, M. Definition of alcoholism. Quart. J. Stud. Alc. **21**: 125–134, 1960.

12. KELLER, M. Psychological psuperstition and physical phallacy. (Guest Editorial.) Brit. J. Addict. **75**: 225–226, 1980.

13. SENECA, L. A. Epistle LXXXIII, On drunkenness. Repr. in: JELLINEK, E. M. Classics of the alcohol literature. Quart. J. Stud. Alc. **3**: 302–307, 1942.

14. CHAUCER, G. The Canterbury tales; the pardoner's tale. In: SKEAT, W. W. The student's Chaucer. New York; Oxford University Press; 1900.

15. HUSS, M. Alcoholismus chronicus eller chronisk alkoholssjukdom; ett bidrag till dyskrasiernas kännedom; enligt egen och andras erfarenhet. Stockholm; 1849.

16. BOWMAN, K. M. and JELLINEK, E. M. Alcohol addiction and its treatment. Quart J. Stud. Alc. **2**: 98–176, 1941.

17. JELLINEK, E. M., ed. Effects of alcohol on the individual; a critical exposition of current knowledge. Vol. 1. Alcohol addiction and chronic alcoholism. New Haven; Yale University Press; 1942.

18. GABRIEL, M. Essaie sur l'alcoolisme, considérée principalement au point de vue de l'hygiène publique. Montpellier thesis; 1866.

19. NATIONAL CONFERENCE ON NOMENCLATURE OF DISEASE. A standard classified nomenclature of disease. Logie, H. B., editor. New York; Commonwealth Fund; 1933.

20. U.S. PUBLIC HEALTH SERVICE. FEDERAL SECURITY AGENCY. Manual for coding causes of illness according to a diagnostic code for tabulating morbidity statistics. (Miscellan. Publ. No. 32.) Washington, D.C.; U.S. Govt Print. Off.; 1944.

21. THOMPSON, E. T. and HAYDEN, A. C., eds. Standard nomenclature of diseases and operations. 5th ed. (Published for the American Medical Association.) New York; McGraw-Hill; 1961.

22. AMERICAN PSYCHIATRIC ASSOCIATION. Diagnostic and statistical manual of mental disorders. 3d ed. (DSM-III.) Washington, D.C.; 1980.

23. WORLD HEALTH ORGANIZATION. Manual of the international statistical classification of diseases, injuries and causes of death; based on the recommendations of the Ninth Revision Conference, 1975, and adopted by the twenty-ninth World Health Assembly. (International Classification of Diseases, ICD-9.) Geneva; World Health Organization; 1977.

24. STRAUS, R. Alcohol and society. Psychiat. Ann. **3** (No. 10): 8–107, 1973.

25. JELLINEK, E. M. The disease concept of alcoholism. Highland Park, N.J.; Hillhouse Press; 1960.

26. JELLINEK, E. M. Alcoholism, a genus and some of its species. Canad. med. Ass. J. **83**: 1341–1345, 1960.

27. JELLINEK, E. M. Phases in the drinking history of alcoholics; analysis of a survey conducted by the official organ of Alcoholics Anonymous. Quart. J. Stud. Alc. **7**: 1–88, 1946.

28. RUSH, B. An inquiry into the effect of ardent spirits upon the human body and mind, with an account of the means of preventing and of the remedies for curing them. Brookfield, Mass.; Merriam; 1814. [Orig. 1785.]

29. TODD, J. E. Drunkenness a vice, not a disease. Hartford, Conn.; Case, Lockwood & Brainard; 1882.

30. CAHALAN, D. Problem drinkers. San Francisco; Jossey-Bass; 1970.

31. CAHALAN, D. and ROOM, R. Problem drinking among American men. (Rutgers Center of Alcohol Studies, Monogr. No. 7.) New Brunswick, N.J.; 1974.

32. KELLER, M. The disease concept of alcoholism revisited. J. Stud. Alc. **37**: 1694–1717, 1976.

33. BACON, S. D. The process of addiction of alcohol; social aspects. Quart. J. Stud. Alc. **34**: 1–27, 1973.

34. ISRAEL, Y. and MARDONES, J., eds. Biological basis of alcoholism. New York; Wiley-Interscience; 1971.

35. MENDELSON, J. H. and LA DOU, J. Experimentally induced chronic intoxication and withdrawal in alcoholics. Pt. 2. Psychophysiological findings. Quart. J. Stud. Alc., Suppl. No. 2, pp. 14–39, 1964.

36. KINGHAM, R. J. Alcoholism and the reinforcement theory of learning. Quart. J. Stud. Alc. **19**: 320–330, 1958.

37. KEPNER, E. Application of learning theory to the etiology and treatment of alcoholism. Quart. J. Stud. Alc. **25**: 279–291, 1964.

38. LUDWIG, A. M. and WIKLER, A. "Craving" and relapse to drink. Quart. J. Stud. Alc. **35**: 108–130, 1974.

39. LUDWIG, A. M., WIKLER, A. and STARK, L. N. The first drink; psychological aspects of craving. Arch. gen. Psychiat. **30**: 539–547, 1974.

40. KELLER, M. Concepçoes sobre o alcoolismo. Rev. Ass. Brasil. Psiquiat. **2**: 93–100, 1980.

41. HAGGARD, H. W. Foreword. Pp. xi–xii. In: Alcohol, science and society; twenty-nine lectures with discussions as given at the Yale Summer School of Alcohol Studies. New Haven; Quarterly Journal of Studies on Alcohol; 1945.

42. KELLER, M. Preface to the sixth printing. In: Alcohol, science and society; twenty-nine lectures with discussions as given as the Yale Summer School of Alcohol Studies. New Haven; Quarterly Journal of Studies on Alcohol; 1954.

43. BACON, F. Novum organum [orig. 1620], I, 59.

44. R' Tryphon [fl. 1st century B.C.] in Babylonian Talmud, Tract. Aboth, ch. 1, v. 11.

45. KOPPE, R. Nochmals die Errichtung von Lehrstühlen der Alkohologie. Int. Mschr. Alksm. **13**: 188, 1903.

46. BACON, S. D. Alcohol research policy; the need for an independent,

phenomenologically oriented field of study. J. Stud. Alc., Suppl. No. 8, pp. 2–26, 1979.

47. KELLER, M. Afterwords. Quart. J. Stud. Alc., Suppl. No. 8, pp. vii–x, 1979.

8

The Social Costs of Alcohol in the Perspective of Change, 1945–1980

Robert Straus

It is not possible to relate this review of the social "costs" and social problems to any single chapter of the original *Alcohol, Science and Society* because discussions of costs and problems are found throughout that volume. This is the way it should be. Just as alcohol, because of its low molecular weight, can permeate all systems and virtually all tissues of the human body, so alcohol problems and the "costs" of alcohol permeate all major social systems and are related to most issues of contemporary social concern.

The theme of this chapter is change; change in knowledge about alcohol, change in our perceptions, change in our assumptions about what we know and what we do not know and what we expect to learn. There have been changes in the contexts of drinking; in the social contexts of who drinks what, where, when, with whom and why; and in the chemical contexts of other drugs, chemicals or substances that are present in the human body when alcohol is consumed. There have been changes in the liability of intoxication, in the labeling of people with alcohol problems, in what we perceive to be the social costs and problems of alcohol, and in our social responses.

There have been several attempts to express the social costs of alcohol in terms of dollars. Although such estimates may have utility in raising levels of social concern and stimulating social response, they have little meaning, being subject to vagrancies of exclusion and inclusion and to such factors as variations in the value of the dollar. For example, in the 1974 report to Congress (1), the economic costs of alcohol misuse and alcoholism was 25.37 billion dollars; in the 1978 report (2), the estimate was 42.75 billion. The appropriate framework for a consideration of social costs is not one of economic measures, but, as Bacon (3) so effectively indicated in 1945, must incorporate a consideration of numerous concomitants of complexity that characterize contemporary societies.

134

In considerations of change, great caution must be attached to comparisons of numerical estimates. The authors of *Alcohol, Science and Society* assumed that 40 to 45 million people in the U.S. were users of alcoholic beverages in 1945—28 to 32% of the total population. Today, we assume that there are 100 to 110 million users or 45 to 49% of the total population. Our current estimates are based on different measuring instruments, and different criteria including an acceptance rather than a denial of the reality of drinking among young people. Similarly, the 1945 estimate of 2 to 2½ million "excessive" drinkers represents different criteria than today's commonly used estimates of 9 to 13 million "problem" drinkers.

What can we say about who drinks? It is generally assumed that there has been a rise in the proportion of alcohol users, especially among youth under the age of 18 and probably among adult women. Of more significance has been a rise in the per capita consumption of alcohol that has been evident in most Western industrialized societies since 1960 and has increased by about a third in the U.S. during this 20-year period. Only a modest part of the rise can be accounted for by increases in the relative number of drinkers; a more significant factor is an increase in levels of consumption on the part of people who drink.

In 1980, there is special interest in the drinking problems of youth, women, minorities and the elderly. Although, in each group, there are indeed distinct aspects of drinking, and distinct problems, the special concern for their drinking "problems" has arisen primarily because of society's more general contemporary interest in the problems of youth, women, minorities and the elderly.

Changing Perceptions

Whatever the social motives may be, focus on the drinking behavior and drinking problems of identifiable segments of the population is timely in relation to our changing perceptions about alcohol and the human body. Since 1945, we have shifted our perceptions from an emphasis on the similarities in the way that people respond to alcohol to an emphasis on differences.

In 1945, we were using our limited knowledge as a basis of generalizations that could help us understand and predict in terms of averages how alcohol would affect people and how people would react to alcohol. We were extending our insights to recommendations for public policy regarding how measures of blood alcohol concentration might be used to establish norms for the legal liability for driving while intoxicated. We were suggesting that people be educated so that they could use norms to determine their own safe or dangerous limits for drinking.

Individual Variability in Response to Alcohol

In 1980, we have a greater understanding of individual variability in the relationship between alcohol and the human body. We expect soon to identify factors of a genetic nature that affect the sensitivity to alcohol by different organs and tissues of the body. We assume that genetically determined potentialities and limitations interact with changes induced by alcohol or its metabolites to produce responses that represent the unique combined impact of nature and nurture. We recognize that alcohol will affect different individuals in varying ways. Some may be so sensitive that they cannot or choose not to drink at all. Some may be able to drink with little apparent harm amounts that would incapacitate most others. We now assume that all people do not experience exactly the same metabolic process; that in some alcohol finds alternate metabolic pathways and produces metabolites that may have uniquely toxic characteristics. We recognize that different organs of the body and even different areas of an organ such as the brain may vary in sensitivity to alcohol. Thus, in response to apparently similar exposures to alcohol, some people may develop cirrhosis of the liver, some pancreatitis, some cardiac myopathy and some peptic ulcer. Or, in response to similar blood alcohol concentrations, some brains may respond with aggression, some sleep, some crying, some with little visible change in brain function. We recognize that many individuals will respond differently under varying circumstances to similar exposures to alcohol. They will react differently in the morning than in the evening, when fatigued than when rested, when they become older than earlier in their lives, when sick than when well. We recognize that the mood of the drinkers before drinking and the social setting in which drinking takes place can also significantly affect the impact of alcohol on the drinker.

All of these factors of variability between and within individuals point to the dangers of employing simplistic formulas in determining the effects of alcohol. They stress the need for more sophisticated conceptualizations that stress variations rather than averages in the alcohol–human relationship.

Changing Definitions and Measurements

In 1945, the problems of alcohol were defined primarily in terms of intoxication and alcoholism. The former was defined as a problem when it occurred in public, or interfered with task performance, or led to accidents harmful to self or others. The latter was seen as a problem largely for men who had experienced many years of increasingly heavy drinking and frequent intoxication culminating in a state of compulsive uncontrollable addiction to alcohol. Alcoholism had been studied largely in captive populations such as inebriates in jails, disturbed alcoholics in mental hospitals, or in conspicuous populations such as the men of Skid Row. The traits of social and emotional instability that characterized these groups provided the prevailing stereotypes of "alcoholism." Although most people had friends or relatives who repeatedly drank too much, because they were different from the stereotypes and still had jobs or family or community stability they were protected and were often supported in denying their alcohol problems until or unless they too eventually "hit bottom."

A number of physical problems associated with excessive drinking were recognized in 1945 (cirrhosis, several brain syndromes, polyneuropathy, cardiomyopathy), but they were assumed generally to be nutritional diseases. Alcohol has a high caloric value and many observed that people who obtained a high proportion of their caloric needs through alcohol did not eat enough of the foods which provided other essential nutrients. In 1980, scientists are prepared to incriminate alcohol itself or its various metabolites, or both, more directly in an increasing number of major diseases. In fact, a recent report of the Surgeon General (4) suggested that excessive alcohol consumption was a contributing factor to virtually all of the major causes of death and disability in the U.S.

In 1945, Jellinek (5) reported an estimate of three million inebriates in the United States of whom 2.6 million (87%) were men; and of the men, 2.1 million (81%) were aged between 30 and 60. These numbers demonstrate the impact that countability, visibility and classification have had on the way in which alcohol problems were measured. The vast majority of "captive" problem drinkers— those in jails, mental hospitals and on Skid Row—were middle-aged men. Of these, only Skid Row men and inebriates prior to arrest were visible yet all were easily countable. There were other groups of heavy drinkers: men and women whose problems were very visible to their families, fellow workers or neighbors but were hidden from the rest of society. Because they did not fit the stereotype, they were not classified, nor were they readily countable. Still another group of frequently heavy drinkers in 1945 were young men, between the ages of 18 and 25. They were often visible, but they too did not fit the stereotype because they had not been drinking heavily for very long. Also, they were able to drink with a certain degree of social license and their problems were seen as those of "sowing wild oats" rather than of alcohol per se. Not until the epidemics of illicit drug use of the 1960s and other manifestations of youthful protest focused society's attention on more general problems of adolescents and young adults did the reality of drinking problems among youth come into sharp focus. Changes in public preparedness to reclassify excessive drinking by young people as an alcohol problem were supported at the scientific level by major changes in how the epidemiology of problem drinking was studied.

Who Are Problem Drinkers

Two developments, already underway in 1945, changed the studies of the characteristics of excessive drinkers somewhat: The rapid growth of Alcoholics Anonymous and the emergence of community alcoholism treatment programs. These developments made possible new studies based on the characteristics of people who sought the aid of A.A. (albeit anonymously) and those who sought treatment in community clinics or residential programs (6). The new resources attracted a very different socially stable clientele than had been seen at jails or mental hospitals, but the clients were still primarily men and for the most part in middle adulthood. They were also highly selected because they had generally had to suffer for a fairly long period of time before they were motivated enough to seek help.

At that time, it was assumed that the stigma of alcoholism was so great that only people who were "captive" or motivated to seek help would be willing to participate in studies of drinking. With rare exceptions it was not until the mid-1960s that social scientists expanded their studies to drinking patterns and drinking problems in the population-at-large. Following the pioneer work of Genevieve Knupfer (7) in California, Don Cahalan, Ira Cisin, Robin Room and their associates have provided a whole new prospective of drinking in American society through their studies based on the probability sampling of households (8–10). Of special significance to this review are their findings of a high prevalance of problem drinking among young men, of a significant amount of spontaneous (without treatment) remission, and of movement into and out of problem drinking. They also found that the problems of drinking were related to a wide variety of social problems including difficulties with marriages, relatives, friends, neighbors, jobs, police, accidents, health, aggressive behavior, and economics.

A similar challenge to long held beliefs about the natural history of problem drinking emerged from Fillmore and Bacon's 27-year follow-up study (11) of the college student population originally studied by Straus and Bacon in 1949–1951 (12). Of particular note is the extent to which drinking patterns changed over time in accordance with changes in various life situations.

The "Wettening" of Society

In 1945, American society was still rebounding from a period of national prohibition and a devastating depression and was still embroiled in a massive war effort. The powerful temperance movement that had been active for more than a century was still vibrant and local option prohibition still prevailed in one state, and hundreds of counties. Advocates of the "wet" cause were still gingerly protective of their gains and by legal or tacit agreement most communities limited the number and nature of public settings where drinking could occur.

By contrast, in 1980, the U.S. has become a much "wetter" society. There is less local prohibition, more liberal controls on age of drinkers, hours and places where drinking can occur or alcohol can be sold and many more social situations which offer, call for or even "require" drinking. For example, most bowling establishments in 1945 included the bowling alleys, a soft drink machine and possibly a snack bar. Most bowling establishments in 1980 include bars and the assumption that alcohol consumption is a desirable concomitant of bowling. Similar arrangements have developed in many theaters and in connection with most sports events. The advent of television has brought another occasion or setting in which drinking is seen as appropriate or desirable. Television advertising has provided a new media for promoting the use of wine and beer. Although the advertising of distilled spirits is prohibited, there are no apparent limits on the promotion of heavy drinking in television dramas. Studies of prime-time television programs have found that alcohol consumption is depicted many times more frequently than beverages such as coffee or water. The drinking often depicts amounts and speeds of consumption that would produce intoxication or incapacity in most people, but in television dramas such drinking is usually associated with highly successful outcomes in activities such as consummating love, apprehending criminals, winning business contracts and even driving automobiles.[1]

In 1980, as in earlier periods of our history, at both the societal

---

[1]DILLIN, J. Reported in the *Christian Science Monitor*, 30 June, 1 July, 11 July and 26 December, 1975.

and the individual levels, there are conflicting values about alcohol. Governments are expressing increasing concern about the problems of alcohol, yet they continue to protect the alcohol industry and to benefit from alcohol as a major source of tax revenue. Individuals, too, tend to support measures designed to control drinking problems as long as they do not impinge on personal drinking practices. Whenever decisions about alcohol are made, there is an unwritten and unexpressed but overriding mandate that policies directed at drinking problems must not impinge on the policy-maker's own access to alcohol.

As previously noted, in 1945 the problems of alcohol were defined as (1) intoxication and its consequences; (2) some diseases that were then thought to be only indirectly caused by alcohol; and (3) alcoholism, then thought to be an addictive consequence of prolonged excessive drinking occurring primarily in alienated and disturbed segments of the male population. By 1980, there were numerous changes in both the nature and perceptions of these problems.

The Chemical Context of Drinking

In 1945, it was generally assumed that most people who became intoxicated were reacting to alcohol. Except for barbiturate use, which was limited to a relatively small group of prescription recipients, there were few other intentionally used intoxicants. Although, in retrospect, many common industrial pollutants were capable of interacting with alcohol their possible contributing role to drinking behavior was not seriously considered. In any event, alcohol was indeed the primary intoxicating substance in common use.

In 1980, alcohol no longer has such a distinction. The number and variety of industrial intoxicants, especially the polyvinyls, has increased significantly. The deleterious interaction of some of these substances with alcohol are suspected. More significant, a wide variety of new intoxicating medicines have been introduced since 1945 and are now used by a large segment of the population, especially the antihistamines and antianxiety drugs. Antihistamines have a mild sedative action for most people. They are prescribed for allergies and are a major component in most nonprescription cold remedies and sleeping pills. It is estimated that some 25 to 40 million Americans are using one or more forms of antihistamines at any time. Although relatively innocuous by themselves, when taken coincidentally with alcohol, antihistamines can significantly enhance the intoxicating effect. In fact, the combined reaction may be synergistic (i.e., greater than the sum of the parts).

Antianxiety drugs such as Valium (diazepam) and Librium (chlordiazepoxide) are now the most widely prescribed medicines in this country. It is estimated that about 20 million Americans are using such drugs at any particular time. Antianxiety drugs and alcohol share several functions; both have sedative and tranquilizing effects; it is suspected that they are metabolically competitive and when combined they can also produce synergistic effects. Although the directions clearly warn against the concomitant use with alcohol, several studies have revealed that most prescribing physicians do not warn their patients and that most patients are oblivious to the danger.

There are numerous additional drugs that interact deleteriously with alcohol either because they can enhance the dangers of intoxication or because alcohol can interfere with their intended purpose.

In addition, the last 35 years have witnessed a vast increase in the nonmedical use of intoxicating substances, most of which interact deleteriously with alcohol. Of particular significance is marijuana because estimates suggest that as many as 20 million people may be using the drug. When the modern wave of marijuana use in the U.S. began in the 1960s, there was a major focus on marijuana per se and users sought to experience its unique sensations. Gradually, over the last 15 years, marijuana use has become more casual and more associated with other activities, including drinking. Although, as yet, relatively little is known about alcohol–marijuana interactions, both drugs produce complex central nervous system responses that can affect motor control, reaction time, perception and judgment. Marijuana has a particular impact on time and space perception. Until the last few years, the type of marijuana most common in this country was of relatively low toxicity and its potential dangers were not considered serious. The importing of much more potent products, however, has greatly increased the toxicity of the marijuana now available. Since most people are now using more powerful marijuana products, more frequently, in association with alcohol, the potential for serious problems of marijuana–alcohol interaction appears substantial.

In summary, there have been many factors of change that have significantly altered the chemical context of alcohol use in the U.S., including the pollution of the environment with chemicals, the introduction and widespread use of many mood modifying medicines and the nonmedical drug movement. All of these developments contribute to a vastly increased potential for intoxication when alcohol and other chemical interactions occur.

Liabilities of Intoxication

During the same time period that rising rates of alcohol consumption and the advent of many new medicinal and nonmedicinal intoxicants have increased the probability for intoxication in our society, other changes in the society-at-large have increased the liability of being intoxicated.

Technological changes in industrial processes have shifted the requirements of more and more jobs from brawn to brain. In jobs requiring the expenditure of energy through hard physical labor, alcohol with its high caloric value was functional. But in jobs requiring precision in motor control, keen perception, quick reaction time, reasoning and wise judgment, alcohol, to the extent that it compromises central nervous system responses, is dysfunctional. At the same time, with the emergence of huge conglomerates, the magnitude of corporate decisions provides less and less tolerance for errors of judgment on the part of management personnel.

Changes in transportation provide a vivid example. The driver of a horse and wagon was protected by a well-trained horse. The driver of a motor vehicle is operating a complex and potentially lethal piece of machinery. As the density of traffic and the speed of automobiles have increased the demands for efficient brain function, the liability of intoxication in drivers has increased correspondingly.

Military organizations throughout the world have a history of using alcohol for its anti-inhibitory, antianxiety and high caloric properties and have valued men who could hold their liquor and drink like a soldier, a sailor or a marine. Yet, since World War II, the nature of warfare has changed drastically. The skills of personal combat, for which alcohol was thought to be functional, have been replaced by those required to operate complex weapons and weapon systems for which intoxication is strikingly dysfunctional. Today, our armed forces face a dilemma in replacing century old norms that supported heavy drinking with those that support moderation.

Alcohol and Aggressive, Violent or Criminal Behavior

Virtually all social policy directed towards controlling both drunkenness and drinking has been supported by assertions that alcohol "releases" aggression and is a major cause of crime, violence and other antisocial behavior. In 1945, psychiatrist Ralph Banay (13) concluded that alcohol might be related to aggression in a number of different ways. It could impair judgment, minimize the anticipation of punishment, lessen inhibitions, or provide a deliberately selected excuse for an act of violence or crime and affect victims as well as perpetrators. Banay concluded that while inebriety was undoubtedly a contributing factor in crime, the role of alcohol was overestimated and the contribution of alcoholism was greater in minor than in major crimes.

In its 1978 report to Congress (2), the National Institute on Alcohol Abuse and Alcoholism concluded that "violence, accidental or intentional, constitutes a substantial part of all mortality, illness and impairment in the United States" and that "alcohol often plays a major role in such violent events as motor vehicle accidents; home, industrial and recreational accidents; crime; suicide; and family abuse." However, efforts to quantify these relationships have produced widely varying estimates.

The issue was also examined in 1980 by a panel of scientists for the Institute of Medicine of the National Academy of Sciences (14). While recognizing that many people who commit or are victims of acts of violence, accidental injuries or crime are found to have been drinking or to have histories of problem drinking, the panel raised a number of basic questions for future research. For example, if violence is frequently associated with alcohol use, is the violence a result of the drinking, are people who are violent also heavy drinkers, or do people who want to be violent drink to facilitate or provide an excuse for their violence? To what extent do alcohol problems occur as one part of a cluster of difficulties experienced by problem-prone people? The panel also suggested asking whether for some people intoxication might, on occasion, prevent some problems such as beating a spouse or committing suicide.

It is apparent that we have not progressed very far towards a better understanding of alcohol's role in violent, aggressive and criminal behavior. Indeed, some of the questions that have been raised in 1980 are remarkably similar to those posed by Banay in 1945.

Of Social Costs and Social Problems

As noted previously, the problems of alcohol so permeate the behavior of society that it is difficult to measure the costs in any precise or economically meaningful way. What is the cost of a damaged fetus, or a disabled or disturbed child, or a desperate wife or husband? What is the cost of a defective automobile—the time and money spent in repairs or the consequences of the accident its defects cause? What amount of the cost of treating heart disease, cancer, diabetes, emphysema, ulcer, cirrhosis, accidental injuries and other "diseases of alcohol" can be charged to alcohol itself?

The problems of alcohol stem essentially from its excessive use. Excess is relative to the sensitivities, vulnerabilities and capacities of the individual and the norms of the social setting. The problems are manifested in three major ways—intoxication, diseases, and dependencies. Alcohol can produce intoxication that compromises the capacity of the individual to function effectively and may lead to acts of commission or omission that endanger the well-being of the drinker, other people or both. Intoxication can be acute or prolonged, incidental or repetitive. Although there is a general relationship between the amount of alcohol consumed and the degree of intoxication experienced, there is wide and significant individual variation. The term intoxication usually refers to alcohol's effect on brain function, although technically the toxicity of alcohol can affect the entire body.

People who repeatedly drink heavily, whether or not they manifest brain intoxication, run the risk of developing a variety of diseases that are alcohol-related. These diseases, such as cirrhosis, pancreatitis, cardiac myopathy and peptic ulcer, can occur without alcohol but the probability of their occurring in people who are heavy drinkers is much greater. The term "heavy" is relative and there is wide variation in organ vulnerability to alcohol. We do not yet know why some people develop cirrhosis, others pancreatitis, others an ulcer, but it is clear that alcohol consumption increases some people's vulnerability to many diseases.

People who repeatedly use alcohol quickly develop a kind of dependence, in that they rely on alcohol to alter their mood. For most people, small amounts of alcohol produce effects that they believe are beneficial or functional. They may never experience apparent problems from drinking but they will develop a kind of "low-level" dependence on alcohol to the extent that they will miss it and even feel uncomfortable if their usual drinking is omitted. Generally, problems of dependence on alcohol occur only in people who repeatedly feel the need to drink amounts that for them are "too much" and will produce undesirable effects. Problem drinking or dysfunctional dependence on alcohol can be generalized (the need is felt all or much of the time) or it can be situationally specific (the need is felt only in particular situations, such as facing a difficult task or an anxiety-provoking interpersonal confrontation, or only on weekends or in association with certain activities such as sex). Dependency needs can also stem from within the individual or from outside. Inner-felt needs are psychologically perceived. They may or may not include a physical aspect in which the absence of alcohol from body tissues produces a withdrawal or abstinence syndrome. Many people however feel that they need to drink, or to drink what for them is "too much," not because of their own psychological or physical reliance on alcohol, but to meet the powerful and important expectations of their social world. Responses to perceived requirements for excessive drinking that emanate outside of the individual constitute another kind of dependence on alcohol—social situational dependence (15).

There is in American society a conflict between the reality of individual variability in capacities for drinking and a lack of recognition of or respect for this variability in the sociocultural norms that govern the way people perceive they are supposed to drink in social settings. Because the drinking norms in many social situations are compatible with the comfortable capacity for alcohol of people who are able to drink larger than average amounts, people with limited capacities who feel that they need to meet the drinking expectations of such situations are repeatedly drinking what for them is too much.

Examples of problem drinking that develop from a dependence on meeting social expectations or demands can be found wherever we turn. Young people who seek the approval of their peers or acceptance into social groups are particularly vulnerable both because their capacties for drinking may be quite limited and because drinking is an important symbolic behavior among adolescents. Many fraternities, sororities and other peer groups evaluate aspirant members in terms of their drinking performance. The adult society is little different. Group practices such as bar hopping (having a drink in each bar) and rounding (taking turns paying for a round of drinks) put pressure on the person who can comfortably handle only one or two drinks to keep up with those who can handle more. Whether in a car pool of workers who stop for beer on their way home or in a martini-drinking lunch group of corporate executives, or a guest at a friend's home, there is often pressure on moderate drinkers to drink "more." I suggest that social dependence on drinking is the basis for a large amount of problem drinking in the United States, especially among young people who often drink too much in response to social needs before they have had a chance to recognize or appreciate the psychologically reinforcing qualities of moderate drinking.

The Costs of Alcohol Permeate Society

We have suggested that the costs of alcohol involve the consequences of drinking too much, that what constitutes too much is a matter of individual variation and that these costs are manifested in intoxication, in medical complications and in dysfunctional dependencies on alcohol. Costs include our still tentative assumptions about the teratological properties of alcohol and the fetal alcohol syndrome. They include the problems of infant and child care that occur when one or both parents drink too much, marital problems, the role of alcohol in the expression of violence against children, spouses or others, and the emotional trauma of living with problem drinkers. They include school children who underachieve, are absent, are unhappy or get into trouble because their parents or they themselves drink too much. They include the combination of factors—genetic, psychological, sociocultural—associated with the greater risk that the children of problem drinkers will become problem drinkers. They include the fact that people who drink too much greatly increase the probability that they will experience serious illness or injury, will occupy a disproportionate number of hospital beds, and will utilize a disproportionate share of other health resources. They include the damage that drinkers can do to themselves and to others when they operate automobiles, airplanes or other complex machinery while intoxicated. They include the lost productivity, wasted training, costly rates of turnover and absenteeism of problem-drinking workers. The costs of alcohol involve the family, business and industry, churches, health care, welfare, criminal justice, recreation, transportation, communication, government and politics.

In concluding this chapter on social costs of alcohol in the perspective of change, let us consider the implications of certain changing perceptions for future social efforts to contain or reduce "costs."

With the exception of a simple dichotomy between "social" or "normal" drinking and "problem" drinking or alcoholism, education and public information about alcohol in the past has tended to emphasize similarities in the way people react to alcohol. Guidelines for safe drinking have been based on averages that have exceeded the safe boundaries of many people. Numerous factors that we now know provide significant variability in the way people respond to alcohol generally have been unrecognized or ignored. Problem drinkers have been held individually responsible for drinking too much. Social interventions have been designed to treat, control or influence individual behavior.

Two changes in our current perceptions about drinking suggest the need for modified social policy and social responses. On the one hand we now recognize that individual variability in responses to alcohol is more significant than we believed in the past. On the other, we recognize that sociocultural norms and group pressures play a more significant role than we had acknowledged in supporting or pushing individuals to use what for them is too much alcohol. We also recognize increasingly the role of social situations in reinforcing the pharmacological meaning of alcohol.

Thus, it is important that future strategies for the containment of the social costs of alcohol include *(1)* an educational approach that enables individuals to recognize their own unique responses to alcohol as a basis for wise personal decisions about drinking, and *(2)* a program of prevention which emphasizes the importance of social forces in supporting excessive drinking and which acknowledges social as well as individual responsibility for reducing the costs of alcohol problems.

References

1. U.S. NATIONAL INSTITUTE ON ALCOHOL ABUSE AND ALCOHOLISM. Second special report to the U.S. Congress on alcohol and health, June 1974. Mark Keller, editor. (DHEW Publ. No. ADM 75–212.) Washington, D.C.; U.S. Govt Print. Off.; 1974.
2. U.S. NATIONAL INSTITUTE ON ALCOHOL ABUSE AND ALCOHOLISM. Alcohol and health; third special report to the U.S. Congress from the Secretary of Health, Education and Welfare, June 1978; technical support document. Ernest P. Noble, editor. (DHEW Publ. No. ADM-79-832.) Washington, D.C.; U.S. Govt Print. Off.; 1979.
3. BACON, S. D. Alcohol and complex society. Pp. 179–200. In: Alcohol, science and society; twenty-nine lectures with discussions as given at the Yale Summer School of Alcohol Studies. New Haven; Quarterly Journal of Studies on Alcohol; 1945.
4. U.S. SURGEON GENERAL'S OFFICE. Healthy people; the Surgeon General's report on health promotion and disease prevention. (DHEW Publ. No. PSH 79–55071.) Washington, D.C.; U.S. Govt Print. Off.; 1979.
5. JELLINEK, E. M. The problems of alcohol. Pp. 13–29. In: Alcohol, science and society; twenty-nine lectures with discussions as given at the Yale Summer School of Alcohol Studies. New Haven; Quarterly Journal of Studies on Alcohol; 1945.
6. STRAUS, R. and BACON, S. D. Alcoholism and social stability; a study of occupational integration in 2023 male aocoholism clinic patients. Quart. J. Stud. Alc. **12**: 231–260, 1951.
7. KNUPFER, G. Epidemiologic studies and control programs in alcoholism. V. The epidemiology of problem drinking. Amer. J. publ. Hlth **57**: 973–986, 1967.
8. CAHALAN, D., CISIN, I. H. and CROSSLEY, H. M. American drinking practices; a national study of drinking behavior and attitudes. (Rut-

gers Center of Alcohol Studies, Monogr. No. 6.) New Brunswick, N.J.; 1969.

9. CAHALAN, D. Problem drinkers; a national survey. San Francisco; Jossey-Bass; 1970.

10. CAHALAN, D. and ROOM, R. Problem drinking among American men. (Rutgers Center of Alcohol Studies, Monogr. No. 7.) New Brunswick, N.J.; 1974.

11. FILLMORE, K. M., BACON, S. D. and HYMAN, M. The 27-year longitudinal panel study of drinking by students in college, 1949–1976; final report. (Report No. PB 300–302.) Springfield, Va.; National Technical Information Service; 1979.

12. STRAUS, R. and BACON, S. D. Drinking in college. New Haven; Yale University Press; 1953.

13. BANAY, R. S. Alcohol and aggression. Pp. 143–150. In: Alcohol, science and society; twenty-nine lectures with discussions as given at the Yale Summer School of Alcohol Studies. New Haven; Quarterly Journal of Studies on Alcohol, 1945.

14. INSTITUTE OF MEDICINE. DIVISION OF HEALTH PROMOTION AND DISEASE PREVENTION. Alcoholism, alcohol abuse and related problems; opportunities for research. (Publ. No. IOM 80-04.) Washington, D.C.; National Academy Press; 1980.

15. STRAUS, R. The challenge for reconceptualization. J. Stud. Alc., Suppl. No. 8, pp. 279–288, 1979.

9

An Animal Model of Alcoholism

David Lester

Species other than man were not mentioned in the progenitor of the present volume. Despite the facts and ideas presented there, many of which owed their being to the lecturer's own dependence in research on the use of infrahuman species (the rat and dog, for example), the audience gained no inkling of animals' importance in our knowledge and understanding of the issues involved with the pharmacological, physiological, nutritional, psychological and behavioral effects of alcohol. This discussion focuses on one such use of animals.

Let me begin by restating a thought expressed in the Foreword of the first *Alcohol, Science and Society*: Alcoholism—its definition(s) you will gather elsewhere in these chapters—is an especially complicated problem because the great majority of users of alcoholic beverages do not become alcoholics. Are the experimental animals we use also subject to these constraints? Are we limited as to species, perhaps only to homo sapiens, and to animals with specific characteristics within those species?

In the study of reproductive physiology, cancer, teratology and gerontology, nutrition, drug metabolism, and hereditary and other disease processes, the list of mammalian species used in research is long indeed, ranging from the cat to (probably) the zebu, the latter a bovine mammal, but certainly including cattle, chickens, dogs, ducks, hamsters, horses, mink, mice, primates, rabbits, rats, sheep and swine (1). The processes modeled in the assorted species also range from albinism to the all-encompassing zoonoses, including anemias, cataracts, diabetes, dwarfism, glaucoma, hydronephrosis, jaundice, obesity and yes, even systemic lupus erythematosus!

The studies use infrahuman species in which to model a process, not wholly because it is cheaper and easier to do so, but because it is possible by using experimental animals to exercise a profound degree of control. Naturalistic observations, on the other hand, are uncontrolled. The complexity of most processes also demands that they be dissected and examined in isolation for greater control and especially for such simplification as may make understanding them less difficult. It is not only in the medical sciences that models are widely used. In engineering, models are more economical to construct than the prototype; more important, the physics underlying the model may be understood better than the complex process of which it is a part. The model reduces the number of variables by introducing concepts of dimensional analysis and similarity, concepts familiar to most of us who have debated whether saccharin is a carcinogen or not, and thus makes the problem easier to handle (not necessarily of political issues such as the saccharin one) and the results more generally applicable. Let me reiterate that many models may be needed to gain understanding of a complex process, each model being studied representing a part of the process, each to be studied independently and interactively.

As I have indicated, models are, of course, used in biology and medicine. It should be underscored that the model, though similar, is not the same as the actual process. This point has been well made by Kac (2) for mathematical models, and is certainly applicable to biological models, that, "Models are, for the most part, caricatures of reality, but if they are good, then, like good caricatures, they portray, though perhaps in distorted manner, some of the features of the real world." Let there, however, not be too much distortion. In the present controversy (3), for example, about primate "language," the point is well-made that a study of sign language taught to apes can be expected to cast as little light on human language as a study of human jumping can cast light on the mechanism of bird flight: the trap is that flying is more than rising into the air without the aid of special equipment!

With the exception perhaps of the excessive consumption of alcohol, differences among alcoholic individuals are numerous. Although descriptions of real alcoholics in their own life settings are necessary, the real world of the alcoholic is not easily amenable to manipulation. Even experiments inducing intoxication in alcoholics (4), while approximating drinking in the natural setting, still depend on subject selection and lack control of a host of other variables. The use of animal models provides an opportunity to study etiological parameters, symptoms, prevention and cure of a disorder in a variety of controlled situations not possible with humans either in a natural or laboratory setting.

Here I draw, to an extent, on certain arguments which my long-time friend and colleague, Dr. Earl X. Freed, and I set out some years ago (5), suitably altered to correspond to the passage of time. The criteria for an animal model of alcoholism must, of course, correspond to essential dimensions and topology of the original. In human addiction to alcohol, it is evident that more than one variable is etiologically influential and that the multiple variables interact. Many genetic, physiological, social and cultural influences have been described and the phenotype we denote as alcoholism is probably expressed in those individuals where the required constellation of environmental factors acts on a suitably disposed substrate. Since social and cultural influences are not likely to be duplicated in an animal model, except perhaps trivially and tangentially, it is the biological matrix—in a global sense—on which the burden falls for representation (6).

Should the biological contribution carry little weight in the configuration of forces producing alcoholism, it may well be that an animal model of alcoholism is unattainable. The criteria for an animal model may thus not be possible to meet although certain components of the complete model are assuredly amenable to study and have, in fact, received considerable attention.

Let me now enumerate six successive criteria for an animal model. These are:

1. Substantial oral ingestion of alcohol without deprivation of food and in the presence of other competing fluids;

2. Ingestion directed to the intoxicating character of alcohol, with intoxicating levels of circulating blood alcohol reached;

3. Work performed, even in the face of aversive consequences, to obtain alcohol;

4. Intoxication sustained over a lengthy period; sufficient for the

5. Production of a withdrawal syndrome and the development of physical dependence; and

6. After an abstinent period, reacquisition of drinking to intoxication and reproduction of the alcoholic process.

It is easiest to dispose of stages 4 and 5. There is ample evidence that, under a variety of involuntary circumstances, intense intoxication with alcohol sustained over even a short period (7) will produce many signs corresponding to the withdrawal syndrome in man, ranging from tremors and hyperactivity to hallucinations and elicited and spontaneous convulsions.

The involuntary circumstances to which I refer are the many ways by which alcohol can be introduced into an animal. These include oral intubation, intraperitoneal injection, respiration of air containing alcohol, and intake of an alcohol-containing liquid diet as the sole source of food, fluid, and calories. The amount and rate of alcohol ingestion can be manipulated and various quantitative features of the production of the development of tolerance and of the withdrawal syndrome in the rat and the mouse are now well-described (8). It is clear from this work that alcohol acts on many organ systems and that the temporal course of withdrawal symptomatology reflects the waxing and waning of activity of more than one portion of the central nervous system. The ability to produce the withdrawal syndrome means that the neurochemical mechanisms underlying the various symptoms can be studied: through manipulation by drugs, for example, of the variety of neurotransmitters that may be involved. Not too far distant in time then should be our ability to inhibit, rationally, such life-threatening aspects of withdrawal as delirium tremeus, even before alcohol ingestion terminates.

Before turning to the first criteria for an animal model, I should point out that the sixth is related to the widely held belief that physical dependence on alcohol drives the alcoholic to begin drinking after a previous episode which has had withdrawal as its sequel. Yet in other species, it has not been possible, even after long-sustained periods of alcohol intake, where intoxication was a prominent feature, to show a turn to the voluntary ingestion of alcohol in the quantity and at a rate which would produce a second episode of withdrawal.

As in man, the effects of alcohol in animals are related to the rate at which alcohol is drunk. The amount of alcohol ingested by an animal increases to a degree with the length of time the alcohol is available. The greater the rate of intake and the more prominent the central effects of alcohol, the more likely is tolerance to be acquired and the withdrawal syndrome to appear once access to alcohol is terminated. These consequences of alcohol use show only that its potency as an addictive agent is substantially less than that of heroin. None of this information, unfortunately, tells us how, in the first instance, one induces an animal, on its own volition, to ingest alcohol in a dosage substantial enough to produce the results we seek for our model. If the first three criteria can be met then the fourth and fifth will inevitably be met.

To meet the first criteria, alcohol ingestion must result from directed behavior: the animal must work for the alcohol. When the laboratory animal drinks an alcohol solution in competition with response tendencies to behave in other ways, such behavior is akin to the compulsive component of drinking by alcoholics; the alcohol-seeking behavior of the animal is a measure of its motivation for alcohol. Although motivational aspects of alcoholism appear to be the most significant criterion for an animal model, they have received the least attention, perhaps because they are also the most elusive requirement to attain.

A variety of anecdotal, clinical and experimental evidence argues that alcohol intake is a strongly reinforcing behavior in some humans, based wholly, it would appear, on the positive central nervous system benefits which alcohol confers. In these humans, drinking alcohol becomes a compulsive need, resistant to change, and a dominant theme of life. The behavior seems to underlie motivations conceptualized as tension reducing (9). Human beings know that alcohol gives a lift, that it is an intoxicant and tranquilizer, and that it exercises an anxiety-attenuating effect on behavior even before the first drink because the effects of alcohol have been described by the written word and its actions portrayed in the public media. This knowledge of alcohol lore precedes actual experience and undoubtedly plays an important role in the genesis of alcohol-seeking behavior. It contributes to the positive reinforcement which accrues when, one day, the individual does drink alcohol and experiences at first hand alcohol's palliative, ameliorative and other properties. A discomforting feeling, depressive mood, or disquieting internal state such as conflict is modified favorably or attenuated as a consequence of drinking. Learning occurs and in similar future circumstances one can understand how the individual might again resort to alcohol. To what extent the effects of alcohol are preconditioned by genetic or cultural susceptibility, or are psychopharmacological and primarily induced by alcohol per se, or represent psychodynamic expectations by the drinker, are questions remaining to be answered. In any event, the fact seems to be that the act of drinking becomes a powerful reinforcer in some individuals.

But how does the experimental animal learn about alcohol? What does it detect when it drinks an alcohol solution? Besides taste, the animal usually learns that alcohol has caloric value: a large number of experiments using rats attests to this. Whether the sine qua non of the central effects of alcohol are appreciated by the animal and then connected to the previous alcohol ingestion seems a quite different matter; the gap in time between drinking and awareness of alcohol's central effects seems sufficiently long to preclude such an association. Of course, the search for the animal's response specifically to alcohol would have little meaning if alcohol possessed no important pharmacodynamic attributes—indeed there would be no problem. But alcohol does have important central actions which any animal model must take into account, and until it can be shown that the animal drinks alcohol for the same kind of central reinforcement that human beings seek, then so-called animal models will remain mere analogues bearing only a superficial resemblance to alcoholism.

Thus, some attempts at studying animals' drinking behavior

only simulate or mimic, rather than model, human drinking. If an animal model is feasible (and you may even believe now that I have considerable doubts about this), then the model should be one in which the internal phenomenal states of the animal closely approximate those of the alcoholic. In a simulation, it is only the external manifestations, the outward appearances which are represented. It is no difficult feat to achieve high blood alcohol concentrations in, for example, rats constrained to drink alcohol by fluid or food deprivation, but where the alcoholic drinks from strong inner motivations, the rat is constrained to do so by equally compelling external manipulations alien to man. The net result for both rat and man is the excessive consumption of alcohol without a model of functional relationships having been attained.

A wealth of literature demonstrates how alcohol affects experimental neurosis and conflict. Absent, however, are demonstrations that animals select alcohol to attentuate tension states. There have been several failures (10, 11) to induce rats to emit operant responses for alcohol or to drink an alcohol solution in response to stress; rats more readily drank larger amounts of alcohol when it was freely available as the only fluid than when its availability was contingent on operant behavior. On the other hand, it is also reported that stressed rats drink alcohol in a choice situation (12, 13), and operant performance for alcohol has been found in rats (14); in a more recent study with monkeys, however, addictive-type drinking was not produced in a stressful avoidance situation (15).

These findings suggest that alcohol can, and does, reduce tension when its administration is controlled by the experimenter, but it also suggests either that animals have not learned this or that self-selection of alcohol by animals may not be based on this need. If animals select alcohol for "reasons" unrelated to tension reduction, it constitutes an important drawback to research involving animal models of psychopathology because an underlying hypothesis for such paradigms has always been that conflict, stress and anxiety should result in increased alcohol intake. I have indicated elsewhere (16) that: "If stress and anxiety are involved in the mechanism of addiction, then the ingestion of alcohol to reduce these states seems eminently reasonable. In the absence of any showing that enough alcohol has been ingested to produce such a reduction, however, the nexus of alcohol ingestion and degree of stress is nebulous. Experiments directed to the production of various levels of stress or anxiety and the concomitant measurement of alcohol selection might well illuminate this relationship." Such experiments have not been performed. There are barriers to be overcome for its successful accomplishment. What makes for a necessary and sufficient psychological or behavioral stressor? How do we acquaint animals a priori with alcohol's tension-reducing character, as in human social learning, without at the same time producing a learned aversion to the taste of alcohol?

There are a plethora of reports in the scientific literature of the last 40 years on the selection of alcohol by animals. By and large, animals choose to drink water rather than an alcohol solution, but when alcohol is selected, the amount drunk rarely, if ever, produces intoxication. In the usual experiment, the choice of fluids is restricted to water and a solution of alcohol. When the choice is broadened, the essential unimportance of any alcohol choice is highlighted by the fact that a third fluid choice of, for example, a sugar solution (or other sweet or caloric fluid) eliminates what might have been regarded as a preference for a solution of alcohol (17). Obviously, those humans who choose to drink an alcoholic beverage do so in the presence of a limitless variety of other fluids. In a way, therefore, self-selection experiments in animals, which are limited to a choice of two fluids, increase the likelihood that an alcohol solution will be chosen, thus decreasing the generalizability of such a model. Nonetheless, experiments of this nature continue to be carried out; although some elements of value unquestionably will result, it is well to recognize the restricted nature of any findings.

Two groups of investigators have, in fact, selectively bred lines of rats which exhibit a greater or lesser degree of "preference" for an alcohol solution. In Finland (18), these lines have been designated, fancifully and optimistically, AA for alcohol-addicted, and ANA for alcohol-nonaddicted. It does not appear that AA rats drink alcohol at a rate productive of frank intoxication; ANA rats choose virtually none of the alcohol solution. In the United States, P (for preferring) and NP (for nonpreferring) lines of rats have been bred (19). Although only slightly elevated blood alcohol concentrations are ever found in the P line, the daily intake of alcohol approaches that maximally metabolized in 24 hr by the rat, and after a period of many months of such intake, minimal signs of a withdrawal syndrome are apparently noted when alcohol intake is terminated. Although certain features of these experiments may be important for understanding various effects of chronic intake of alcohol (also attainable in other ways) they do not appear to meet the decisive criteria for an animal model of alcoholism.

One of the most astonishing methods reported to increase choice of an alcohol solution is the injection of minute amounts of alcohol into the cerebral ventricles (20, 21). Curiously, a lengthy, and surprising, list of other compounds also have been reported (22), by essentially one group of investigators, to produce this increase. In some cases, the rat reportedly chooses to reject water entirely and drink only the alcohol solution profferred, for long periods of time. Most recently, the compounds injected (23) have been compounds which can be formed by the reaction of acetaldehyde, arising from alcohol metabolism within the body, and one or another of neurotransmitter materials occurring in the brain (24). Given the obvious speculations and hypotheses that can arise, it is a matter of some dismay that, with one exception (25), the work (i.e., using alcohol and other tested compounds) has not received independent confirmation, either in the rat (26), or in other species (27). Although the reasons for the disagreement are, to say the least, controversial, it seems, nonetheless, to be clear that even if alcohol is chosen, it is chosen in a restricted two-fluid situation; moreover, there has been no rationale advanced for focusing on an alcohol solution as the second fluid choice: why should the ventricular stimulus not be directed, say, to a solution of sodium lactate? And finally, there is a conceptual block: if alcohol (or the condensed product of acetaldehyde and a neurotransmitter) injected into the ventricles gives rise to a "preference" for alcohol, why should not orally ingested alcohol (and acetaldehyde from the normal metabolism of such alcohol and its condensed product) not also yield the same result? Yet, as I have already implied, even the largest of long-term alcohol intakes, with alcohol also appearing in the ventricles, does not thereafter lead to any increase in the voluntary intake of alcohol.

What appeared to some investigators (28) as the dead-end nature of most "preference" experiments has led to the selective breeding of mouse and rat lines differing in their pharmacological response to alcohol. The rationale for such breeding is that human beings who become addicted to alcohol appear in many cases to respond differently towards alcohol than those not so addicted; at the very least, they do not find alcohol ingestion to be aversive. If there are biological differences between alcoholics and nonalcoholics, it appears sensible to begin to study animals which show disparate reactions to alcohol. Thus, two lines of mice (29), LS (for "long-sleeping") and SS (for "short-sleeping") exhibit a wide range of differences, in addition to the phenotype of a 3-fold difference in the duration of anesthesia which is the basis for the breeding; the dose of alcohol is, of course, a nontrivially elevated single dose of 3.5 to 4.0 g of alcohol per kg of body weight, about 30% of the alcohol which a mouse can metabolize in 24 hr. Here at Rutgers (30), we have been selectively breeding rats for more than 20 generations using a moderately intoxicating dose of alcohol (1.5 g per kg) and selecting lines which exhibit greater (MA for most affected) and lesser (LA for least affected) degrees of motor impairment after alcohol. The group in Finland, who bred AA and ANA rats, have also begun to breed two lines of rats on the basis of differing degrees of motor impairment. In neither the mice nor the rats do there appear to be differences in the rates at which alcohol is metabolized; the differences are then at the level of central nervous system responses to alcohol.

Whether these lines of mice or rats will prove useful in meeting the first criteria of an animal model is not yet clear. One can as easily argue, for example, that a line with enhanced sensitivity to alcohol (LS, MA) would be more easily addicted to alcohol as that a more resistant line (SS, LA) would exhibit a lesser aversion to alcohol. Such a decision, now impossible to make, will undoubtedly appear, pragmatically, as the result of experiment, not conjecture. In any event, the advantage of such hereditary or genetic models is great: although the neurochemical processes underlying alcohol's effects are hardly simple, the availability of mouse and rat lines representing extremes of response to alcohol should make the task easier. Let me note, too, that the differences in the effects of alcohol emphasize the role of biological constitution. One should, however, note the caveats drawn against the use of albino animals, especially the laboratory rat, as being unrepresentative, biologically, of pigmented mammals (31).

Alcohol has, as you know, a distinctive taste and odor. Under ordinary circumstances, animals in different species show a broad range of responses to varying concentrations of alcohol. Its taste is without question recognized by most animals, even though drinking it in quantity, in solution, is only appreciated by few. It is, however, a relatively simple matter (32) to establish an aversion to an alcohol solution, especially in the rat. In this species, associating the ingestion of a fluid, not previously experienced, with a following episode of sickness, separately induced, produces a long-lasting and intense aversion to the conditioned solution (33). The conditioning is impressive and it occurred to us that one might as easily, and as strongly, condition not aversion, but preference.

By associating *recovery* from sickness with alcohol ingestion, we have been able to increase, substantially and significantly, the voluntary alcohol intake by rats in a choice situation (34). This was accomplished by making rats severely deficient in Vitamin B_1 (thiamine), thus producing sickness, and then alleviating the sickness abruptly by an intramuscular injection of thiamine, and associating the recovery from this sickness with the drinking of an alcohol solution. The increase thereafter in selection of an alcohol solution persisted for at least 10 days. Numerous variations on this theme can be played, including how sickness is induced, all designed to convince the rat that alcohol is good for what ails it! Since the association would be uniquely with an alcohol solution, the procedure seems likely to produce a situation in which the rat might choose an alcohol solution not only over water, but in preference to sugar, saccharin and other "naturally" preferred flavors or sources of calories.

I daresay that I have left you all in a quandary, with more questions than answers. We have traveled far in the years since the lectures in 1945. No one talked then of an animal model of alcoholism. And it is not at all certain that when the lectures reappear 35 years hence that we will have an animal model, but we may then be in a position either to assert or to reject the possibility of such a model. We can be sure, however, that the availability of such a model will be of help in the real world of human beings.

How will such a model, or models, help? I have already indicated that our present models of the withdrawal syndrome in rats, whose convulsions are the counterpart of delirium tremens, allow the testing in animals of a variety of treatments designed to inhibit the appearance of the delirium as intoxication ebbs. Of no less interest will be the definition of the biochemical and physiological attributes which determine sensitivity to the effects of alcohol, as well as those attributes which divide animals with lesser or greater desires for alcohol when given a choice. Because of the biological unity of mammalian species, we believe that such infrahuman models will allow manipulations whose goal is a rational therapy of the afflictions which alcohol bestows on some men.

There are thus two aspects of successful animal models: the simpler will allow us to study remediation of the effects of alcohol, whether of delirium tremens or fatty liver, for example; the other aspect is prevention: can we elucidate those constitutional and biological factors which predispose to a heightened need for alcohol, and then intervene to reduce the likelihood that they will be manifested in addictive drinking behavior?

I am reminded of the temperance lecturers who once exhorted their audiences, with eloquence, on the subject of drink, occasionally even demonstrating the agonies of a worm immersed in a glass of whisky—assuredly a model you should now find unacceptable as a model of the effects of alcohol in humans. Sometimes, however, the lecturers would go up the evolutionary ladder—and shades of the precursors of laboratory experiments on self-selection—inquiring of their hearers: "If I put a pail of water and a pail of whisky in front of a hard-working donkey in the fields, which would he drink?" From a not wholly convinced audience would come the lusty bellow: "The water." "That's true, my friend. And why would he drink the water?" "Because he's a jackass!" was the immediate reply.

It's not at all sure that we can have an animal self-selection model like the human one, unless—remember our selectively bred lines—we devote our attention to rats who do not behave like jackasses!

REFERENCES

1. MITRUKA, B. M., RAWNSLEY, H. M. and VADEHRA, D. V. Animals for medical research; models for the study of human disease. New York; Wiley; 1976.

2. KAC, M. Some mathematical models in science. Science **166**: 695–699, 1979.

3. WADE, N. Does man alone have a language? Apes reply in riddles, and a horse says neigh. Science **208**: 1349–1351, 1980.

4. MENDELSON, J. H., ed. Experimentally induced chronic intoxication and withdrawal in alcoholics. Quart. J. Stud. Alc., Suppl. No. 2, 1964.

5. LESTER, D. and FREED, E. X. Criteria for an animal model of alcoholism. Pharmacol. Biochem. Behav. **1**: 103–107, 1973.

6. FRIEDMAN, H. J. and LESTER, D. A critical review of progress towards an animal model of alcoholism. Pp. 1–19. In: BLUM, K., BARD, D. L. and HAMILTON, M. G., eds. Alcohol and opiates; neurochemical and behavioral mechanisms. New York; Academic Press; 1977.

7. MAJCHROWICZ, E. Induction of physical dependence upon ethanol and the associated behavioral changes in rats. Psychopharmacologia, Berl. **43**: 245–254, 1975.

8. MELLO, N. K. A review of methods to induce alcohol addiction in animals. Pharmacol. Biochem. Behav. **1**: 89–101, 1973.

9. CAPPELL, H. and HERMAN, C. P. Alcohol and tension reduction; a review. Quart. J. Stud. Alc. **33**: 33–64, 1972.

10. FREED, E. X. Failure of stress to increase alcohol consumption by rats. Newslett. Res. Psychol. **9** (No. 3): 24–25, 1967.

11. MYERS, R. D. and HOLMAN, R. B. Failure of stress of electric shock to increase ethanol intake in rats. Quart. J. Stud. Alc. **28**: 132–137, 1967.

12. ADAMSON, R. and BLACK, R. Volitional drinking and avoidance learning in the white rat. J. comp. physiol. Psychol. **52**: 734–736, 1959.

13. CASEY, A. The effect of stress on the consumption of alcohol and reserpine. Quart. J. Stud. Alc. **21**: 208–216, 1960.

14. MELLO, N. K. and MENDELSON, J. H. Operant performance by rats for alcohol reinforcement; a comparison of alcohol-preferring and nonpreferring animals. Quart. J. Stud. Alc. **25**: 226–234, 1964.

160   David Lester

15. MELLO, N. K. and MENDELSON, J. H. Evaluation of a polydipsia technique to induce alcohol consumption in monkeys. Physiol. Behav. **7**: 827–836, 1971.
16. LESTER, D. Self-selection of alcohol by animals, human variation and the etiology of alcoholism; a critical review. Quart. J. Stud. Alc. **27**: 395–438, 1966.
17. LESTER, D. and GREENBERG, L. A. Nutrition and the etiology of alcoholism; the effect of sucrose, fat and saccharin on the self-selection of alcohol by rats. Quart. J. Stud. Alc. **13**: 553–560, 1952.
18. ERIKSSON, K. Rat strains specially selected for their voluntary alcohol consumption. Ann. Med. exp. Fenn. **49**: 67–72, 1971.
19. LUMENG, L., HAWKINS, T. D. and LI, T.-K. New strains of rats with alcohol preference and nonpreference. Pp. 537–544. In: THURMAN, R. G., WILLIAMSON, J. R., DROTT, H. and CHANCE, B., eds. Alcohol and aldehyde metabolizing systems. Vol. III. Intermediary metabolism and neurochemistry. New York; Academic Press; 1977.
20. MYERS, R. D. Alcohol consumption in rats; effects of intracranial injections of ethanol. Science **142**: 240–241, 1963.
21. MYERS, R. D., VEALE, W. L. and YAKSH, T. L. Preference for ethanol in the rhesus monkey following chronic infusion of ethanol into the cerebral ventricles. Physiol. & Behav. **8**: 431–435, 1972.
22. MYERS, R. D. and VEALE, W. L. Alterations in volitional alcohol intake produced in rats by chronic intraventricular infusions of acetaldehyde, paraldehyde or methanol. Arch. int. Pharmacodyn. **180**: 100–113, 1969.
23. MYERS, R. D. and MELCHIOR, C. L. Alcohol drinking; abnormal intake caused by tetrahydropapaveroline in brain. Science **196**: 554–556, 1977.
24. COLLINS, M. A. Identification of isoquinoline alkaloids during alcohol intoxication. Pp. 155–166. In: BLUM, K., BARD, D. L. and HAMILTON, M. G., eds. Alcohol and opiates; neurochemical and behavioral mechanisms. New York; Academic Press; 1977.
25. DUNCAN, C. and DEITRICH, R. A. A critical evaluation of tetrahydroisoquinoline induced ethanol preference in rats. Pharmacol. Biochem. Behav. **13**: 265–281, 1980.
26. FRIEDMAN, H. J. and LESTER, D. Intraventricular ethanol and ethanol intake; a behavioral and radiographic study. Pharmacol. Biochem. Behav. **3**: 393–401, 1975.
27. KOZ, G. and MENDELSON, J. H. Effects of intraventricular ethanol infusion on free choice alcohol consumption by monkeys. Pp. 17–24. In: MAICKEL, R. P., ed. Biochemical factors in alcoholism. New York; Pergamon; 1967.
28. FREED, E. X. and LESTER, D. Schedule-induced consumption of ethanol; calories or chemotherapy? Physiol. & Behav. **5**: 555–560, 1970.
29. McCLEARN, G. E. and KAKIHANA, R. Selective breeding for ethanol sensitivity in mice. Behav. Genet. **3**: 409–410, 1973.

30. RILEY, E. P., FREED, E. X. and LESTER, D. Selective breeding of rats for differences in sensitivity to alcohol; an approach to an animal model of alcoholism. I. General procedures. J. Stud. Alc. **37**: 1535–1547, 1976.

31. CREEL, D. Inappropriate use of albino animals as models in research. Pharmacol. Biochem. Behav. **12**: 969–977, 1980.

32. LESTER, D., NACHMAN, M. and LEMAGNEN, J. Aversive conditioning by ethanol in the rat. Quart. J. Stud. Alc. **31**: 578–586, 1970.

33. NACHMAN, M., LESTER, D and LEMAGNEN, J. Alcohol aversion in the rat; behavioral assessment of noxious drug effects. Science **168**: 1244–1246, 1970.

34. BASS, M. B. and LESTER, D. Increased alcohol selection in rats after alcohol drinking paired with recovery from thiamine deficiency. Physiol. & Behav. **19**: 293–302, 1977.

10

Alcoholism and Heredity: Update on the Implacable Fate

Donald W. Goodwin, M.D.

> Question: You referred to a sample of over 4,300 inebriates of whom 52% came from alcoholic parents. What was the extent of the alcoholism of the parents?
>
> Answer: The alcoholism in those parents was real honest-to-goodness inebriety. . . . The data do not mean the alcoholism was transmitted biologically. It was transmitted socially (1, *p. 113*).

This no-ifs-ands-or-buts reply was furnished by E. M. Jellinek to a question put by a student at the Yale Summer School of Alcohol Studies in 1944. How did this most undogmatic of alcohologists come to such a dogmatic conclusion?

He never really said. From his other comments, however, I can suggest three possibilities. Although Jellinek recognized that alcoholism was transmitted in families, he was equally impressed by the number of families in which it was not. Somehow this suggested to him that "if a hereditary constitutional factor is present it does not become operative without intercurrent social factors."

Second, Jellinek was indeed antidogmatic. One of the dogmas of the day, held by almost everybody, was that alcoholism was hereditary. Since there was almost no evidence for this except for the familialness of the condition, Jellinek naturally rebelled, although his rebellion might have gone a bit too far.

Third, Jellinek fell back on a non sequitur that one still hears on the alcoholism lecture circuit. It goes like this: If alcoholism is hereditary, it cannot be treated. If it cannot be treated, it must not be hereditary. Jellinek surely perceived the illogic of this, but still believed therapists should not be faced by "impacable fate," i.e., heredity.

This is silly, of course, since many hereditary illnesses are

162

treatable and treatability is not relevant to the issue of etiology in the first place.[1]

Still, Jellinek knew what questions to ask. If anything is inherited, What is it? Does it involve tolerance? Are there internal "musts" about alcoholism which have to be present for the illness to appear? Is a predisposition to alcoholism, a "readiness to acquire the disease," reflective of a specific proneness to excessive drinking or a manifestation of general psychological proclivities which favor the development of alcoholism if no other "escape" is available? This chapter will discuss these questions in light of recent evidence.

The past 35 years have produced even more evidence that alcoholism is a family disorder. One reviewer (2) found some 140 studies reporting on the prevalence of alcoholism in families of alcoholics; all showed a greatly increased risk of alcoholism in both male and female relatives of alcoholics compared to relatives of nonalcoholics.

Not everything transmitted in families is hereditary. Speaking French is transmitted by families but is not hereditary. There are several ways "nature" and "nurture" can be separated; all have now been applied to alcoholism. One is the genetic marker approach, another involves studying twins, and a third uses the "experiment of nature" of adoption to separate genetic or at least congenital influences from those apparently due to upbringing.

When Jellinek discussed the subject in 1944 there existed only one nature–nurture study, the one described by Anne Roe in the chapter just after Jellinek's in the original book with this title (3). Roe studied children of heavy drinkers who had been raised by foster parents, by then in their early thirties. None was an excessive drinker. Roe concluded that heredity was not a factor in alcoholism. Jellinek no doubt knew about the Roe study—both were at Yale— but did not refer to it in his 1944 Summer School talk. Since Roe's findings differ from those in subsequent reports, I will return to them later.

There have been many genetic marker studies (4). A genetic marker is a trait known to be inherited, usually in a straight-forward

---

[1]That Jellinek had mixed feelings about alcoholism and heredity was evident when he came up with a second non sequitur of even greater beauty: "The only permissible conclusion is that not a disposition toward alcoholism is inherited but rather a constitution involving such instability as does not offer sufficient resistence to the social risk of inebriety" (1). No better example of having your cake and eating it can be found in the alcoholism literature.

Mendelian fashion, such as blood groups, colorblindness, and certain diseases. If more alcoholics have the "marker" than would be expected by chance, than this would indicate that alcoholism is influenced by heredity.

Many studies show an association of alcoholism with some marker; however, for almost every study showing an association, another study shows none. Studies of colorblindness have been most consistent. A group in Chile first reported that alcoholics often were colorblind. Other groups had similar observations, but found that the colorblindness went away in time. In short, the colorblindness apparently was of nutritional or other nongenetic origin.

There has been a number of studies of twins. In perhaps the most ambitious, identical twins were found to be more concordant for alcoholism than fraternal twins (5). Other studies showed some difference in drinking habits between identical and fraternal twins, implying a genetic factor, but alcoholism as usually defined seemed under little or no genetic control (6).

Three adoption studies have been published in the last few years. The first, conducted in Denmark (4), interviewed children of alcoholics raised in nonalcoholic foster families. By their late twenties the boys were four times more likely to be alcoholics than adopted boys of nonalcoholic parentage. By their mid-thirties the girls were four times more likely to be alcoholics than nonadopted controls. But the adopted girl controls also had a high rate of alcoholism, raising the possibility that adoption itself (for women) may favor the development of alcoholism (there is no other evidence supporting this finding).

The interviews were performed by Danish psychiatrists and the data were analyzed in St. Louis. The interview consisted of a complete psychiatric and mental status examination and obtained a wealth of information about the adoptive parents and the subjects' life experiences.

Only two factors distinguished sons of alcoholics from controls: divorce and alcoholism. Sons of alcoholics were three times more often divorced than controls. The reason for this was not clear, since the divorces did not appear to be related to drinking.

Alcoholism was defined by operational criteria similar to the American Psychiatric Association's DSM III. There also were categories of "heavy drinking" and "problem drinking." There were no differences between subjects and controls in either group. Nor were the sons of alcoholics likely to have a psychiatric illness other than alcoholism. Alcoholism had developed during their teens and twenties and was severe enough to require treatment. Drug misuse did not distinguish the sons of alcoholics from controls, a finding relevant to Jellinek's question about substance use specificity.

Subsequently two other adoption studies have been published. Bohman (7) found that Swedish adoptees with a biological parent with many alcohol problems were much more likely to have alcohol problems than were adoptees of nonalcoholic parentage. If their biological parents were criminals, they were no more likely to be criminal or alcoholic. In other words, transmission appeared to be specific for alcoholism, at least regarding criminality.

Cadoret (8) studied a group of adoptees in Iowa and found that those with a biological parent who was alcoholic tended to be alcoholic themselves and lacked any other diagnosable psychiatric disorder.

Based on these studies, all conducted in different countries with different methodologies, the following tentative conclusions can be drawn: *(1)* Children of alcoholics are about four times more likely to become alcoholic than are children of nonalcoholics, whether raised by their alcoholic biological parents or by nonalcoholic foster parents; *(2)* Their alcoholism develops at a rather early age, almost exposively in some cases; *(3)* The alcoholism is particularly severe, requiring early treatment such as disulfiram; *(4)* They are not more prone to other psychiatric disorders, including drug misuse, than are sons of nonalcoholics.

These conclusions apply only to men, since the studies of women (4) have produced ambiguous results.

"Familial alcoholism" might be a useful subtype of alcoholism, although perhaps more useful in research than in treatment. The subtype would have the following distinguishing characteristics: *(1)* There would be a positive family history of alcoholism; *(2)* The alcoholism would start early in life and have a florid course; *(3)* The likelihood of an "underlying" or second diagnosable psychiatric disorder would be no greater than expected in nonalcoholics.

Separating alcoholics into familial and nonfamilial types has advantages for research. To begin with, patients in most alcoholism wards are almost evenly divided into those with a positive family history of alcoholism and those without one. This provides equal size subsets, which is useful for statistical analysis. There is usually no question about the family history. Almost always, when a alcoholic patient has one family member who is alcoholic, he has two or more who are alcoholic. In comparing familial and nonfamilial alcoholics one could examine distinguishing characteristics such as age of onset and severity of alcoholism, and a wide variety of social, psychological, and biological factors.

Such studies have indeed been conducted in the past two or three years. Several indicate that alcoholics with a family history of alcoholism are younger and have a particularly extreme form of alcoholism. Severity appears to be a particularly important parameter, which brings us back to the study by Roe (3).

As noted, in the early 1940s Roe studied a group of adoptees, half of alcoholic parentage and half of nonalcoholic parentage. Neither group drank very much. Why did Roe's findings differ from those of other studies?

One possibility was that the biological parents were not "truly" alcoholic. They were described as "problem drinkers with a syndrome"; the syndrome included a variety of antisocial behaviors. There was no history of treatment for alcoholism. In short, Roe may have been dealing with a group of sociopathic alcoholics whose alcoholism is one expression of a range of antisocial behavior. The more recent studies examined nonsociopathic alcoholics.

In the Danish study, the alcoholic parents were severely alcoholic, because the diagnosis of alcoholism in Denmark is rarely made when an alternative is possible, and because the clinical record described most of the alcoholic parents as being in a state of severe withdrawal. Because of this, I feel severity should be taken into account when studies are conducted. Evidence from the twin studies also points to severity as an important factor. The younger the twins, and the more severe the alcoholism, the greater the difference in concordance rates between identical and fraternal twins.

Jellinek said it would be impossible to separate biological from social inheritance in familial alcoholism. He may not be entirely correct. To be sure, twin and adoption studies have inherent flaws. Since adoptions, for example, do not occur until after the baby is born, the baby has a minimum of nine months of intimate exposure to the mother. Thus, it is premature to use terms like "genetic" or "hereditary" to describe the adoption study findings; "constitutional" or "congenital" might be more accurate. Perhaps the safest thing to say is that there is now evidence suggesting that exposure to an alcoholic parent apparently is not a prerequisite for the development of alcoholism. Nevertheless, some observers are convinced that "biological" means of transmission should be explored.[2]

[2]For followers of the fetal alcohol syndrome story it will be of interest that the issue was as alive in 1944 as today. Jellinek was asked, "Is there evidence that excessive use of alcohol increases the likelihood that offspring be moronic or stunted?" Jellinek replied, "Much has been written on this question. There have been many experimental investigations on animals and there have been many statistical investigations on humans. . . . Those who have reviewed the material critically say that

Jellinek (1) asked, "If anything is inherited, what is it? Does it involve tolerance?" We can now say, unequivocally, that it involves tolerance—or at least *lack* of tolerance. It has only been appreciated recently how many millions of people possess a profound intolerance for alcohol. Orientals, in response to very little alcohol, often develop a cutaneous flush and develop a strong disinclination to continue drinking.

There is a low rate of alcoholism in the Orient, usually attributed to cultural sanctions against drunkenness and a preference for derivatives of the poppy plant. It now appears that a physiological intolerance for alcohol may be just as important a deterrent. Unquestionably, this intolerance is inherited; it is genetic. Little Oriental infants also flush after tiny amounts of alcohol.

Many non-Orientals also have a physiological intolerance for alcohol. There is some evidence that more women than men are intolerant. There is just a little evidence that perhaps more Jews than Gentiles are intolerant. In other words, the Jewish mother or her equivalent may not be the sole determinant in producing that paragon of our times, the moderate drinker.

With the development of pharmacogenetics, it has become apparent that the metabolism of most drugs is under strong genetic control. For example, if you give drug X to identical twins, they metabolize the drug almost identically. If you give drug X to fraternal twins, the rate of elimination varies greatly. The same is true of alcohol. Alcohol is metabolized at almost identical rates by identical twins and at very different rates by fraternal twins. If rates of elimination are controlled genetically, it is reasonable to believe that effects from drugs also are genetically determined to some extent.

When, however, we come back to the question, "If anything is inherited, what is it?" we are as much in the dark as was Jellinek. We have a few clues, all tenuous. Alcoholics are said to have higher blood levels of acetaldehyde than nonalcoholics after drinking the same dose of alcohol, usually attributed to effects of alcohol on the liver (9). One study also reported that nonalcoholic family members of alcoholics have higher blood levels of acetaldehyde after alcohol ingestion (10). High blood levels of acetaldehyde are usually associated with intolerance, rather than tolerance, but the story may be more complicated than that. There is one line of speculation that alcoholics lose control of drinking not because of the pleasant effects

---

it is inconclusive . . . and that one cannot make any positive statement at this time" (1). Announcement of the "discovery" of the fetal alcohol syndrome 30 years later betrayed no awareness of a very large literature on the subject.

of alcohol but because of unpleasant effects, and the latter conceivably may be related to acetaldehyde levels.

Another intriguing clue relates to that formidable enzyme, monoamine oxidase (MAO). One study reports that alcoholics with low levels of MAO have more alcoholic relatives with low levels than those with high levels (11). MAO degrades a number of neurotransmitters, including serotonin and norepinephrine. Transmitters are associated with mood states. Conceivably, decreased MAO activity may result in accumulation of these transmitters in various proportions which influences alcohol's effects on mood differently in different people.

"The etiology of inebriety is baffling," Jellinek said in 1944, and the situation is not much improved today. We still do not know why some people become alcoholics and most do not. My colleagues in Denmark and America and I are currently involved in a long-term study of children of alcoholics who are just entering a period when serious drinking begins. It is assumed that one of four of the boys will become alcoholic. Collecting a tremendous amount of information about these children before they begin drinking may provide clues to the mystery of why one becomes an alcoholic and the other three do not.

There was a tendency throughout the original *Alcohol, Science and Society* to view alcoholism as a product of many forces: biological, sociological, psychological. To some extent, this may be a device to make all the students of alcoholism feel useful—the biologists, sociologists, psychologists. There is a kind of unspoken gentlemen's agreement that since experts from diverse backgrounds study alcoholism, alcoholism must have diverse origins.

It may be so. In certain ways, it clearly is so. Genes give us enzymes to metabolize alcohol; society gives us alcohol to metabolize; and our psyches respond in wondrous ways to these combined gifts. Nevertheless, beyond this obvious level, the evidence for multiple causes of alcoholism is no better or no worse than the evidence for a single cause. When one does not know, it is best to say one does not know.

This point needs amplification, particularly since almost everybody still talks about the "multifactorial" nature of alcoholism. I close by quoting Lewis Thomas, President of the Sloan–Kettering Cancer Institute and the Montaigne of Medicine:

> The record of the past half century has established, I think, two general principles about human disease. First, it is necessary to know a great deal about underlying mechanisms before one can really act effectively. . . .

Second, for every disease there is a single key mechanism that dominates all others. If one can find it, and then think one's way around it, one can control the disorder. This generalization is harder to prove, and arguable—it is more like a strong hunch than a scientific assertion—but I believe that the record thus far tends to support it. The most complicated, multicell, multitissue, and multiorgan diseases I know of are tertiary syphilis, chronic tuberculosis, and pernicious anemia. In each, there are at least five major organs and tissues involved, and each appears to be affected by a variety of environmental influences. Before they came under scientific appraisal each was thought to be what we now call a "multifactorial" disease, far too complex to allow for any single causative mechanism. And yet, when all the necessary facts were in, it was clear that by simply switching off one thing—the spirochete, the tubercle bacillus, or a single vitamin deficiency—the whole array of disordered and seemingly unrelated pathologic mechanisms could be switched off, at once.

I believe that a prospect something like this is the likelihood for the future of medicine. I have no doubt that there will turn out to be dozens of separate influences that can launch cancer, including all sorts of environmental carcinogens and very likely many sorts of virus, but I think there will turn out to be a single switch at the center of things, there for the finding. I think that schizophrenia will turn out to be a neurochemical disorder, with some central, single chemical event gone wrong. I think there is a single causative agent responsible for rheumatoid arthritis, which has not yet been found. I think that the central vascular abnormalities that launch cononary occlusion and stroke have not yet been glimpsed, but they are there, waiting to be switched on or off (12).

Who knows? Maybe alcoholism also has a single switch. Maybe it could be turned off if we knew how.

References

1. JELLINEK, E. M. Heredity of the alcoholic. Pp. 105–114. In: Alcohol, science and society; twenty-nine lectures with discussions as given at the Yale Summer School of Alcohol Studies. New Haven; Quarterly Journal of Studies on Alcohol; 1945.
2. COTTON, N. The familial incidence of alcoholism; a review. J. Stud. Alc. **40**: 89–116, 1979.
3. ROE, A. Children of alcoholic parents raised in foster homes. Pp. 115–127. In: Alcohol, science and society; twenty-nine lectures with discussions as given at the Yale Summer School of Alcohol Studies. New Haven; Quarterly Journal of Studies on Alcohol; 1945.
4. GOODWIN, D. W. Alcoholism and heredity; a review and hypothesis. Arch. gen. Psychiat. **36**: 57–61, 1979.
5. KAIJ, L. Alcoholism in twins; studies on the etiology and sequels of abuse of alcohol. 2 vol. Stockholm; Almquist & Wiksell; 1960.

6. PARTANEN, J., BRUUN, K. and MARKKANEN, T. Inheritance of drinking behavior; a study on intelligence, personality, and use of alcohol of adult twins. (Finnish Foundation for Alcohol Studies, Vol. 14.) Helsinki; 1966.
7. BOHMAN, M. Some genetic aspects of alcoholism and criminality; a population of adoptees. Arch. gen. Psychiat. **35**: 269–276, 1978.
8. CADORET, R. J. and GATH, A. Inheritance of alcoholism in adoptees. Brit. J. Psychiat. **132**: 252–258, 1978.
9. LIEBER, C. S. Metabolism of ethanol. Pp. 1–29. In: LIEBER, C. S., ed. Metabolic aspects of alcoholism. Baltimore; University Park Press; 1977.
10. SCHUCKIT, M. A. and RAYSES, V. Ethanol ingestion; differences in blood acetaldehyde concentrations in relatives of alcoholics and controls. Science **203**: 54–55, 1979.
11. SULLIVAN, J. L., CAVENAR, J. O., JR., MALTBIE, A. A., LISTER, P. and ZUNG, W. W. K. Familial biochemical and clinical correlates of alcoholics with low platelet monamine oxidase activity. Biol. Psychiat. **14**: 385–394, 1979.
12. THOMAS, L. Medical lessons from history. In: THOMAS, L. Medusa and the snail; more notes of a biology watcher. New York; Viking; 1979.

11

The Role of Blood in Alcohol Intoxication and Addiction

Edward Majchrowicz

In many behavioral, clinical and medicolegal situations it is accepted that the concentrations of alcohol in the blood accurately reflect the concentrations prevailing at a particular site of action. The knowledge of the blood alcohol concentrations (BAC) is a very important and valuable diagnostic indicator in a number of conditions: BACs correlate quite well with the severity of intoxication and associated behavioral changes. There is a quantitative relation between the BACs, and the severity of the impairment of body functions, the physiological and biochemical changes and the behavioral responses. Changes in such biobehavioral parameters as emotion, mood, attention and motivation all appear to correlate positively with the concurrent blood ethanol concentrations (1–8).

The development of methods for accurate determination of ethanol in blood and body fluids is of great importance in the diagnosis and treatment of acute alcoholic states, in forensic medicine, and traffic safety as well. Usually the direct determination of alcohol in the blood is accepted as a reliable indicator of the severity of intoxication, however, this technique involves accurate sampling and processing of the blood. The determination of ethanol in the expired air obviates the blood handling procedure. The concentration of ethanol in the expired air remains in a constant ratio to that in the blood, and therefore can be used to assess the concentrations of alcohol in the blood. This method of ethanol determination is widely used to assess BACs in traffic offenses or in detoxication centers. One of the instruments based on this method is the breathalyzer. It depends on rapid distribution of ethanol between the alveolar air and blood in the lungs. The ratio of distribution, determined under standard conditions in several laboratories, has the value of 1 to 2100, i.e., one volume of blood contains as much ethanol as 2100 volumes of air exhaled under standard conditions (9).

Although ethanol affects all organs of the body to a varying

Blood Alcohol Concentration (%)	Effect

0.020–0.099	A. Impaired sensory function
	  1. Reduced visual acuity (flicker-fusion test)
	  2. Decreased sense of smell and taste
	  3. Elevated threshold for pain
	    (A) Decreased sensitivity of cornea of eye
	    (B) Decreased sensitivity to local heating of skin
	B. Muscular incoordination
	  1. Spontaneous and induced nystagmus
	  2. Decreased steadiness while standing (Romberg test)
	  3. Impaired performance on tests of skill (ring test, finger-to-finger test, target practice, typing)
	  4. Slight impairment of ability to drive an automobile
	C. Changes in mood, personality, and behavior
	  1. Dizziness
	  2. Reduced sense of fatigue
	  3. Mild euphoria
	  4. Self-satisfaction
	  5. Release of inhibitions
	  6. Loud, profuse speech
	D. Impaired mental activity
	  1. Subtraction test
	  2. Reading comprehension tests

0.100–0.199	A. Staggering gait B. Marked impairment on mental tests C. Marked impairment of driving ability D. Lengthened reaction time
0.200–0.299	A. Nausea and vomiting B. Diplopia C. Marked ataxia D. Extreme clumsiness
0.300–0.399	A. Hypothermia, cold clammy skin B. Loss of ability to speak C. Amnesia D. Anesthesia E. Heavy breathing

CHART 1.—*Relationship between Blood Alcohol Concentrations and the Signs and Symptoms of Intoxication.* This table was originally published in Maling (10, *p. 277*), and is reproduced from Majchrowicz (11) with the permission of the author and publisher.

extent, the brain and the liver are two major centers which deserve special attention, because the interaction of these two organs with ethanol results in the immediate and overtly expressed effects of alcohol. Since brain is heavily vascularized and therefore richly provided with blood, the behavioral effects of ethanol can be experienced subjectively by a person or observed objectively by others within 10 to 15 min after the consumption of even a single drink.

Chart 1 shows the most common signs and symptoms of alcohol intoxication in relation to the concurrent BACs in human subjects. These responses include the most commonly observed signs and symptoms of muscular incoordination, sensation, performance, skill, behavior and judgment which reflect the severity of the central nervous system (CNS) depression observed at increasing of blood concentrations of alcohol. The values are compiled on the basis of a number of reports published between 1943 and 1963 (10, 12–14).

Although a number of routes of alcohol administration are possible (15–20), the predominant, if not exclusive route in nonclinical conditions in humans is by drinking. Such routes as intramuscular, intravenous, intracerebral or intraperitoneal injection are used almost exclusively for pharmacological studies in animals and only occasionally for biomedical purposes in humans. Ethanol can also be absorbed by inhalation in the lungs, throughout the skin and through the epithelia of the mouth. However, these routes of administration are only occasionally used for either scientific or clinical purposes in human subjects or experimental animals.

Measurements of the rates of ethanol disappearance from the blood have been used to determine the rates of ethanol metabolism in the body (21, 22). After an alcoholic beverage or a defined dose of ethanol has been consumed, a time period of 2 to 4 hr is allowed for the equilibration of ethanol between the blood and body fluids. After equilibration the BAC is determined by a number of sequential blood samples taken at measured periods of time. The decrease in BAC is an indication of the rate or speed of elimination of alcohol by the body. Usually BACs are expressed as the weight of ethanol removed from a unit volume of blood in a defined period of time. BACs are often expressed as the number of milligrams (mg) of ethanol contained in 1 deciliter (dl) of blood, or as a percentage; for example, 100 mg per dl = 0.1%.

Absorption and Distribution

Alcohol entering the body via the gastrointestinal tract is distributed throughout the body by means of blood similarly to a variety of nutrients and fluids. Whereas most foodstuffs must be rendered soluble and absorbable by prior digestive processes, alcohol is absorbed directly (at different rates) from essentially all parts of the gastrointestinal tract; most rapidly from the small intestine. The rates of absorption vary somewhat between individuals; however, major factors influencing the uptake of ethanol from the gastrointestinal tract depend on the presence or absence of food, the volume and the concentration of ethanol in the beverage and the rate of drinking. Since ethanol is soluble in water in all proportions, and its small molecule has only a weak electric charge, it is rapidly distributed throughout the body water by simple diffusion. Consequently, in contradistinction to nutrients, ethanol requires no predigestion and no energy-facilitated transport across the biological membranes and absorptive epithelia. In addition, there is no evidence that ethanol is bound to tissues or is stored in any organ. The rate of ethanol uptake depends largely on the water content of a given tissue, the density of vascularization and the rate of blood flow. Consequently, ethanol is equilibrated throughout the body water within a relatively short period of time. Optimum concentrations of ethanol in alcoholic beverages for the most rapid absorption from the gastrointestinal tract is in the range of 10 to 30%.

Following the consumption of alcoholic beverages, ethanol induces two sets of reactions which are essentially similar to the reactions observed after consumption of any other substance whether a nutrient, a drug, a medicine or a poison. First, ethanol induces a number of direct and indirect behavioral, physiological and biochemical effects in the body. Second, the body responds to the presence of ethanol as it does to any foreign substance, by disposing of it either by using it in normal metabolic processes or by detoxication systems. In the case of ethanol, the organism removes it from the body by changing its molecule in a specific enzymatic mechanism in the liver (reviewed by Roach in this volume).

Blood Concentrations of Ethanol during Short and Prolonged Administration of Alcoholic Beverages

After equilibration, the BAC curve accurately represents the concomitant status of various physiological and behavioral aspects of intoxication. However, after both short-term and prolonged consumption the blood contains, in addition to ethanol, a large variety of substances which are derived directly from ethanol or appear as a consequence of alcohol's interaction with the body. For example, immediately after the metabolism of ethanol is initiated, there are significant increases in blood concentrations of acetaldehyde and acetate. Lactate, free fatty acids, various enzymes and other compounds are also released into the blood from different parts of the body.[1] The kinetics of ethanol absorption, distribution and elimination have also been discussed by Roach in this volume and elsewhere (1–8, 10–14, 21–24), therefore, the following sections will be directed towards the basic aspects of concentrations of ethanol and related substances which are present in the blood during short and prolonged periods of drinking.

Figure 1 depicts typical blood concentrations of ethanol, acet-

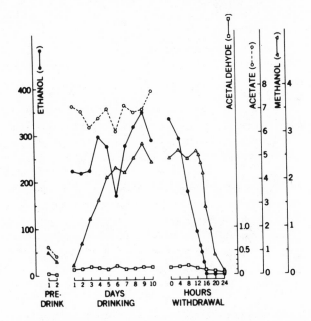

FIGURE 1.—*Blood Concentrations of Ethanol, Acetaldehyde, Acetate and Methanol in an Alcoholic Subject during an Experimental Drinking Episode.* The concentrations are shown during three characteristic time periods related to a drinking episode: Predrinking period; continuous drinking (10 days); and alcohol withdrawal, which is conventionally depicted from the time of administration of the last dose of an alcoholic beverage. The concentrations of all substances are expressed in milligrams per deciliter (100 mg per dl=0.1%). The determinations were made every morning throughout the drinking period and at designated time intervals during the early stage of the withdrawal period (prodromal detoxication phase). Reproduced from Majchrowicz (11) with permission of the publisher.

---

[1]In her chapter in this volume, Roach reviews the mechanisms and consequences of the release of some of these compounds.

aldehyde, acetate and methanol in volunteer alcoholic subjects during an experimental drinking period lasting up to 10 days (11, 25–27). As a result of essentially continuous consumption of alcoholic beverages, elevated BACs are sustained throughout the entire period of drinking and at the same time the subjects display various signs and symptoms of intoxication. After the termination of the drinking period BACs returned to zero within 24 hr. Usually, when BACs descend, after long-term drinking, to about 100 to 200 mg per dl (0.1 to 0.2%), typical signs and symptoms of the alcohol withdrawal syndrome develop (Figure 2). A similar pattern in the onset of withdrawal is observed in experimental animals (28–30). A few salient points on the relationship of the withdrawal syndrome and BACs will be introduced later in this chapter.

Alcohol Metabolites

The first step in the metabolism of ethanol is the formation of acetaldehyde. Since acetaldehyde at unphysiologically high concentrations suppresses a number of metabolic reactions in isolated organs and tissue preparations in vitro, it has been suggested that several physiological responses observed after either acute or chronic consumption of alcoholic beverages may be mediated by acetaldehyde.

It was only relatively recently (25, 31, 32) that the concentrations of acetaldehyde were found to be around 0.2 to 0.3 mg per dl (0.0002-0.0003%) of blood (Figure 1). Regarding brain concentrations of acetaldehyde there is still a lack of reliable data, but most likely they do not exceed those in the blood. In fact, there are a number of reports suggesting that acetaldehyde in the brain is either considerably lower than in the blood or present in concentrations that are almost undetectable. In general, the concentrations of acetaldehyde in blood seem to be too low to induce significant changes in the general metabolism of the body or in any particular organ. The possible exception may be the liver, where acetaldehyde is formed from ethanol and consequently there might be increased local concentrations. Some of the acetaldehyde seems to be able to escape destruction in the liver and subsequently is released into the blood. However, a number of scientists posed the question whether there should be any acetaldehyde present in the blood, since there are several enzymes in the liver which can metabolize acetaldehyde before it enters the blood (33).

The most important and at the same time paradoxical situation is that there is presently no reliable method available for accurate determination of acetaldehyde at low concentrations in the blood

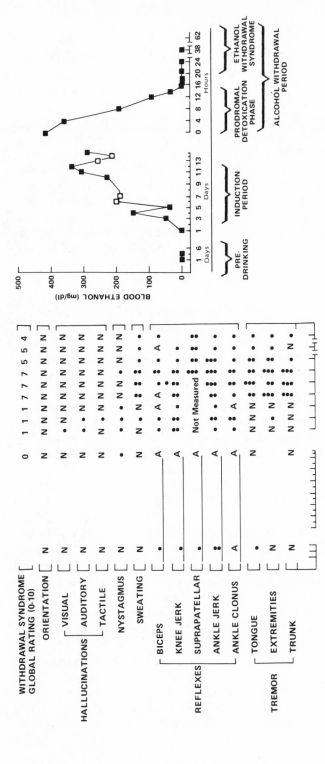

FIGURE 2.—*Relationship between the Onset of the Alcohol Withdrawal Signs and Symptoms and BAC after a 14-day Drinking Period in Human Subjects.* BACs were determined every morning throughout the drinking period and at designated time intervals during the early stages of the withdrawal period (prodromal detoxication phase). The intensity of a response was rated subjectively on a decreasing scale of 3 to 1, and is represented in the upper portion of the figure with closed circles. A=absent, N=normal. Slightly modified and reproduced from Majchrowicz (11) with permission.

and other body fluids. To resolve the controversies concerning the role of acetaldehyde in the actions of ethanol, it is essential to have dependable methods for the accurate determination of acetaldehyde in blood, body fluids, and tissues.

It might be of interest to point out that there are numerous suggestions that the toxicity of acetaldehyde is greater than that of ethanol by a few hundred or even a few thousand times, and consequently, acetaldehyde might play a significant if not dominant role in the actions of ethanol. However, the examination of rats which had ethanol or acetaldehyde administered intragastrically indicates that the ratio of these toxicities is about 7 (34). There are also a number of biochemical studies indicating rather low inhibitory effects of acetaldehyde on a number of metabolic pathways of several tissues.

Acetate: Its Formation and Distribution

Although acetate is one of the major substrates of liver metabolism, the total amount of acetate produced in the liver during alcohol metabolism exceeds the capacity of the liver to metabolize it further. Therefore, a large portion of ethanol-derived acetate is released into the circulation system and distributed throughout the body where it is metabolized to carbon dioxide and water (11, 35). A small fraction of acetate may be incorporated into fatty acids, cholesterol and other components of the body. The increase in blood acetate concentrations occurs immediately after the metabolism of ethanol is initiated in the liver and remains on a plateau until all ethanol is eliminated from the body. The blood acetate concentrations increase about 10 to 20 times above the normal levels after short and prolonged administration of alcohol. Figure 1 shows typical blood acetate concentrations in alcoholic subjects during a free choice drinking period. Upon the initiation of the drinking period the blood acetate concentrations ranged up to 7 to 10 mg per dl. The mean BACs at the same time ranged between 50 to 400 mg per dl (0.05 to 0.40%). The apparent dose–response relationship between blood alcohol and blood acetate concentrations exists only at relatively low concentrations of ethanol. A similar pattern of dose–response relationship also exists between blood acetaldehyde and blood alcohol concentrations (25, 32).

Acetate is devoid of any toxic properties and is compatible with life, in the usual physiological concentrations. Acetate is an important precursor of the citric acid cycle (Krebs cycle) in which the energy-yielding metabolic substrates are processed to transform the energy stored in their chemical bonds to biologically useable forms of energy. This suggests that ethanol might be considered a nutrient. However, the oxidation of ethanol is independent of any homeostatic mechanism; its metabolic transformation proceeds at a constant rate until all ethanol is removed from the system. In various body organs and tissues, the ethanol-derived acetate merges with the endogenous pools of acetate and subsequently saturates most pathways concerned with the metabolism of endogenous substrates. Since this ethanol-derived acetate is produced regardless of the homeostatic body requirements, it plays the role of an aberrant nutrient. It should be noted that most of the principal biosynethic processes in the body are interconnected to acetate or its derivative, acetyl coenzyme A, in a variety of pathways. Furthermore, the intermediary metabolism of acetate derived from ethanol is mediated by the same metabolic pathways and uses the same enzyme and coenzyme systems that mediate the metabolism of the principal energy-yielding substrates (carbohydrates, fats and proteins) under carefully balanced homeostatic control mechanisms (1–8).

Methanol and its Assimilation

A number of inhibitory effects of ethanol on liver metabolism are reflected in the release into the blood of a variety of metabolic products. One compound that appears in the blood, urine and other body fluids during prolonged alcohol consumption is methanol (26). It was reported about 20 years ago that traces of methanol are found in the breath and urine of healthy human subjects (36). However, methanol does not usually accumulate because it is rapidly metabolized by the same enzyme systems that metabolize ethanol. It is well established that a number of similar chemical compounds which belong to the same homologous series are metabolized in the living organism by the same enzyme system. Since methanol and ethanol are members of the alcohol homologous series, they are oxidized by the same enzyme system, i.e., alcohol dehydrogenase and aldehyde dehydrogenase. Another characteristic of the homologous series is that the higher members have greater affinity for the enzymes that metabolize them than the lower members of the series. Thus ethanol will displace methanol in the competition for the enzyme-active sites, leaving the methanol unmetabolized. This principle has been used in the treatment of methanol poisoning. Under normal conditions the methanol is oxidized by alcohol-metabolizing enzymes into formaldehyde and subseqeuntly into formic acid, both of which are very potent poisons. (It has been reported that even an ounce of two of methanol may render a person blind, and a few ounces are fatal.)

Therefore, a rationale in the treatment of methanol poisoning is to prevent the formation of poisonous metabolites. This can be achieved by allowing the patient to drink an alcoholic beverage. The ethanol in the beverage will competitively inhibit the oxidation of methanol to the toxic metabolites, forcing it to be eliminated from the system intact in the urine and breath (37, 38).

During long-term consumption of alcoholic beverages, the enzymes metabolizing alcohol are saturated with ethanol. Consequently, the trace amounts of endogenous methanol are not metabolized thus increasing the methanol concentration in the blood. When, in our studies, volunteer alcoholic subjects were allowed to consume large quantities of bourbon daily for up to 10 or 14 consecutive days, the blood concentrations of methanol had increased progressively to 2 to 4 mg per dl at the end of the experiment (Figure 1). After the subjects stopped drinking blood ethanol clearance was completed within 10 to 18 hr (depending on the existing BAC at the cessation of drinking). The highest methanol levels were found at the termination of the drinking period. After the initiation of alcohol withdrawal, blood methanol levels remained relatively stable for about 10 to 18 hrs, but when the BACs decreased to approximately 70 to 20 mg per dl (0.07 to 0.02%), methanol levels began to decline, coinciding with the emergence of the withdrawal signs and symptoms. The blood methanol clearance lagged behind the linear disappearance of ethanol by approximately 6 to 8 hrs.

Blood Alcohol, Intoxication, and Withdrawal

As discussed above the accurate measurement of BACs provides a reliable index of the biobehavioral and clinical status of acute effects of single doses or short-term binges, and can provide extensive and valuable information about the potential onset of the alcohol withdrawal reactions that follow prolonged drinking. The prediction of the onset, duration and potential severity of the withdrawal syndrome was made possible by numerous studies under emergency ward conditions and in experimental research wards (39–41).[2]

The development of physical dependence (or addiction) is related to the pattern of alcohol intake. Critical blood ethanol concentrations and the sustained presence of ethanol in the body are necessary during the periods of time preceding the onset of the with-

[2]Animal models of physical dependence on alcohol (see the chapter in this volume by Lester) also enable the exploration of the neurochemical determinants of intoxication, tolerance and addiction in relation to concomitant BAC (e.g., 11, 15–20, 23, 30, 38, 39, 42–44).

drawal syndrome. This presence of ethanol in the body can be best monitored by measuring the BACs at selected time intervals. As can be seen in Figure 2 there are three consecutive periods in the natural history of alcohol dependency in humans under experimental laboratory conditions. These periods can be clearly distinguished on the basis of temporal relationships, BACs and a characteristic sequence in the onset and disappearance of the corresponding symptoms and responses either in the presence or absence of alcohol in the blood.

The onset of the withdrawal reactions is closely related to the pattern of the blood alcohol disappearance curve (Figure 2). The predrinking period, characterized by the absence of ethanol in the blood, usually serves as a control or reference period against which to compare a number of biobehavioral responses that develop later during the subsequent phases of intoxication and withdrawal. The drinking period (pharmacologically defined as the induction period) is the time when the subjects continuously consume alcoholic beverages. The drinking period is characterized by the presence of variable amounts of alcohol in the blood, represented by a seesaw pattern in the BAC curve. It is important to remember that the continuous presence of ethanol in the blood results in the concurrent flooding of the brain with ethanol at similar concentrations. This in turn results in prolonged CNS depression. Under the experimental ward conditions the volunteer subjects consumed a variety of alcoholic beverages as frequent small doses. The induction period is characterized by continuous depression of the CNS. Under experimental conditions the rate of alcohol administration and the degree of resulting intoxication are closely dependent on the concentration of alcohol in the blood.

Although in alcoholism a state of mild to moderate CNS depression is sustained for long periods of time, it is usually a binge of excessive drinking that precipitates withdrawal reactions. When, for a variety of reasons, a person stops drinking, the binge is interrupted resulting in the development of typical signs and symptoms of the alcohol withdrawal syndrome. To alleviate the onset of the withdrawal reaction recognized as morning tremulousness, an alcoholic occasionally replenishes the system by taking additional drinks. In severe cases alcoholics may develop a "full-blown" withdrawal syndrome. The dependence on alcohol cannot be revealed until drinking is interrupted and the BAC declines to a certain threshold level, when the dependency is manifested by the characteristic signs and responses of withdrawal hyperexcitability.

The CNS depression, initiated during the drinking period (or induction period), continues after the last drink of alcohol, i.e., after the initiation of the withdrawal period. In humans (and experimental animals) the first phase of withdrawal is characterized by the rectilinear rate of blood ethanol clearance (Figure 2). As the BAC is declining there is a concurrent lessening of the CNS depression which can be observed as relative sobering. The sobering is essentially the reversal of gradual intoxication process when BACs rise after the initiation of the drinking episode (Chart 1). When the BAC declines to approximately 200 to 100 mg per dl (0.2 to 0.1%) there is a short-lived period of pseudonormalcy, during which the patients attempt eating, drinking nonalcoholic beverages, washing, shaving and resuming social contact. At approximately the same time there is a gradual onset of the withdrawal signs and responses which mark the initiation of the second phase of alcohol withdrawal, i.e., the overt withdrawal syndrome. It is important to note that the onset of the withdrawal syndrome occurs at relatively high BACs. Therefore, it is important to establish whether alcohol is still present in the blood before undertaking administration of therapeutic drugs. The typical signs of the withdrawal syndrome include reflex and motor changes, tremors of hands, tongue, body and head and general hyperactivity. A number of symptoms may be reported by the patient, including changes in mood and affect, mental activities, delusions, disorientation in time and space, etc. Convulsive seizures constitute one of the most dramatic and dangerous signs of the withdrawal reactions and may develop while ethanol is still present in the blood or after its complete elimination.

REFERENCES

1. THOMPSON, G. N., ed. Alcoholism. Springfield, Ill.; Thomas; 1956.
2. WALLGREN, H. and BARRY, H., ed. Actions of alcohol. 2 vol. Amsterdam; Elsevier; 1970.
3. TRÉMOLIÈRES, J., ed. Alcohol and derivatives. 2 vol. (International Encyclopedia of Pharmacology and Therapeutics, Sect. 20.) Oxford; Pergamon; 1970.
4. ROACH, M. K., McISSAC, W. M. and CREAVEN, P. J., eds. Biological aspects of alcohol. (Advances in Mental Science, No. 3.) Austin; University of Texas Press; 1971.
5. KISSIN, B. and BEGLEITER, H., eds. The biology of alcoholism. 5 vol. New York; Plenum; 1971–
6. LIEBER, C. S., ed. Metabolic aspects of alcoholism. Baltimore; University Park Press; 1977.
7. MAJCHROWICZ, E. and NOBLE, E. P., eds. Biochemistry and pharmacology of ethanol. 2 vol. New York; Plenum; 1979.

8. JACOBSEN, E. The metabolism of ethyl alcohol. Pharmacol. Rev. **4**: 107–135, 1952.

9. HARGER, R. N., RANEY, B. B., BRIDWELL, E. G. and KITCHELL, M. F. The partition ratio of alcohol between air and water, urine and blood; estimation and identification of alcohol in these liquids from analysis of air equilibrated with them. J. biol. Chem. **183**: 197–213, 1950.

10. MALING, H. M. Toxicology of single doses of ethyl alcohol. Pp. 277–299. In: TRÉMOLIÈRES, J., ed. Alcohol and derivatives. Vol. 2. (International Encyclopedia of Pharmacology and Therapeutics, Sect. 20.) Oxford; Pergamon; 1970.

11. MAJCHROWICZ, E. Reversal in the central nervous system function during ethanol intoxication in humans and experimental animals. Fed. Proc. **40**: 2065–2072, 1981.

12. GOLDBERG, L. Behavioral and physiological effects of alcohol on man. Psychosom. Med. **28**: 570–595, 1966.

13. MARDONES-R., J. The alcohols. Pp. 99–182. In: ROOT, W. S. and HOFMAN, F. G., eds. Physiological pharmocology. Vol. 1. The nervous system. Part A. Central nervous system drugs. New York; Academic; 1963.

14. VICTOR, M. and ADAMS, R. D. The effect of alcohol on the nervous system. Res. Publ. Ass. nerv. ment. Dis. **32**: 526–573, 1953.

15. FREUND, G. Induction of physical dependence on alcohol in rodents. Pp. 311–325. In: MAJCHROWICZ, E., ed. Biochemical pharmacology of ethanol. (Advances in Experimental Medicine and Biology, Vol. 56.) New York; Plenum; 1975.

16. PIEPER, W. A. Induction of physical dependence upon alcohol in non-human primates. Pp. 327–337. In: MAJCHROWICZ, E., ed. Biochemical pharmacology of ethanol. (Advances in Experimental Medicine and Biology, Vol. 56.) New York; Plenum; 1975.

17. POHORECKY, L. A. Animal analogues of alcohol dependence. Fed. Proc. **40**: 2056–2064, 1981.

18. GOLDSTEIN, D. B. Pharmacological aspects of physical dependence on alcohol. Life Sci., Oxford **18**: 553–561, 1976.

19. MELLO, N. K. A review of methods to induce alcohol addiction in animals. Pharmacol. Biochem. Behav. **1**: 89–101, 1973.

20. LESTER D. and FREED, E. X. Criteria for an animal model of alcoholism. Pharmacol. Biochem. Behav. **1**: 101–107, 1073.

21. LI, T.-K. Enzymology of human alcohol metabolism. Adv. Enzymol. molec. Bio. **45**: 427–483, 1977.

22. KHANNA, J. M. and ISRAEL, V. Ethanol metabolism. Pp. 275–315. In: JAVITT, N. B., ed. Liver and billiary tract physiology. (International Review of Physiology, Vol. 21.) Baltimore; University Park Press; 1980.

23. KALANT, H., LeBLANC, A. F. and GIBBINS, R. J. Tolerance to, and dependence on, some non-opiate psychotropic drugs. Pharmacol. Rev. **23**: 135–191, 1971.

24. HIGGINS, J. J. Control of ethanol oxidation and its interaction with other metabolic systems. Pp. 249–351. In: MAJCHROWICZ, E. and

NOBEL, E. P., eds. Biochemistry and pharmacology of ethanol. Vol. 1. New York; Plenum; 1979.

25. MAJCHROWICZ, E. and MENDELSON, J. H. Blood concentrations of acetaldehyde and ethanol in chronic alcoholics. Science **186**: 1100–1102, 1970.

26. MAJCHROWICZ, E. and MENDELSON, J. H. Blood methanol concentrations during experimentally induced ethanol intoxication in alcoholics. J. Pharmacol. **179**: 293–300, 1971.

27. MAJCHROWICZ, E. Metabolic correlates of ethanol, acetaldehyde, acetate and methanol in humans and animals. Pp. 111–140. In: MAJCHROWICZ, E., ed. Biochemical pharmacology of ethanol. (Advances in Experimental Medicine and Biology, Vol. 56.) New York; Plenum; 1975.

28. MAJCHROWICZ, E. Comparison of ethanol withdrawal syndrome in humans and rats. Pp. 15–23. In: GROSS, M. M., ed. Alcohol intoxication and withdrawal—IIIb. Studies in alcohol dependence. (Advances in Experimental Medicine and Biology, Vol. 85B.) New York; Plenum; 1977.

30. MAJCHROWICZ, E. Induction of physical dependence upon ethanol and the associated behavioral changes in rats. Psychopharmacologia, Berl. **43**: 245–254, 1975.

31. TRUITT, E. B., JR. and WALSH, M. J. The role of acetadehyde in the actions of ethanol. Pp. 161–195. In: KISSIN, B. and BEGLEITER, eds. The biology of alcoholism. Vol. 1. Biochemistry. New York; Plenum; 1971.

32. KORSTEN, M. A., MATSUZAKI, S., FEINMAN, L. and LIEBER, C. S. High blood acetaldehyde levels after ethanol administration; difference between alcoholic and nonalcoholic subjects. New Engl. J. Med. **292**: 386–389, 1975.

33. WESTERFIELD, W. W. The metabolism of alcohol. Texas Rep. Biol. Med. **13**: 559–577, 1955.

34. MAJCHROWICZ, E. and HUNT, W. A. Neurobiological correlates of intoxication and physical dependence upon ethanol; introductory remarks. Fed. Proc. **40**: 2048–2050, 1981.

35. LUNDQUIST, F., TYGSTRUP, N., WINKLER, K., MELLEMGAARD, K. and MUNCK-PETERSEN, S. Ethanol metabolism and production of free acetate in the human liver. J. clin. Invest. **41**: 955–961, 1962.

36. ERIKSEN, S. P. and KULKARNI, A. B. Methanol in normal human breath. Science **141**: 639–640, 1963.

37. ROE, O. The metabolism and toxicity of methanol. Pharmacol. Rev. **17**: 399–412, 1955.

38. KOIVUSATO, M. Methanol. Pp. 465–505. In: TRÉMOLIÈRES, J., ed. Alcohol and derivatives. Vol. 2. (International Encyclopedia of Pharmacology and Therapeutics, Sect. 20.) New York; Pergamon; 1970.

39. MENDELSON, J. H., ed. Experimentally induced chronic intoxication and withdrawal in alcoholics. Quart. J. Stud. Alc., Suppl. No. 2, 1964.

40. VICTOR, M. Treatment of alcoholic intoxication and the withdrawal syndrome; a critical analysis of the use of drugs and other forms of therapy. Psychosom. Med. **28**: 636–650, 1966.

41. ISBELL, H., FRASER, H. F., WIKLER, A., BELLEVILLE, R. E. and EISEN-MAN, A. J. An experimental study of the etiology of "rum fits" and delirium tremens. Quart. J. Stud. Alc. **16**: 1–33, 1955.

42. MAJCHROWICZ, E. and HUNT, W. A. Similarities in some neurological, physiological and neurochemical aspects of the ethanol withdrawal syndrome in humans and experimental animals. Pp. 419–424. In: ERIKSON, K., SINCLAIR, J. D. and KIIANMAA, K., eds. Animal models in alcohol research. New York; Academic; 1980.

43. GOLDSTEIN, D. B. Animal studies of alcohol withdrawal reactions. Pp. 77–109. In: ISRAEL, Y., GLASER, F. B., KALANT, H., POPHAM, R. E., SCHMIDT, W. and SMART, R. G., eds. Research advances in drug and alcohol problems. Vol. 4. New York; Plenum; 1978.

44. WALKER, D. W. and ZORNETZER, S. F. Alcohol withdrawal in mice; electroencephalographic and behavioral correlates. Electroenceph. clin. Neurophysiol. **36**: 233–243, 1974.

12

Alcoholism: Psychological and Psychosocial Aspects

Edith Lisansky Gomberg

If one searches *Alcohol, Science and Society* for those lectures which deal with some of the concerns of psychology, the roots of three areas of interest may be delineated. First, there is the area of experimental or laboratory psychology in Jellinek's chapter in the 1945 book on the effects of alcohol on psychological functions (1). Second, there is the area of personality theory and empirical study discussed by Landis in the "Theories of the Alcoholic Personality" (2). Third, lectures by Fleming (3) and by Baker (4) considered treatment and rehabilitation.

It is interesting to note how Jellinek described the whole work. In his Introduction, he indicated that the curriculum dealt with the effects of alcohol on the individual. Of the five lectures on effects, only one—a kind of afterthought—dealt with effects on "psychological functions" and Jellinek qualified it as belonging to physiology as well. The section of the Introduction which is labeled "Personality and Alcoholism" includes summaries of lectures on heredity, on alcohol and agression, and "the alcoholic personality." From this point on Jellinek uses the term *inebriety*, and summarizes the lectures as dealing with social factors in inebriety, the effect of inebriety on society, the controls of inebriety, and the treatment of inebriety. Jellinek used the term inebriety as we use the current term, alcoholism. It is clear that, from the beginning of organized work on alcohol studies, it was difficult to distinguish between the interest in alcohol studies in general and the interest in alcoholism per se. With a substance so widely used as alcohol it was inevitable that interest would focus in the decades following the publication of *Alcohol, Science and Society* on alcoholism and alcohol-related problems. For one thing, it was the beginning of the era in which the social sciences seemed to hold a promise for the solution of many social problems. For another, alcoholism was a disruptive manifestation and alcoholics were troublesome to their families and to society. Thus the drift of research and study in the years that followed *Alcohol, Science and Society* went to issues relating to alcoholism and alcohol problems. More recently, perhaps because of a burgeoning interest in prevention, policy makers have begun to see the need for study of ordinary, moderate, everyday drinking behavior. Studies of the proportions of the population who drink, the amount and frequency, the contexts and social customs, the effects of alcohol on cognitive and affective and social response, have been increasing.

In this chapter I will confine the discussion to the problems of heavy drinking–problem drinking–alcoholism, and the search for an answer to the question, Why do some people manifest such behavior and others do not? It is not a trivial question as the key to effective prevention as well as effective treatment may lie in the response. I will distinguish between two sets of theories, those I term "psychological" seek explanation in individual terms, and those I term "psychosocial" view individuals as embedded in social contexts so that explanation must involve a combination of psychological and social variables.

Psychological Views of Alcoholism

Psychological theories, as a rule, reject genetic explanations although some theorists may note the possibility of a genetic component or a biochemical predisposition.

Psychoanalytic Views

Based on Freudian theory of psychosexual development which posited growth from early, infantile oral stages to adult maturity and the capacity, as Freud said, "to work and to love," the view of alcoholism which emerged early on was of *developmental failure*. All addictive dependencies, whether on substances, persons, activities, etc., result from such failure and are attributable either to fixation at different stages of development or to regression to previously outgrown stages (5). The earlier the stage of fixation, the poorer the prognosis.

Fenichel (6) described "the oral and narcissistic premorbid personality" which developed from "specific oral frustrations in childhood." The denial of unpleasant reality and the projection of blame onto others, the most primitive of ego-defenses, was cited as evidence of the *oral* level fixation of many alcoholic persons.

There are also individuals, described as "reactive" or "neurotic" alcoholics, "fixated at or regressed to the *anal* stage" (5). In such persons, love relationships have moved beyond the self-love of the oral level to love for others of the same sex, i.e., homosexuality. And a third psychodynamic constellation involves fixation at or regression to the *phallic-oedipal* state: these persons have moved to heterosexual relationships but experience a good deal of conflict, "unreconciled impulses of tenderness and sexuality which they are unable to direct toward one and the same woman" (5).

A valuable distinction has been made between "essential" and "reactive" alcoholics. Knight (7) described the essential alcoholic as egocentric, hedonistic and incapable of maintaining intimate relationships with others; the reactive alcoholic is a more compulsive individual who has achieved some aspects of adult maturity and adjustment. A number of research studies (8–11) suggest that this is a useful, valid distinction, a dimension of social competence. It is one viable way of classifying alcoholic persons.

Another psychoanalytic contribution of great potential worth is the description by Menninger (12) of the self-destructiveness of alcoholics. This form of self-destruction, the excessive use of alcohol, Menninger believed to be a manifestation of attempts to control aggressive impulses. There has been some experimental support (13) but the Menninger hypotheses have not produced much research interest. Perhaps one of the best and most succinct statements about etiology was made by Lolli (14): "Although physiological events may facilitate its origin, [alcoholism] is rooted in a distortion of the early mother–child relationship. It represents the abnormal survival in the adult of a need for the infantile normal experience of unitary pleasure of body and mind. The alcoholic rediscovers this experience in the course of intoxication. He cannot resist its gratification, however illusory and temporary it turns out to be, and this is the basis of alcoholic addiction" (*p. 106*).

Psychological-Trait Views

Landis, in his lecture in *Alcohol, Science and Society*, put the question very clearly: "Is there a personality type, a relatively constant combination of psychological traits, which, appearing in an individual, renders him especially susceptible to the intemperate use of alcohol; or is there no such special personality type, but only alcohol acting on any or all of the varieties of personality?" (2, *p. 129*).

It was a legitimate question for its time although now it surely seems to be naive. In 1950 and again in 1957, reviews (15, 16) of the published findings from psychological test studies of alcoholics showed no consistent findings. The studies reviewed were often methodologically inadequate but a more important challenge is whether the question they sought to answer was a legitimate one, Is the researcher looking for single, unitary traits? Could characteristic behaviors found common to alcoholics be a result of the alcoholism rather than an antecedent? It was generally agreed that there might be "a constellation of personality traits carried into the onset period" (17) and, after the alcoholism developed, "patterns of personality traits common to most alcoholics" (18). The same distinction has been made in the most recent review of psychological test research (19), i.e., that we must distinguish between the prealcoholic personality, and the picture presented by alcoholics entering treatment.

Understanding prealcoholic traits is possible only in longitudinal study and there have been a number of these (20–23). Robins et al.'s (20) follow-up study 30 years after children were seen at a psychiatric clinic suggested the importance of low family social status and parental inadequacy as antecedents but, in the subjects themselves, "serious antisocial behavior" was significantly related to the development of alcoholism in later life. The McCords' report (21) is more concerned with personality traits; in this study of more than 200 boys, the description of those who later became alcoholic includes the following: "apparently self-contained, outwardly self-confident, undisturbed by abnormal fears, indifferent toward their siblings, and disapproving of their mothers. Also, they more often evidenced unrestrained aggression, sadism, sexual anxiety, and activity rather than passivity" (*p. 427*).

Interestingly enough, as adult alcoholics the same persons who were self-confident now felt victimized by society, the same persons who had emphasized their independence now were often highly dependent people. Finally, a longitudinal study of a nonclinical population by Jones (22, 23) showed that boys who later became problem drinkers were, as youngsters, "undercontrolled, impulsive and rebellious." Again the description is one of attention-seeking, aggressive, acting out youngsters resentful of authority.

The similarity of prealcoholic behaviors reported in three separate longitudinal studies suggests the importance of such research. Jones (22) noted: "The results indicate that alcohol-related behavior is to some extent an expression of personality tendencies which are exhibited before drinking patterns have been established."

The question whether there is an alcoholic personality is therefore misleading. What should be asked, instead, are two questions: First, what do people who become alcoholic in their twenties, thirties and forties have in common as behavioral traits when they are children and adolescents? Second, when people come to treatment facilities with alcohol problems or when people with serious alcohol problems are identified in community and national surveys, what behavioral traits do they have in common at this moment?

Behavioral Views

The origins of behaviorism lie with Pavlov and his pioneer work in conditioned reflexes and conditioned responses. In the United States, Watson and Thorndike were the first to build a theory around "the law of effect" and they were followed by Guthrie who wrote of associative learning and by Hull who constructed his learning theory around the concepts of habit strength and "the experimental extinction" of learned habits. Application of these learning theories to explain disordered or unacceptable behavior came later.

Interestingly enough, aversion treatment for alcoholism goes back as far as Roman civilization. The associative pairing of alcoholic beverages and an unpleasant experience may indeed be as old as man. The application of principles of learning and unlearning to alcoholism and to various forms of psychopathology is relatively recent, although a few treatment facilities offered conditioned response therapy decades ago. Modern learning theory's application to therapy is based on the work of Skinner, Wolpe, Eysenck and others.

The principles involved are stated by Ullman and Krasner (24): "Alcoholic overindulgence depends upon a specific operant behavior: the ingestion of alcoholic beverages. ... The behavior becomes a matter of concern to others, and hence the amount consumed and the specific physiological effects are of secondary concern to the target behavior of drinking under conditions considered inappropriate. ... The first step ... is to determine what the appropriate alternative behaviors are and under what conditions they may be emitted. ... The person must be trained and capable in the emission of certain acts, must emit them, and must be reinforced for their emission" (p. 451).

Theoretical and experimental work in the 1950s attempted to link learning theory and various forms of disordered behavior. Dollard and Miller (25) presented evidence for viewing alcohol as a reducer of fear from experiments with animals in which alcohol reduced fear and thereby acted as reinforcer, and a cross-cultural survey suggested that the amount of insobriety in a given society was "reliably related" to the degree of subsistence hazard, i.e., it was a measure of anxiety. This theorizing about the relationship between alcohol consumption and reduction of fear, anxiety, conflict and tension became known as the tension-reduction hypothesis and resulted in a good deal of experimental work, primarily with laboratory animals. A review of this work in 1972 (26) concluded that much of the evidence was "negative, equivocal and often contradictory." To complicate matters further, experimental study of the effects of alcohol in alcoholics suggested that a drinking bout often increased tension and anxiety; and Mello (27) reported, "Alcoholics most frequently show an increase in anxiety and depression and further impairment of an already fragile self-esteem during the course of a drinking episode" (*p. 282*).

Learning theorists have also described alcohol consumption as a negatively reinforced operant, a positively reinforced operant, and as a discriminative operant response. Results of experimental work suggest that the search for a simple definition of alcohol's effects is a will o' the wisp. The understanding that alcohol is a complex drug with complex effects whose impact may vary with the subject, the dose, at different times, in different contexts, with varying physiological states of the person drinking, and with other sources of variability, is slow in coming.

Current developments in learning theory seem to be placing more emphasis on the social contexts of drinking. Bandura's "social learning theory" (28), for example, views alcohol as a positive reinforcer and suggests that prolonged excessive drinking produces metabolic alterations which serve as secondary motivator, the alcohol now serving to stave off the unpleasant consequences of alcohol withdrawal. This view emphasizes the prealcoholic social learning of drinking behavior, i.e., the cultural norms of the group in which a person has membership and the stresses to which a person is subjected. Marlatt's experimental work (see the discussion by Nathan in this volume) has also emphasized the social variables involved in drinking behavior and has examined frustration and anger in interpersonal contexts, expectations of alcohol effects, and the effects of alcohol on cognitive perceptions of personal control (29). The emphasis of some current learning theorists on modeling behavior and the familial–social setting in which drinking behavior is learned is also relevant. One may say that this current thinking is moving in the direction of psychosocial theory.

The behavioral and psychoanalytic–psychodynamic approaches to drinking and alcoholism vary in a number of different ways. First, and probably most important, was the assumption by psychoanalytic theorists that character would be determined by early life experience and that such character formation would determine the person's later drinking behavior and problems. The behaviorists may speak about modeling and behaviors learned in interpersonal contexts but the learning may take place in the here-and-now and not necessarily in past experience. Thus, a young person raised in an abstinent family may do some rapid learning about drinking when he or she moves into permissive peer settings such as college or the military.

Second, dynamic theory assumed that psychotherapy or "talking out" therapy with its attendant insights would produce the desired goal, i.e., the alleviation of symptoms. Behavior theory has sought to "shape" behavior toward the desired end by experimental manipulations in experimental settings. The therapies included aversive conditioning, biofeedback, relaxation training, systematic desensitization, cognitive and contingency management techniques.

Third, dynamic theory placed relatively little emphasis on the drinking behavior per se, focusing attention on the patient's early and present life experiences, feelings and interactions. For behaviorists, on the other hand, treatment starts with assessment or analysis of the drinking behaviors, i.e., its frequency, contexts, patterns and consequences.

There have been many reviews of the behavioral literature on alcohol and alcoholism, starting in 1958 (30), up to more recent reviews (e.g., 31), and the work discussed by Nathan this volume.

Theories in Conflict

There is a communication gulf between those psychologists and therapists who have adopted a psychoanalytic–psychodynamic orientation towards alcoholism and those who have adopted a behavioral one. The origins of psychology lie both in philosophy and in psychophysics and the conflict between humanists and experimentalists dates back to these origins. Humanistic psychology views the person as a whole being who experiences, decides and acts, while the research tradition is defined by method and systematic, factual inquiry in which the observable and the measurable is studied (32). Both humanists and experimentalists have advantages and limitations in their approaches. Humanistic psychologists, most often clinicians, are frequently vague in definition, and the concepts they generate are difficult, often impossible, to verify given our present state of methodology. On the other hand, an extreme, pure-science view among experimentalists (e.g., 33) restricts inquiry to questions which can be studied by the most objective, most precise, most replicable methodology. Child (32) discussed the "several virtues" of humanistic psychology and the research tradition; that the conflict still remains central to psychology is evident, for example, in a recent discussion of limits to the integration of psychoanalytic and behavior therapy (34).

Fortunately, there are mutual influences. Generally, clinicians tend to view the world in humanistic terms, and while they may disagree about the effectiveness of one treatment or another, they tend to be holistic in view. The pressures to justify their treatment approach has brought a more recent emphasis on evaluation research as clinicians learn the facts of life about budgets and cost effectiveness. The research tradition in alcohol studies emphasizes precise definition and measurement, but studies do vary in the size of the behavioral units studied and in awareness of social variables.

The most important commonality, it seems to me, of humanistic clinicians and scientific researchers is a common enemy, or at least a strong competitor in explanatory models. Both clinicians and behaviorists agree on the assumption that experience is the major determinant of behavior, not genetics or biochemistry. Whether the basic assumption is about distal life experience or proximal life experience, the common ground is the primacy of experience as the determinant of the behavior under study.

A major limitation to psychological theories has been their person-centeredness. The social environment is often relegated to a secondary role. In its extreme form, the limitation may manifest itself among clinicians treating clients as though they existed in a vacuum. It may manifest itself among experimentalists by laboratory research with individual subjects in which a baseline of performance is established and then remeasured after administration of alcohol. Drinking occurs in social contexts and psychological work has only recently begun to recognize the importance of the social contexts in which people live, learn how to drink, engage in drinking behavior, and experience the consequences of drinking.

In psychopathology the issue of person-centeredness, the focus on the person at the expense of social context, has led to its description as "blaming the victim." Although clinicians and social scientists are concerned with social issues, this concern has not often been reflected in their professional orientation. In clinical work, social issues often intrude themselves, but published research has been heavily person-centered. An analysis of social science research from the years 1936, 1956, and 1976, concentrated in several problem areas, shows a very heavy emphasis on the personal characteristics of subjects (35). The reports were analyzed in terms of person, milieu, and system variables. Over the time period studied, there was little change. Research in alcohol and substance misuse showed the strongest person-centered orientation of the six subject areas investigated; the authors commented that, over the period studied, there was only "a shift in language from morality to psychiatry."

One by-product of the social unrest of the last two decades has been an antithesis of the person-centered view, i.e., an explanation of alcoholism (or any other social problem) as produced by deprived social environment, oppression, discrimination and slum life. Rates of breakdown and rates of social problems are greater among poor people. But explanation in terms of deprivation alone is inadequate, for deviant behaviors are not necessarily endemic among minority persons and among poor people, and the question again returns to, Why some and not others?

There is a point of view which includes personal characteristics, deprived social environment, and other antecedent circumstances in explaining alcoholism. This is a model of alcoholism which encompasses personal characteristics of the person, the patterns of drinking of the various groups to which the person belongs (family, friends, religious, ethnic, social class, community, etc.), as well as the more immediate events or stresses which may "trigger" an alcoholismic episode.

A Psychosocial View of Alcoholism

The antecedents of alcoholism are genetic, biological, psychological and social, and these antecedents interact in a complex, albeit lawful way. For this discussion, I will leave genetics and biology aside, and look at psychological and social antecedents. Psychological antecedents of two sorts will be discussed: those dealing with individual personality and those dealing with individual motives and patterns of drinking. Social variables will also be discussed in two parts: those which remain a part of the constant, long-term environment and those which may be temporary, immediate and situational.

Individual Variables

Personality. Who will and who will not become an alcoholic is not a random matter. Some people are more likely to behave alcoholismically than others and there are personal characteristics of those who are more likely to become alcoholic which may be called predisposing characteristics. An individual develops, from his or her life experience, habitual patterns of dealing with other people, of coping with problems, of reacting to the world; and some of these patterns, I believe, are more likely to be a part of people who will later become alcoholics. This is the prealcoholic person; in a recent review of the published work on the subject, Barnes (19) concludes that what characteristics we have identified so far are impulsivity, nonconformity, gregariousness, antisocial behavior, and poor self concept.

The supporting evidence is somewhat ambiguous. Furthermore, there has been an assumption that alcoholism is a unitary disorder and that, in seeking the prealcoholic patterns of behavior, we are looking for the antecedent individual variables which characterize all alcoholics. We are not going to find the universal antecedent but if we classify alcoholics by gender, social class, age at onset and the like, we may find some common prealcoholic characteristics. I believe, for example, that among women alcoholics, difficulties in impulse control may characterize women problem drinkers but the difficulties will be greater in women with an early-onset. Further, I hypothesize that antisocial behavior may be a concomitant and predictor for early-onset women and depressive behavior a more likely concomitant and predictor for those with onset in middle age.

Predisposition to alcoholism is not a hit-or-miss affair. One learns in early life, in childhood, and in adolescence patterns of behavior and ways of dealing with problems, and it is a safe assumption that some patterns are more likely to occur in persons who will later become alcoholics. Perhaps it is a particular combination of patterns. The behaviors we see prealcoholically are those we see in other diagnostic groupings and it may be that a particular combination of behavior patterns and coping mechanisms is involved. It may be that combination which will eventually be defined as vulnerability.

Patterns and Motives for Drinking. One learns whether to drink and how to drink from the people in the social environment. That does not mean one pattern is learned forever: a young man may grow up in a devout, abstaining family and not drink at all, go away to school or the army and drink heavily with his peers, later marry and get a job and drink moderately, and even later in life, drink heavily and alcoholismically. We know that there are changes in drinking patterns which occur at different points during the life cycle although we have done little research into the reasons or motivations for such changes. Young men, when they drink, tend often to drink heavily and to intoxication. As people age, the proportion of heavy drinkers drops dramatically, among women at about age 50 and among men at about age 65.

A brilliant concept developed by Allport (36) many years ago is the *functional autonomy of motives*. Behavior, at its point of origin, may be motivated by X but when examined 20 years later, although the behavior has persisted, X is irrelevant. Allport says, "dynamic psychology . . . regards adult motives as infinitely varied, and as self-sustaining, *contemporary* systems, growing out of antecedent systems, but functionally independent of them" (*p. 194*). If the conception, the functional autonomy of motives, is applied to alcoholism, we can see that by the time an alcoholic comes into treatment, his or her reasons for drinking may not be the same as those which initially triggered the problem drinking. The functional autonomy of motives is also applicable to social drinking: in adolescence, it may be conformity, and in young adulthood, part of dating–mating behaviors. During younger years, drinking may be motivated by a wish for relaxation, as a solution to social uneasiness, or simply to do what everybody else is doing. In adult life and middle age, drinking may be motivated by social custom or by a wish to escape from depression; by now it is habitual behavior. Among older persons, alcohol may be linked to sociability and leisure time activities and it also acts for some as a relaxant–sedative: a "nightcap" before bedtime.

The point is that the patterns of drinking and the motives for drinking will vary among social drinkers and among alcoholics. The inference is clear for both treatment and research: we need to deal first with what the current patterns and motives are but their history is necessary in the study of the natural history of alcoholism if we are ever to shorten the process or prevent its development.

There is still another way in which time enters the picture. We encounter people who have had episodes of problem drinking and, apparently by spontaneous remission, cease the problem drinking. There is a very high proportion of alcohol-related problems among young male drinkers and we find that the majority "mature out", i.e., they continue drinking but drink less and show fewer alcohol-related problems. Here are patterns of heavy or problematic drinking that change over time. It would be a most profitable line of research if we were to investigate what actually occurs in spontaneous remission or maturing out—it is more than possible that such information would make early intervention and prevention programs much more effective.

Social Variables

The Social Environment. People have families, kin networks and friends. They go to a workplace along with other people, they affiliate with educational, religious and political institutions, and they come into contact with health-concerned and leisure-concerned institutions as well. They belong to groups which influence not only value systems and social attitudes and voting behavior and child-rearing patterns, but attitudes toward alcohol and drinking behavior as well. The sociological aspects of alcoholism are reviewed by White in this volume; the social variables that determine whether or how one drinks are indispensable in a psychosocial theory about drinking and alcoholism. Once again there is a dimension of change over time to consider. Influences from family groups tend to be reasonably stable over time but peer groups, friendships and social networks vary and may have quite different drinking customs. Other things being equal, it is apparently true that a person in a heavy drinking social group is likely to drink more than when in a moderate or light drinking social group. And the social environment includes ethnic and religious identification, socioeconomic status, age group, networks in the workplace, and the influences which come from community, geographical status and location.

The Immediate Situation. To be poor or a member of a low status minority group is to be stressed but such stresses are long term, if not lifetime stresses. There are some situations which occur in the here-and-now and which may influence drinking. Nor are these situations always stressful; people drink more under certain facilitating conditions like "the happy hour" (37). Evidence is accruing which indicates a strong relationship between certain types and periods of stress and heavy drinking. Women under 35 who are experiencing or have recently experienced marital disruption appear to be under some risk for heavy or problem drinking (38). The relationship between economic strains and economic hardship and drinking for stress relief has been explored (39). And there is some evidence of high risk status among elderly men experiencing the double loss of widowerhood and retirement (40).

Heavy and problem drinking which occurs in response to stress and pressures is more of a reactive response. The interaction of individual vulnerability, personality characteristics, drinking habits and attitudes, and ongoing stress needs investigation: people recently laid off from work or older persons under stress are groups with which to start such research.

Psychosocial Views of Alcoholism: Discussion

In a study of women alcoholics, Kinsey (41) proposed a "symbolic-interaction approach" which included the following set of psychosocial conditions or processes associated with alcoholism: "predisposing" or personality factors, and "orienting" factors. The latter include norms which encourage a tension-reducion definition of alcohol and also membership in heavy drinking, hedonistic groups. From these factors, Kinsey stated that a "vicious circle" of dependence on alcohol develops. Working from this theoretical approach, Kinsey derived a number of hypotheses and tested them with data from the interviews of his state hospital respondents. It was a laudable attempt to deal with psychological and social variables in a single study.

A more recent psychosocial approach—"a developmental approach"—has been posited by Zucker (42). Zucker includes intraindividual influences, primary group influences, intimate secondary group influences, community and sociocultural influences.

Although much of the published literature is still empirical, based for the most part on specific and limited hypotheses about specific and limited aspects of drinking, there is an encouraging trend toward research which takes into account social and individual influences on behavior. The problem remains that combining individual and social variables in a single study does not occur as often as desirable.

Progress since 1945?

From the time *Alcohol, Science and Society* was published until the 1960s, the focus on and the preoccupation with individual motives, needs, symptoms, behavior, was intense. Reflecting a general dissatisfaction with such limited explanation, the 1960s and 1970s seem to have shifted emphasis to some degree to social and sociological variables. It is true that psychodynamic explanations stressed the role of the *family* in which adaptive and maladaptive mechanisms were learned, and it is true that the concept of reinforcement which is the core of the behavioral approach is *social* reinforcement. In the last two decades, the concept of "blaming the victim" emerged and the emphasis in the rhetoric of those years was on social stress and societal injustice. As the 1970s ran down, a marked shift toward genetic and biological explanations of alcoholism appeared, and now a search for biological indicators of alcoholism is on. Research dealing with metabolism and biochemical effects of alcohol is currently encouraged by the research establishment, the funding agencies and the foundations. There is a recent shift of favor toward the epidemiological study of alcohol use and misuse.

It is too bad. We do not seem able to focus our attention on several disciplines at once but seem, somehow, ready to hail a savior group which will solve the problems of drinking and alcoholism once and for all. That is all the more regretable because there has been a development of areas of research in which several disciplines are involved. There is a burgeoning area of study of the effects of alcohol on cognitive functioning in which physiology, neurology and psychology meet. There is another rapidly developing area which involves study of the problem drinker in the context of family. If disciplinary turfdoms can be set aside, the time is ripe for multidisciplinary research. And it is clear that no one discipline holds a monopoly on the future clarification of drinking behaviors.

At the beginning of this report, we distinguished three areas of psychologists' concerns which were included in *Alcohol, Science and Society*: the psychological effects of alcohol, personality issues, and questions of treatment and rehabilititation. A brief examination of each area follows:

The Psychological Effects of Alcohol

The effects of alcohol on individual behavior have surfaced in many studies lately which support the old idea of "set" or expectancy, i.e., within reasonable limits, the drinking person responds not only pharmacologically but in terms of what he or she expects or understands about the effects of alcohol. This research is stimulating an interest in how people acquire their beliefs about alcohol.

Another area of research deals with the social conditions under which people drink. Again an old concept has surfaced, "social facilitation," and experiments include differences in amounts consumed with light drinking companions and heavy drinking companions, the effects of the "happy hour" on the amount of drinking, and similar studies.

There has been a growing sophistication around the question of alcohol's effects in the social drinker vs the alcoholic, and investigators are more aware than before of the problems in generalizing from one group to another.

And there is more—but not enough—sophistication in defining the social characteristics of the group studied. It has taken a while to deliver the message that there are gender differences in the effects of alcohol (43). We have known since the early 1940s that effects of alcohol vary with the past drinking experience of the person. We need more information about effects of alcohol as a function of age differences, including its effects on older people. There is some research, primarily biological, on alcohol's effects in different racial and ethnic groupings. Perhaps the baseline of psychosocial research on the effects of alcohol on psychological functioning is a definition of the psychosocial characteristics of the individual or group under study.

Personality Issues

For the most part, this research has been heavily concentrated on alcoholics. Some trend appears to study the psychosocial behaviors of adolescents who are heavy drinkers as opposed to those who abstain or drink moderately but such research has been limited. Most of the research deals with personality or diagnostic features of people in alcoholism treatment centers, e.g., studies involving the Minnesota Multiphasic Personality Inventory. In spite of the fact that the effort to study alcoholics moved from weaker to stronger psychological tests (in terms of their psychometric qualities such as reliability and validity), the results are not impressive. Psychologists have used test data in several attempts to distinguish subgroups of alcoholics by factor analysis; the research using techniques of observation and report has been pretty much left to anthropologists and sociologists.

One current trend among psychological researchers is to move away from the restriction of psychological tests and to incorporate not only the method of participant-observation but of standardized interview as well. Such techniques hold a promise of broader information about psychological and social variables.

Treatment and Rehabilitation

Very shortly after the publication of *Alcohol, Science and Society* the importance of group therapy in working with alcoholics became clear, and alcoholism treatment agencies were among the first to use group therapies. From the beginning the limitations of psychoanalytic treatment of alcoholics and the meteoric rise of a lay, self-help organization (Alcoholics Anonymous) carried the message: to be effective, treatment programs had to recognize the importance of social rehabilitation. Very quickly, alcoholism therapists understood that they had to deal not only with the drinking but with the wreckage of families, social networks, job relationships, etc.

In the last decade, treatment of alcoholics through family therapies has become increasingly popular in rehabilitation facilities. The trial-and-error of traditional therapies suggested that treating alcoholics out of the family context was to ignore the interactional effects of alcoholism in a family. There has been, in fact, such an embarassment of riches as to produce a dozen different styles of family therapy and an increasing need to systematize and organize the information obtained from such family therapies.

The main fact which emerges from any historical review of treatments is the current emphasis on the social therapies. If one examines the treatment programs described in *Alcohol, Science and Society*, they are, as Dr. Sugerman points out in this volume, not very different from the spectrum of treatments available today. Disulfiram (Antabuse) is still used but not as widely as it has been in the past. The major shift has been in the direction of social forms of treatment.

A Final Word

If I select that which I find most discouraging after all the years which have gone by, it would be the continuing parochialism and narrowness of view of the different disciplines and the different schools of thought within the same discipline. There is such a long way to go: we need to know about the significant variables in human biology, in development and behavior, in genetics, and in social conditions. And we need to know not only the significant variables but the combination and interaction of those variables. Yet in dealing with such a complex topic as drinking behavior, normal or abnormal, we seem unable to avoid the tunnel vision which permits us to see only what is in front but not that which is peripheral. There is a profession-centricism which leads each discipline to view itself as the keeper of the Truth. When will we understand that we are like the blind men, all describing the elephant from our different perspectives?

If I select the most encouraging trend, it is the increasing awareness of research investigators that persons must be studied in social environments, whether in a bar, at a party, in tribal rites, or in a laboratory group. We once had a debate as to whether chronic drunkenness offenders lived in a "pseudo-community," but we know that no one functions in a person-less environment, and whether it is "a bottle gang" or a subtle interaction between schizophrenic patients, we live in social environments. This awareness has led to new questions about people's beliefs about alcohol and its effects, about the influences of groups or persons with whom one drinks, about the influence of "happy hours" and advertising and the like. Such questions can lead to greater understanding of the learned drinking patterns, drinking motives, drinking contexts, and this is the understanding we need for effective therapy and for imaginative prevention programs.

REFERENCES

1. JELLINEK, E. M. Effects of small amounts of alcohol on psychological functions. Pp. 83–94. In: Alcohol, science and society; twenty-nine lectures with discussions as given at the Yale Summer School of Alcohol Studies. New Haven; Quarterly Journal of Studies on Alcohol; 1945.

2. LANDIS, C. Theories of the alcoholic personality. Pp. 129–142. In: Alcohol, science and society; twenty-nine lectures with discussions as given at the Yale Summer School of Alcohol Studies. New Haven; Quarterly Journal of Studies on Alcohol; 1945.

3. FLEMING, R. Medical treatment of the inebriate. Pp. 387–401. In: Alcohol, science and society; twenty-nine lectures with discussions as given at the Yale Summer School of Alcohol Studies. New Haven; Quarterly Journal of Studies on Alcohol; 1945.

4. BAKER, S. M. Social case work with inebriates. Pp. 419–435. In: Alcohol, science and society; twenty-nine lectures with discussions as given at the Yale Summer School of Alcohol Studies. New Haven; Quarterly Journal of Studies on Alcohol; 1945.

5. BLUM, E. M. Psychoanalytic views of alcoholism; a review. Quart. J. Stud. Alc. 27: 259–299, 1966.

6. FENICHEL, O. The psychoanalytic theory of neurosis. New York; Norton; 1945.

7. KNIGHT, R. P. The dynamics and treatment of chronic alcohol addiction. Bull. Menninger Clin. 1: 233–250, 1937.

8. DALE, P. W. and EBAUGH, F. G. Personality structure in relation to tetraethylthiuram disulfide (Antabuse) therapy of alcoholism. J. Amer. med. Ass. 146: 314–319, 1951.

9. RUDIE, R. R. and McGAUGHRAN, L. S. Differences in developmental experience, defensiveness, and personality organization between two classes of problem drinkers. J. abnorm. soc. Psychol. 62: 659–665, 1961.

10. SUGERMAN, A. A., REILLY, D. and ALBAHARY, R. S. Social competance and essential–reactive distinction in alcoholism. Arch. gen. Psychiat. 12: 552–556, 1965.

11. LEVINE, J. and ZIGLER, E. The essential–reactive distinction in alcoholism: a developmental approach. J. abnorm. Psychol. 81: 242–249, 1973.

12. MENNINGER, K. Man against himself. New York; Harcourt, Brace; 1938.

13. MATHIAS, R. An experimental investigation of the personality structure of chronic alcoholic, Alcoholics Anonymous, neurotic and normal groups. Ph.D. dissertation, University of Buffalo; 1955.

14. LOLLI, G. Alcoholism as a disorder of the love disposition. Quart. J. Stud. Alc. 17: 96–107, 1956.

15. SUTHERLAND, E. H., SHROEDER, H. G. and TORDELLA, C. L. Personality traits and the alcoholic; a critique of existing studies. Quart. J. Stud. Alc. **11**: 547–561, 1950.
16. SYME, L. Personality characteristics of the alcoholic; a critique of current studies. Quart. J. Stud. Alc. **18**: 288–302, 1957.
17. LISANSKY, E. S. The etiology of alcoholism; the role of psychological predisposition. Quart. J. Stud. Alc. **21**: 314–343, 1960.
18. LISANSKY, E. S. Clinical research in alcoholism and the use of psychological tests; a reevaluation. Pp. 3–15. In: FOX, R., ed. Alcoholism, behavioral research, therapeutic approaches. New York; Springer; 1967.
19. BARNES, G. E. The alcoholic personality; a reanalysis of the literature. J. Stud. Alc. **40**: 571–634, 1979.
20. ROBINS, L. N., BATES, W. M. and O'NEAL, P. Adult drinking patterns of former problem children. Pp. 395–412. In: PITTMAN, D. J. and SNYDER, C. R., eds. Society, culture, and drinking patterns. New York; Wiley; 1962.
21. MCCORD, W. and MCCORD, J. A longitudinal study of the personality of alcoholics. Pp. 413–430. In: PITTMAN, D. J. and SNYDER, C. R., eds. Society, culture, and drinking patterns. New York; Wiley; 1962.
22. JONES, M. C. Personality correlates and antecedents of drinking patterns in adult males. J. cons. clin. Psychol. **32**: 2–12, 1968.
23. JONES, M. C. Personality antecedents and correlates of drinking patterns in women, J. cons. clin. Psychol. **36**: 61–69, 1971.
24. ULLMAN, L. P. and KRASNER, L. A psychological approach to abnormal behavior. 2d ed. Englewood Cliffs, N. J.; Prentice-Hall; 1975.
25. DOLLARD, J. and MILLER, N. E. Personality and psychotherapy; an analysis in terms of learning, thinking, and culture. New York; McGraw-Hill; 1950.
26. CAPPELL, H. and HERMAN, C. P. Alcohol and tension reduction; a review. Quart. J. Stud. Alc. **33**: 33–64, 1972.
27. MELLO, N. K. Behavioral studies of alcoholism. Pp. 219–291. In: KISSIN, B. and BEGLEITER, H., eds. The biology of alcoholism. Vol. 2. Physiology and behavior. New York; Plenum; 1972.
28. BANDURA, A. Principles of behavior modification. New York; Holt, Rinehart & Winston; 1969.
29. MARLATT, G. A. Alcohol, stress, and cognitive control. Pp. 271–296. In: SARASON, I. G. and SPIELBERGER, C. D., eds. Stress and anxiety. Vol. 3. New York; Wiley; 1976.
30. FRANKS, C. M. Alcohol, alcoholism and conditioning; a review of the literature and some theoretical considerations. J. ment. Sci. **104**: 14–33, 1958.
31. MARLATT, G. A. and NATHAN, P. E., eds. Behavioral approaches to alcoholism. (NIAAA-RUCAS Alcoholism Treatment Series, No. 2.) New Brunswick, N.J.; Rutgers Center of Alcohol Studies; 1978.
32. CHILD, I. Humanistic psychology and the research tradition; their several virtues. New York; Wiley; 1973.

33. POMERLEAU, O. Research priorities on alcohol studies; the role of psychology. J. Stud. Alc., Suppl. No. 8, pp. 75–95, 1979.

34. MESSER, S. B. and WINOKUR, M. Some limits to the integration of psychoanalytic and behavior therapy. Amer. Psychol. **35**: 818–827, 1980.

35. GREGG, G., PRESTON, T., GEIST, A. and CAPLAN, N. The caravan rolls on; forty years of social problem research. Knowledge, Beverly Hills **1**: 31–61, 1979.

36. ALLPORT, G. W. Personality; a psychological interpretation. New York; Holt; 1937.

37. BABOR, T. F., MENDELSON, J. H., KUEHNLE, J. C. and GREENBERG, I. Experimental analysis of the "happy hour"; effects of purchase price on alcohol consumption. Psychopharmacology, Berl. **58**: 35–41, 1978.

38. U.S. NATIONAL INSTITUTE ON ALCOHOL ABUSE AND ALCOHOLISM. Alcohol and health; third special report to the U.S. Congress from the Secretary of Health, Education and Welfare, June 1978; technical support document. Ernest P. Noble, editor. (DHEW Publ. No. ADM 79–832.) Washington, D.C.; U.S. Govt Print. Off.; 1979.

39. PEARLIN, L. I. and RADABAUGH, C. W. Economic strains and the coping functions of alcohol. Amer. J. Sociol. **82**: 652–663, 1976.

40. GOMBERG, E. S. L. Drinking and problem drinking among the elderly. Ann Arbor; University of Michigan Institute of Gerontology; 1980.

41. KINSEY, B. A. The female alcoholic; a social psychological study. Springfield, Ill.; Thomas; 1966.

42. ZUCKER, R. A. Developmental aspects of drinking through the young adult years. Pp. 91–146. In: BLANE, H. T. and CHAFETZ, M. E., eds. Youth, alcohol, and social policy. New York; Plenum; 1979.

43. JONES, B. M. and JONES, M. K. Male and female intoxication levels for three alcohol doses or do women really get higher than men? Alc. tech. Rep., Oklahoma City **5**: 11–14, 1976.

13

Sociological Theories of the Etiology of Alcoholism

Helene Raskin White

The original *Alcohol, Science and Society* contained many lectures pertinent to the field of sociology. Most of them covered a specific aspect of alcohol use or alcoholism which have been updated in the present volume by other authors. None of the original lectures, however, directly discussed the sociological theories of the etiology of alcoholism. The classic studies on which many of the theories were developed appeared at about the same time as the original *Alcohol, Science and Society*, and, except for Horton's (1), were not recognized in that volume. Many of these classic studies have withstood the test of time and are still pertinent to sociological theory.

Sociologists are generally interested in group differences. For example, they have attempted to discover why there are differences between men and women in rates of alcoholism and why the differences vary across cultures. In general, sociologists do not try to explain why a specific individual becomes an alcoholic. Instead, they are interested in explaining rates of alcoholism between different groups or cultures. However, the development of a theory that explains a group's use of alcohol or rate of alcoholism can be useful in predicting the probability of individual drinking problems.

Terms such as drunkenness, insobriety, excessive drinking, problem drinking, and alcohol-related damage have been used interchangeably in the literature. One might argue that the conditions which lead to drunkenness or excessive drinking are very different from those that lead to problem drinking or alcoholism. This chapter will focus on the latter type of behavior. Although alcoholism and

The writing of this manuscript was supported, in part, by a grant from the National Institute on Alcohol Abuse and Alcoholism (No. AA 03509-03).

problem drinking are not identical constructs, in this chapter they will be used somewhat interchangeably.[1]

This chapter will not explore all sociological aspects of alcoholism nor sociological aspects of all drinking behavior.[2] Rather, it is concerned with the sociological theories of the etiology of alcoholism which have developed over the last 35 years. It is not possible to review all relevant sociological literature on this topic. Rather, I will selectively report the literature which I find most pertinent.

I propose to first discuss the major sociological theories of etiology from a traditional perspective and then present a model for integrating these theories. In previous reviews (e.g., 4–6), sociological theories have not been consistently categorized. In presenting my categorization reviewing these theories, I hope to clarify the different perspectives rather than merely crowd the field with alternative nomenclature. My framework parcels sociological theory into four categories: Sociodemographic, Sociocultural, Socialization and Social Deviance.

Sociodemographic Theories[3]

Sociological variables such as age, sex, religion, socioeconomic status, ethnicity, and place of residence are known to affect the probability of drinking. Cahalan (7) found that these same variables related to the degree of problem drinking. He also described environmental and personality variables which are related to problem drinking, the most important of which is the attitude towards drinking. Those people who have a highly favorable attitude towards drinking are more likely to be problem drinkers. Other psychosocial variables consid-

[1]While this is not the place to discuss the varying definitions of these concepts, the fact that there is no one accepted definition of alcoholism has created an inordinate amount of confusion in the field of alcohol studies. It has made cross-cultural and comparative research very difficult and has also created problems in the interpretation of results and the development of theory. See Keller's chapter for greater detail.

[2]At the time of the original *Alcohol, Science and Society*, Bacon (2) presented a framework for sociological research on drinking. He complained that there was too much attention on alcoholism although alcoholism only constitutes a fraction of drinking behavior. He argued that there was a wide arena for sociological study in the phenomenon of drinking behaviors. Preoccupation with excessive drinking persisted despite Bacon's plea through the 1960s (3). In the last decade, however, there has been increasing attention to all aspects of drinking by social scientists.

[3]Since this theoretical perspective is the basis for Cahalan's chapter, I will only briefly discuss it here. The reader is referred to his chapter for greater detail and for references to the pioneering work in this area.

ered important are environmental support of heavy drinking, impulsivity, alienation and maladjustment, unfavorable expectations (not expecting to be able to achieve one's goal in life) and looseness of controls (not belonging to a family or a primary group).

Cahalan's data suggest that sociodemographic variables have an impact on an individual's decision to initiate, continue, or cease drinking by determining (1) childhood exposure to drinking behavior of role models, (2) the quantity considered appropriate, (3) the amounts considered safe, (4) the customs surrounding use, (5) the symbolic meaning of alcohol, (6) the activities associated with drinking, (7) the amount of pressure exerted to drink, and (8) the social rewards or punishments for drinking. However, the psychosocial variables are important determinants of the quantity and frequency of drinking that will be maintained (5).

Sociocultural Theories

Sociocultural theories arose out of examination of subcultural and cross-cultural differences in rates of drinking and problem drinking. Subcultural differences suggest that alcoholism is a cultural phenomenon. Cultural norms define alcoholism and contribute to its development or prevention. Jellinek (8) observed that what constituted alcoholism in one country did not constitute alcoholism in another; for example, periodic drunkenness at fiestas which last for several days was considered normal drinking behavior for Andean Indians, yet in the U.S. people who drink in this fashion may be labeled as alcoholics. In France there was alcoholism without open drunkenness (i.e., they maintain a certain concentration of alcohol in their system at all times, yet not enough to be considered intoxicated). In countries like Spain and Finland alcoholism was defined as occasional explosive drinking. Not only are there different rates of alcoholism in different cultures, but there are different definitions of alcoholism (not to mention many different definitions in the same country).

Sociocultural theories attempt to account for the cultural differences in rates of alcoholism. Sociologists and anthropologists have searched for factors which influence drinking practices, believing that the causes of deviant drinking, including alcoholism, will become more apparent by exploring differences in drinking behavior and attitudes toward drinking and drunkenness (4).

Horton's 1943 study (9) of primitive societies marks the beginning of this theoretical approach. He postulated that the degree of inebriety is directly related to the degree of anxiety in a culture. He defined three sources of anxiety (or fear) in primitive societies: (1) the level of subsistence economy, (2) the presence or absence of subsistence hazards, and (3) the degree of acculturation. Horton's work suggested that in primitive societies (and, presumably in modern society) one of alcohol's primary functions is to relieve tension and anxiety.

Field (10) restudied 56 of Horton's tribes. Field argued that the degree of social organization accounts for the differences in rates of drunkenness. He stated that drunkenness in primitive societies is determined less by the level of fear in a society than by the absence of kin groups with stability, permanence, formal structure and well-defined functions.

Field and Horton had comparable findings. Social disorganization could be described as an anxiety-producing situation. In a society with stability and formal structure, options are less and anxiety should also be less. Social organization can also act as an insulator from stress; the kin group can provide a mechanism to help its members deal with their anxiety. (This relationship will be discussed later in greater detail.)

Following in the Horton–Field approach, Bacon et al. (11) analyzed data obtained from 110 preliterate societies. They found that, in addition to reducing anxiety and tension, alcohol also permitted satisfaction of a dependency need. Frequent drunkenness or high consumption tended to occur in cultures where the needs for dependence are deprived or punished in childhood and in adulthood and where a high degree of responsibility, independence and achievement are required.

All three studies have their limitations, especially for etiological theories of alcoholism. The samples are limited to primitive societies (and only those which someone chose to study). In most preliterate societies alcohol serves an integrative function and alcoholism is rarely found. Much of the data were collected on other aspects of behavior and drinking behavior was abstracted from them; hence, data bases are often incomplete. Another problem is that of definition. Horton referred to the degree of inebriety, Field to the degree of drunkenness and Bacon et al. to frequent drunkenness or high consumption. Although the studies concentrated on normal drinking practices and drunkenness rather than alcoholism, these anthropologists set the stage for sociocultural theories by attempting to develop a unifying link to account for the high rate of inebriety or drunkenness in many cultures at once (3).[4]

[4]Anthropological research is reviewed by Marshall (12) and Everett et al. (13), and discussed by Heath in this volume.

In the literature of sociocultural theories, Bales' theory is probably the most comprehensive (6). Bales (14) defined three ways in which culture and social organization can influence the rate of alcoholism and said that the rate of alcoholism is an interaction of the three sets of variables. The first is the dynamic factor, the degree to which the culture creates inner tensions or acute needs for adjustment in its members, that is, the incidence or level of psychic tension or anxiety in the society. The second variable is the normative orientation, that is, the attitudes towards drinking that the culture produces in its members, including the norms, the ideas, and the sentiments which center on the act of drinking. The third factor is the degree to which the culture provides alternatives to satisfy or cope with acute psychic distress.

Bales used the Irish to illustrate the dynamic factor. In Ireland there was a bare existence economy where only one son inherited the farm after the father stopped working. Insecurity, anxiety and frustration were typical because men could not marry until they had an income. Bales hypothesized that the high level of anxiety and frustration in this culture was causally related to the high rate of alcoholism.

Bales' use of the Irish to support his point may be questionable. Although the Irish in Ireland have many social complications resulting from drinking (e.g., arrests, fights, family problems), Ireland has a relatively low rate of cirrhosis mortality. Thus, it has been argued by some researchers (e.g., 15) that the Irish do not represent an example of a culture with a high rate of alcoholism. Statistics for Irish Americans, however, do indicate a relatively high level of consumption and of alcoholism (16). Additional explanations for the postulated high rate of alcoholism among the Irish include: they drink a lot of distilled spirits and beer; they drink heavily; they have conflicting attitudes related to drinking; their children learn to drink outside of the household; and they favor use of alcohol for self-medicating purposes.

Bales suggested four types of normative orientations which influence drinking patterns: (1) complete abstinence, (2) drinking as a ritual act of communion with the sacred, (3) convivial (using alcohol for social or symbolic purposes, creating a sense of social unity or solidarity), and (4) utilitarian (medicinal use of alcohol, use for self-interest reasons such as personal satisfaction, or even use for solidifying business deals).

Bales cited the Moslems as an example of an abstinent subculture. Because there is total prohibition of alcohol and the orientation of the Moslem culture espouses complete abstinence, the rate of alcoholism is low. This orientation can have negative effects, sometimes promoting alcohol problems rather than preventing them. For example, Straus and Bacon (17) reported a relatively high probability of alcohol problems occurring among Mormons who drink.[5] Several reasons have been offered to explain this phenomenon. Mormons have no experience with moderate drinking and no practice learning when to stop. In addition, alcohol is considered the "forbidden fruit," thus making it very tempting. Mormons who do drink learn to do so at a later age and outside the home; this type of learning experience is postulated to create alcohol problems. The fact that any drinking at all is a break with the group norms make extremes in behavior likely. Finally, such drinking will create problems with family or Mormon friends (19). Recently there has been a reexamination of this supposed detrimental effect of an abstinence background on future alcohol problems. It may be true that, among Mormons who drink, a relatively large proportion may develop problems; however, reanalysis of the data seems to suggest that, when abstainers and drinkers are considered together, Mormons have a low proportion of problem drinkers. The rate of alcoholism among all Mormons in the U.S. is relatively low (15).

Bales suggested that those cultures which have a ritual orientation toward the use of alcohol have the lowest rates of alcoholism and he used the Jews as an example of such a culture. Jewish people tend to have very low rates of alcoholism and alcohol problems, although they have one of the highest proportions of drinkers of any ethnoreligious group. Snyder (20) also found that a major reason why drinking problems are minimal among Jewish people is that drinking is integrated into the ceremonies and symbolism of Jewish traditions. Jews are taught to drink in the family situation at home; drinking is incorporated into the lifestyle. Another explanation for the low rate of problem drinkers is that Jews hold a stereotype of the drunken gentile as compared to the sober Jew. Because of a strong sense of ethnocentrism, Jews will maintain the group norms and refrain from excessive drinking. (Interestingly, Snyder found that Jews tend to drink differently when with an out-group such as college or military peers as opposed to when they are with the in-group, their Jewish community.)

It has been argued that as Jews began to assimilate more into American culture, their rates of alcoholism would increase. However, in 1970 Wilkinson (19) reported that, whatever estimates of alcoholism are used, Jews still exhibit fewer problems. Glassner and

[5]Skolnick (18) reported similar findings among Methodist students when reanalyzing the data collected by Straus and Bacon (17).

Berg (21) conducted the most recent systematic study of drinking among Jews. Their large community interview study was comparable to Snyder's (20) and controlled for age, sex, and degree of religiosity. Glassner and Berg concluded that there has been no increase in the rate of alcoholism among Jews. Four processes which act as social controls to prevent deviance in alcohol use among Jews were identified. The first was the association of alcohol misuse as an out-group characteristic. Second, Glassner and Berg found that although orthodoxy has declined, tradition continues. Children are still exposed to the religious and ritual symbolism of drinking in their formative years. The third process is a restriction of adult primary relations to other Jews and, therefore, to other moderate drinkers. These first three processes reconfirm Snyder's findings 20 years later. Finally, it appears that Jews develop a repertoire of techniques to avoid excessive drinking under social pressure (for example, they learn to make one cocktail last for several hours).

Bales' convivial orientation is like the ritual orientation in the sense that it is symbolic of solidarity, but it is also like the utilitarian orientation in the sense that the drinker expects a good feeling. So it is very easy for the convivial to become a utilitarian orientation. According to Bales, the utilitarian orientation is the one most likely to foster heavy drinking and to lead to problem drinking, especially in societies where there is a great deal of inner tension (e.g., the Irish). Bales stated that the normative orientation was quite important in explaining differences between the Jews and the Irish (ritual vs convivial–utilitarian).

Bales' third factor influencing cultural rates of alcoholism, alternative means, is the degree to which the society produces substitute means of satisfaction or release of tension. In those societies in which there are few alternatives there tends to be a high rate of alcoholism; however, in those societies which have other outlets (such as use of opium or meditation) there are low rates of alcoholism. According to Heath (22), a major problem with Bales' theory is that it is based on historical, religious and demographic documents.

Pittman's (23) theory is very similar to Bales. He defined four attitudes towards drinking that a culture may possess: (1) abstinence (negative attitudes and prohibition towards the ingestion of alcohol), (2) ambivalence (a conflict between coexisting value structures), (3) permissive (permissive towards drinking but negative towards drunkenness and drinking problems), and (4) overpermissive (permissive towards drinking, intoxication and drinking pathologies).

Pittman said that the permissive attitude is least likely to lead to alcoholism and the ambivalent attitude most likely. The other two attitudes are paradoxical in that they can lead either to higher or lower rates of alcoholism in a culture. The permissive attitude, which fosters low rates of alcoholism, is exemplified by the Jews, Portuguese, Italians and New York Cantonese. Pittman cited the French as an example of overpermissive culture with a high rate of alcoholism, as measured by deaths from cirrhosis. Yet, on the basis of cirrhosis mortality data, it could be argued that the Portuguese and Italians have high rates of alcoholism like the French rather than low rates like the Jews and Chinese.

It has been posited by Ullman that the presence or absence of ambivalence or culturally consistent stable drinking customs will influence the rate of alcoholism in a society:

> In any group or society in which the drinking customs, values and sanctions—together with the attitudes of all segments of the group or society—are well established, known to and agreed upon by all, and are consistent with the rest of the culture, the rate of alcoholism will be low. . . . [But where] the individual drinker does not know what is expected or when the expectation in one situation differs from that in another, it can be assumed that he will have ambivalent feelings about drinking. Thus, ambivalence is the psychological product of unintegrated drinking customs (24, *p. 50*).

According to Blacker (4), Ullman's hypothesis is supported by the Jews, Chinese, Irish Americans and Protestants from abstinence backgrounds who drink. However, the French have all the characteristics which Ullman expected to lead to a low rate of alcoholism but nevertheless have a high rate. Blacker stated that Ullman failed to attend to the content and quality of the norms and attitudes and only cared about their structural aspects (e.g., strength, consistency).[6] To make up for Ullman's "shortcoming" in addressing the content of the norms, Blacker proposed an elaboration of Ullman's hypothesis which has been labeled the Ullman–Blacker hypothesis by adding the qualification that when the drinking customs, values and

---

[6]Mizruchi and Perrucci (25) also argued that the qualities of norms are important determinants for system maintenance and system integration. They stated that even if the norms are effectively transmitted and collectively controlled, it does not ensure stability or integration of the system. Using data which indicate proportionately greater alcohol problems among drinkers in groups like the Mormons, they claimed that in a culture characterized predominantly by proscriptive norms, deviation from the norms will be related to high rates of drinking problems because of the absence of prescriptive norms about how to behave if one drinks. However, in cultures with predominantly prescriptive norms, deviation will result in low levels of problems.

sanctions *"are characterized by prescriptions for moderate drinking and proscriptions against excessive drinking,"* the rate of alcoholism will be low (4, p. 68).

Myerson (26) also discussed ambivalence, but his views are more from a social-control than a normative-orientation perspective. He used women and Jews as examples of groups with little ambivalence about drinking, saying that their drinking norms are better defined. When the limits are well defined social control or regulation is more possible and probable. Women and Jews will tend to drink less to avoid social disapproval and therefore experience fewer drinking problems. On the other hand, ambivalence (lack of clearly established limits) sets the stage for weak social control and more problems.[7]

A model of the distribution of consumption of alcoholic beverages by populations has also been applied to the study of alcoholism.[8] This model describes the relationship between per capita consumption and prevalence of alcohol-related damage in populations. The model, proposed by Lederman (28, 29), is based on a hypothesis that the distribution of consumption of alcohol in a population follows a lognormal curve—highly skewed, unimodal and continuous. Populations with high levels of per capita consumption will have a larger proportion of drinkers consuming greater amounts of alcohol than populations with low levels of per capita consumption (15).

It has been shown that the rates of alcoholism vary with the level of over-all consumption in a population (30). This finding and in fact the model, however, are predominantly based on a definition of alcoholism as measured by rates of cirrhosis mortality or other alcohol-related health problems. De Lint and Schmidt (31) have shown that in a number of western societies the per capita consumption of alcohol is positively related to cirrhosis mortality rates. They reported that daily consumption of 10 cl of absolute alcohol (approximately 6 drinks) is associated with a risk of physical pathologies. As per capita consumption increases, the proportion of the population which

[7]Room (27) analyzed the use of the term ambivalence in relation to drinking behavior and showed its limits as an explanatory concept. Ambivalence serves only as a catch-all term for individual conflicts with regard to drinking behavior and societal confusion, hence, integrating it with normative belief systems and systems of social control. (Some of these concepts will be discussed again later under the category of social deviance theories.)

[8]Although the distribution of consumption model is not actually a sociocultural theory, it is included in this section of the chapter so that it may be contrasted with sociocultural models.

consumes amounts of alcohol which could be physically damaging also increases.[9]

Frankel and Whitehead (15) compared a sociocultural model (the Ullman–Blacker hypothesis) with the distribution of consumption model. These researchers claimed that a major difference between the two models is the fact that the former does not pay attention to the quantity of alcohol consumed. However, the Ullman–Blacker hypothesis stipulates proscriptions against excessive or heavy drinking, thus it does consider quantity. Another major difference is in their definitions of alcoholism: the sociocultural model incorporates both drunkenness and social problems as in the case of attributing a low rate of alcoholism to the Jews and the Italians and a high rate to the Irish, and it also incorporates physical damage and mortality in attributing a high rate of alcoholism to the French. On the other hand, the distribution of consumption model relies on a definition of alcoholism of physical problems and mortality.

Whitehead and Harvey (34) tested aspects of both models using the data collected by Bacon et al. (11). Their results did not support the Ullman–Blacker hypothesis to any great extent and they found the level of consumption to be the best predictor of alcohol problems. Based on their findings they amended the Ullman–Blacker hypothesis by inserting into the above quote: "... *that keep per capita consumption low enough that few persons in that society will consume in excess of 10 cl of absolute alcohol per day*, the rate of alcoholism will be low" (34, *p.* 63). The addition of a numerical value, that is, 10 cl, to the Ullman–Blacker hypothesis does not considerably change its substance given that it had already considered

---

[9]There have been many criticisms of the distribution of consumption model on the basis of lack of supporting evidence (e.g., 32). One problem is the determination of average consumption cross-culturally: production or sales records are not particularly reliable on an international basis and do not account for illegal manufacturing or unrecorded legal home brewing (33). If the model is based only on drinkers, then a problem of determining their number exists; very few countries have truly valid and reliable estimates of the number of abstainers (if they even define abstainers similarly). Pittman (33) argued that, for per capita measures of consumption to have significance, consumption statistics must be related to the drinking patterns within a society, and they rarely are. Consumption statistics should, but often do not, take into account the age or sex by age distribution of the population. If the majority of drinkers in a population are between 20 and 30 years old, alcohol may produce differential effects, especially in terms of physical damage, than if the majority are between 60 and 70 years old. I would add that per capita measures of consumption should incorporate the factor of individual weight. Six drinks daily could have vastly different effects in a population in which most members are tall and heavy as compared to one in which they are short and slim.

quantity. In fact, it is curious that Whitehead and Harvey retained concepts from the Ullman–Blacker hypothesis, such as "consistent with the rest of the culture" even though their analysis did not strongly support these concepts. I am impressed, however, with their initiative and originality in empirically testing these theoretical models.

Frankel and Whitehead (15) reanalyzed the sociocultural model, the distribution of consumption model and the Whitehead–Harvey reformulation model using path analysis statistics with data from 68 societies from the Bacon et al. (11) compilation. The sociocultural model accounted for 20% of the variance (variation) in the extent of alcohol-related damage (social and physical problems). Neither the degree to which the drinking practices were integrated into the culture nor the existence of prescriptions for moderate drinking had any significant direct influence on the extent of damage. Only proscriptions for drunkenness had a significant direct effect on the extent of alcohol-related problems.

The distribution of consumption model explained 36% of the variance and showed a significant relationship between over-all consumption and extent of damage. Frankel and Whitehead noted the model only contains one variable, over-all consumption, thus it does not help to understand relationships between other factors that may be related to alcohol-related damage.

The Whitehead–Harvey reformulation model explained 48% of the variance in alcohol-related damage and demonstrated that level of over-all use was the most powerful predictor of extent of damage. By elaborating the model, Frankel and Whitehead were able to explain 57% of the variance in damage. They found the clearest relationships between the level of consumption and damage and between proscriptions against drunkenness and damage, while the relationships between both integration of drinking practices and prescriptions for moderate drinking with damage were not clear.

Frankel and Whitehead's analyses have several methodological flaws such as being based on somewhat unreliable data, unexplained relationships, and problems with operational definitions. However, they should be commended for this excellent attempt to empirically test theory.

The sociocultural theories assume a relationship between cultural attitudes towards, purposes for, motivations for or norms of alcohol use and the rate of alcoholism. Several of the problems with studies testing this approach have been discussed above. The definitions of alcoholism, alcohol problems or alcohol damage vary from study to study making comparisons difficult. There is a tendency to use either social or physical complications as indicators of alcoholism rather than both. Thus, there are inconsistencies; for example, Blacker cited Italy as an example of a culture with a low rate of alcoholism while de Lint and Schmidt used Italy as an example of a culture with a high rate. The indicators even differ within analyses; for example, Blacker used the lack of social problems to account for the low rate of alcoholism among Italians and the existence of physical problems to account for the high rate among the French.

Several of the theories were developed in an ad hoc manner from inaccurate generalizations and stereotypes (22). As Heath (22) pointed out, even those theories which were derived deductively (such as the large cross-cultural comparisons) have relied on data not collected explicitly as part of a study on drinking practices. Also, the data were collected during field work in alien societies, through participant observation or nondirect interviewing, with very little attention given to systematic sampling. The rest of the data were collected in surveys of usually unrepresentative or small samples. In addition, the validity and reliability of the data must be questioned given the nature of the topic and the dependence on self-report. Heath also claimed that there is a tendency to compare apples with bananas in much of the sociocultural research. He suggested that there are different kinds of norms, ethnic groups, drinking problems, and cross-cultural studies, which are not necessarily comparable with each other.

Lemert (3) said that the research in this area is marked by inconsistency and lack of clarity in many key concepts. There is also skepticism about the validity of data, choice of units for comparison, and representatives of cultures sampled.

Whether one holds that Bales' theory or the Ullman–Blacker hypothesis represents the sociocultural model, one is struck by the fact that both are quite dated. That is, Bales developed his theory around 1946 and Ullman developed his around 1958. Blacker made the addition to Ullman's hypothesis in 1966. Now, 15 years later, very little progress has been made in the development of a socio-cultural model.

Socialization Theories

The sociocultural theories described above suggest that the attitudes or belief system surrounding alcohol use espoused by a subculture will determine the level of drinking problems experienced. Social-ization is the process by which society's ideational system in the form of values, perceptions, beliefs and norms—the when, why, how, what, where, how much to drink—is passed on to the individual members. It occurs primarily in the family, at least through late childhood. After that other systems, such as the schools and peer networks, serve to reinforce, direct or contradict the internalized system. Several investigators have studied these family influences on drinking.

The McCords (35) conducted an extensive longitudinal analysis examining family characteristics related to alcoholism. The most important characteristics found to be related to alcoholism were: maternal alteration between active affection and rejection, parental (especially maternal) escapist behavior in a crisis, a deviant mother (e.g., criminal, alcoholic, promiscuous), maternal resentment of her role in the family, incest or illegitimacy in the family, overt paternal rejection, absence of high parental demands for sons, lack of supervision, an outsider with conflicting values of the parents living in the home, and parental antagonism toward one another.

Zucker (36) described five levels of influence on a child's drinking. The first level consists of family status, life-style and community involvement and represents the interrelationship between the cultural milieu and drinking behavior of persons living in that culture. The second level comprises the family environment or family interaction, that is, parental conflict, the father–mother interactions concerning drinking, and the family structure (family size and the birth order of the children). The third level constitutes individual parent behaviors: belief structures regarding drinking, drinking behavior, personalities, and child-rearing behaviors. Zucker stated that these family characteristics are tied in with the socialization process and will determine whether and how a child will drink. The other two levels of influence include peer effects and the child's personality. The peer effects are related to the previous levels in that parents' value systems and background and a person's residence will influence choice of peers. Zucker has shown the connection between the first three levels and the child's personality, since the manner in which the child is brought up will definitely affect his or her need and behavior systems.

Alcoholism is often transmitted in families.[10] The family in which there is an alcoholic will be more likely to create an emotionally unstable child: inconsistent relationships with both parents, who may alternate between being kind, considerate, and overbearingly affectionate and being hostile and withdrawn, affect the sense of security and self esteem. Alcoholics are often poor role models

---

[10]See Goodwin's chapter in this volume for a discussion of the heredity–environment issue.

and inadequate parents, creating adjustment problems in the child. The child may become socially isolated from peers. Thus, the environment of an alcoholic home can contribute to the eventual development of alcoholism in the child.

Attitudes about and motivations for alcohol use are important determinants of alcohol problems. Children will learn these attitudes and motivations in the home and will model their parents' behavior. When a child sees alcohol being used as a means of escape, the child will model this behavior and learn problem-producing drinking habits. The child might also learn to use escape to deal with all problems rather than to learn more positive coping mechanisms.

The emotion surrounding an alcoholic's drinking is another way in which parental attitudes can affect the children. Alcohol is rarely a neutral topic and therefore the child will probably develop a sense of ambivalence about alcohol which may also contribute to the development of alcohol problems.

Social Deviance Theories

Sociologists have applied theories of deviance to alcohol problems and alcoholism, hence, combining elements of social and psychological spheres to explain alcoholism. Three perspectives of deviance can be logically applied to alcoholism: anomie, labeling, and cultural transmission.

Anomie

The two leading sociologists in the area of anomie are Durkheim and Merton. Because their theories focus on very different aspects of the concept of anomie, their theories relate to the etiology of alcoholism in distinct ways.

Durkheim's (37) theory, which defines anomie as a state of normlessness or deregulation, was applied to alcoholism by Cheinisse (38) in 1908. Cheinisse observed that alcoholism was very rare among Orthodox Jews who had recently emigrated to Paris, while the incidence of alcoholism was higher among Jews who had been living in Paris for a while. He explained the difference in terms of the solidarity and cohesion of the traditional Jewish community: as ties to the community loosen and consensus breaks down (a state of anomie), rates of alcoholism increase. Snyder (39) stated that Cheinisse's ideas carry over to present day society. He noted that there has been an increase in drinking pathologies as a result of secularization of Jews; secular Jews have the highest rate of alcoholism among Jews, reformed Jews the next highest, then conservative, and finally Orthodox. Snyder claimed that the differences are not related to differences in incidence or frequency of alcohol use nor to changes in generational or social class status but, rather, to the lessening of solidarity.[11]

Snyder suggested that the low rate of problem drinking among Jews, especially Orthodox Jews, may, in fact, be a combination of eunomie (absence of anomie) and the ritual attitude toward alcohol. The quality of the drinking norms due to the ritual use of alcohol reinforces the general eunomia which in turn minimizes alcoholism and inebriety.

Thus, in a Durkheimian sense, a moral community which is cohesive regulates the individual members and protects them from anomie, thereby insulating them from inebriety. An eunomic society implies that people know their purpose in life, know their roles, have fewer options, have tradition, and have social order. Conversely, when there is a breakdown in the moral guidelines and regulatory functions of life, deviant behavior (e.g., alcoholism) is more likely to occur. This analysis brings us back to Field (9) and Horton (10). Field found that inebriety was high in societies where there was social disorganization and absence of a formal kin structure, a state which may be the equivalent of anomie. Likewise, applying Horton's theory, the characteristics of a eunomic society imply that individuals are bound to the group and have a purpose in life; hence, they may suffer less from fear and anxiety.

One may also argue that a cohesive regulatory community has certain norms concerning proper behavior which are shared by all. Since the group has strong social control, a deviant (e.g., a problem drinker) would be ostracized quickly. Thus people conform because of knowledge of the rules and strong social control. Based on the ambivalence theories discussed previously, we would expect the rate of alcoholism to be low in such a society.

Merton's (40) theory of anomie has to do with the disjunction between the goals and means, that is, the strain produced between the societal goals and the institutional means of achieving those goals. Five different types of behavior result from this strain: con-

---

[11]Those familiar with Durkheim's *Suicide* might argue that Cheinisse and Snyder were using Durkheim's concept of "egoism" (egoistic suicide) rather than anomie. Egoism would be defined as a lack of integration and would be contrasted to concepts such as solidarity and cohesion. On the other hand, anomie was related to how the society regulates the individual rather than the individual's attachment to society. It is possible, however, that alcoholism may be related to both the degree of integration into and regulation by the community.

formity, ritualism, innovation, retreatism and rebellion. Alcoholics fit into the retreatist category as do drug addicts, vagrants, and social isolates. Retreatists cannot attain societal goals through either legitimate or illegitimate means.

Merton's use of the term anomie suggests alienation. The alcoholic retreats from society because he or she cannot succeed. If we look at strain as a source of tension and anxiety, then we are back to Horton again. Alcohol is one method of relieving the anxiety caused by the strain of failing in society. Cahalan (7) found that alienation and maladjustment, as well as unfavorable expectations, were related to problem drinking. The reason why one person chooses alcohol and another drugs as a means of retreat may depend on the person's attitudes concerning alcohol use or initial experiences with alcohol.

Labeling

Labeling is a social process that transforms one's conception of self as normal to self as deviant. Labeling theory as it relates to alcoholism states that an alcoholic is a person who, through circumstances, becomes publicly labeled as deviant and by society's reaction is compelled to play a deviant role. Labeling theory is characterized by a two-part process beginning with an act of primary deviance and developing into secondary deviance (41). Primary deviance is the actual behavior that causes someone to become labeled (e.g., arrested for driving while under the influence of alcohol). Secondary deviance is the behavior produced by being placed in a deviant role (e.g., continuous intoxication) and can be seen as self-fulfilling prophecy. Secondary deviance is the social role which becomes a means of defense or adaptation to the problems created by societal reaction to primary deviation. Once a person is labeled it leads to a change in self concept and changes in others' definitions; in the case of alcoholism, the individual begins to drink more because the action is expected by the alcoholic label.[12]

Robins (43) questioned the use of labeling theory to explain alcoholism. According to her, the best predictor of later alcoholism in a person seems to be early deviant behavior, although the nature of the early deviance is unimportant. The second best predictor is having a parent with a history of deviance, although neither the

[12]Labeling the alcoholic as sick also has had negative consequences for alcoholics. As Roman and Trice noted (42), the mere process of labeling serves to aggravate and perpetuate a condition which is initially under the individual's control—"the disease label has disease consequences" (pp. 247–248).

nature of the deviance nor whether the person lives with the parent are very important in determining the power of the predictor. Social group membership runs a distant third place as a predictor.

Robins argued that labeling theory and these findings contradict each other. The predictors are powerful whether or not the behavior has been labeled. Also the decreases in deviance often evidenced as adolescents mature suggest an opposite progression than labeling theory predicts; labels should increase as time goes on, and thus deviant behavior should increase. The fact that early deviance predicts later deviance regardless of the type of deviance does not fit with labeling theory.

Robins did not negate labeling theory as an explanation of alcoholism, she merely suggested that it is probably not a very important explanation. The label is not usually applied until late in the progression toward alcoholism. By that time physical or psychological dependence may perpetuate heavy drinking whether or not the person has been labeled. Robins also pointed out that the label may actually be beneficial to treatment as acceptance of the label is a prerequisite for success in Alcoholics Anonymous (and other programs). Thus the label may actually lead to a positive change in self concept, not a negative change as labeling theory predicts. Also, the label is often removed as heavy drinking decreases, thus it is not irreversible as the theory would predict.

Cultural Transmission

Another major deviance perspective which has been related to alcohol use is cultural transmission (or social and cultural support). There are many different subcultures in society. What one subculture defines as normal behavior may be defined by another as deviant behavior. Cultural transmission theories postulate that the subculture socializes its members into behavior which "outsiders" label as deviant. Members perform these "deviant" acts because their reference group supports and rewards such behavior.[13]

One such cultural transmission theory is Sutherland's (45) differential association, a learning theory applied principally to crime. It postulates that crime—the techniques of committing the crime and the motives and attitudes concerning the behavior—is learned in interactions with other persons within primary groups. The person becomes deviant because of an excess of definitions favorable to the

---

[13]Cultural transmission theories have been categorized as socialization theories in some textbooks, e.g., Clinard and Meier (44).

violation of law over definitions unfavorable to violation of law among significant others.

In terms of heavy drinking and alcoholism, the theory is quite applicable. If the reference group or subculture is supportive of heavy drinking then that drinking norm will be internalized. One is taught how to drink and the norms and values concerning drinking are set. Continued heavy drinking can lead to alcohol problems and alcoholism.

Trice's (46) theory of alcoholism in America brings together labeling theory and cultural transmission theory. In addition, he included aspects of personality—so, in a sense, he transcended the confines of sociology and moved toward an interactive (multidisciplinary) theory. Trice stated that alcoholism in America is a mixture of "prone personalities who regularly imbibe in drinking groups that reflect the functional value of alcohol in complex society, but which exercise widely varying norms about what is deviant drinking" (p. 2). As a result, the drinker experiences weak social controls because he or she can move to groups more tolerant of heavier and deviant drinking. "Finally, cultural values stressing the importance of self control justify a pattern of segregation of those who regularly become intoxicated."

By "prone personalities," Trice meant people who experience dependency–independency conflict in childhood and feel insecure and afraid of failure. There is stress on individualism and independence, yet middle-class values encourage dependency on parents and for some the conflict remains unresolved. It is the fit between vulnerable personalities and drinking groups that Trice felt is the key to alcoholism.

By drinking-centered groups he meant those in which people drink together whether it be at cocktail parties, taverns, etc. Trice claimed that repeated exposure to these groups tends to satisfy emotional needs of the prone personality and to set the pattern of the use of alcohol as a primary coping mechanism. The group is actually a pseudo-primary group, that is, there is no demand for repeated long-term emotional relations, yet it meets the immediate demands of group feeling.

Social ambivalence and weak social controls in American society are a result, in part, of the Temperance Movement which created fractured attitudes about alcohol. Also, as a result of ethnic pluralism, there are many norms about alcohol's meaning, no uniform social control and more than one set of rules.

According to Trice, the American value system stresses self control and will-power. Because of these values, once the heavy drinker begins to drink more than the other people in his drinking group, ostracism occurs. The heavy drinker turns to a more tolerant, heavier drinking group. Often, the move to a heavier drinking group means a decrease in status and the process continues. (It has been suggested that the Jews or Chinese, in order to switch drinking groups, have to sever community ties; presumably they would rather conform to the community's drinking norms than lose its emotional support.) The movement from group to group leads to unalterable social segregation. Exclusion becomes permanent and finally the heavy drinker hits bottom. By this point, the person has possibly been formally labeled if arrest, divorce, job loss, or alcoholism treatment has occurred. After being labeled he or she is totally segregated from the main stream of society.

A Socioenvironmental Model

A socioenvironmental model of alcoholism can tie together all of the information presented in this chapter. The socioenvironmental model is an expansion of Roebuck and Kessler's (6) ideas and is based on Bales' (14) theory with the addition of a social control element.

The first component of the socioenvironmental model is the degree of stress, anxiety, or tension (cf. Bales' dynamic factor, psychic anxiety) in the society. In societies in which people feel the most stress, the members are going to be most in need of relief from that stress. This stress could result from: acculturation or subsistence anxieties as Horton (9) hypothesized or social disorganization as Field (10) suggested. This component takes into account modern society with its greater complexity including increased stratification and intergroup competition. It therefore includes the alienation of modern man and the disjunction between social goals and the legitimate means available for people to achieve them. In addition, the normlessness, purposelessness and rootlessness found in modern society contribute to the high level of stress.

The second component in the model is the normative belief system. This component encompasses much more than Bales' normative orientation factor. Of course, it does encompass people's purposes for use of alcohol and their motivations for drinking. Attitudes toward drinking and especially those conducive to alcoholism can have an important influence on the development of alcoholism. These conducive attitudes include Bales' utilitarian orientation and Pittman's (23) overpermissive attitude, that is, those attitudes which condone drinking to deal with personal problems (or for personal pleasure). In addition, this component includes the norms concerning appropriate quantities, frequencies, and occasions for use. In broader terms, it encompasses the larger normative belief system, including values concerning escaping from one's problems and appropriate ways of dealing with stress.

Bales' alternative factor is the third component of the socioenvironmental model. If a society offers means or organized occasions for dealing with stress, tension, depression, anomie, and alienation either through religious or mystical rituals, recreation, other drugs, or psychoanalysis, etc., then the number of members choosing alcohol as their coping response will be fewer. The choice of alcohol to relieve culturally induced tensions is determined by both the normative belief system (attitudes toward alcohol) and the available alternative means of tension release.

Finally, the additional component of the socioenvironmental model is the element of control. This factor includes the concept of having ties to a group or community. The control component encompasses anomie in that it takes into account the degree of regulation and cohesion within the community. The more unified the society and the greater the consensus on norms among members, the greater the social regulation. Anomic societies will have higher rates of alcohol problems than eunomic societies.

In societies with greater degree of social control, members' behavior is more observable and can be condoned or condemned. Thus the control element encompasses the concepts of labeling and societal reaction. If labeling leads to isolation and exclusion of the problem drinker from significant others and the development of an alcoholic self concept and secondary deviance, then it may be responsible for perpetuating further alcoholism. The isolated alcoholism-prone individual will then avoid social regulation of alcohol use and become anomic. Myerson (26) explained the lower rates of drinking problems among women as compared to men as partly due to the norms about and controls on appropriate drinking for women. Likewise, it has been suggested that the Jewish or Chinese person who exceeds customary drinking practices would be excluded from the group. The control concept encompasses group controls over all behavior, including belonging to a group and not being a loner, in addition to specific controls over drinking behavior.

The socioenvironmental model of alcoholism may make more sense if it is described on the individual level. Take, for example, the case of an individual who experiences stress, anxiety or depression. Stress can be caused by many sources: an inability to achieve one's goals, unfavorable expectations, a series of life failures, childhood trauma, etc. Hence, the individual's life stress may be a result of background characteristics (sociodemographic), childhood experiences (socialization) or life experiences, particularly failures (strain). Or it might also be the result of some psychological or physiological handicap. Whatever the causes, there is a need to escape.

The individual who values the need to escape from problems and believes that alcohol is a viable means of escape or a mechanism for coping, is likely to use alcohol. In addition, if this individual's reference group (family, friends, co-workers, community, etc.) condones or encourages heavy drinking and the person adopts this behavior then he or she may develop an alcohol problem. The results of social learning and socialization, including differential association, affect the normative belief system. The individual learns through continued and reinforcing use of alcohol and perhaps from significant others that alcohol can relieve tension and anxiety. Therefore, the use of alcohol as a pain reliever is internalized. The normative belief system must incorporate values which view alcohol use, particularly heavy use, as positive and beneficial to the individual.

Behavior must also be considered here. That is, not only must the individual have certain motivations, purposes and reasons for using alcohol, he or she must drink at a certain level. Social psychologists continue to debate the question of whether attitudes shape behavior or behavior determines attitudes. Both are necessary to explain the development of alcoholism.

The element of alternative means of satisfaction is tied in with the normative belief system, socialization and social learning. Everyone meets in life some traumas, or at least some disturbing events. We all must learn to cope with these events. If this individual never learns "socially acceptable" coping mechanisms (e.g., rational thinking, prayer, or seeking counsel) then he or she may turn to detrimental means, such as aggression, drugs or alcohol, to express frustrations and alleviate pain. The individual may not have learned appropriate coping mechanisms because of inadequate socialization or because the reference group passed on inappropriate skills.

The final element, social control, includes societal reaction and labeling. An individual tied to a group cannot act out (drink to excess, go on binges, etc.) without being identified by the group as having a drinking problem. Of course, if the reference group members (significant others) drink in the same fashion then an individual would not be labeled until the behavior was truly excessive relative to group norms. So the individual in our example either did not belong to a group or has already drunk to excess, been labeled and been excluded from a group. Either way, the behavior is no longer under group regulation.

Therefore, according to the socioenvironmental model, the type of individual likely to become an alcoholic is one who (1) experiences a great deal of stress, (2) has learned that alcohol is an appropriate means of relieving stress and uses alcohol often for this reason, (3) has no other, or at least no better, means of relieving this stress, and (4) is anomic in the sense of not belonging to or having been excluded from all groups (especially a primary group) which regulate his or her behavior.

I believe this model offers a framework for compiling the sociological literature on the etiology of alcoholism and alcohol problems. Our next task is to empirically test aspects of the model to determine which are the most powerful predictors of the problem behavior and what are necessary combinations that result in alcoholism.

An Interactive Model

Sociologists are interested primarily in group behavior. Their theories attempt to explain and predict alcoholism rates for a society (or subcultural group) rather than for an individual. Trice (47) is one of the few sociologists in the alcohol field who bridges the gap between sociology and psychology by describing a fit between prone personalities (individuals with a psychological vulnerability) and drinking (reference) groups which allow for heavy drinking. Cahalan (7) has also identified an interaction between sociological and psychological variables in the process of becoming a problem drinker.

The findings from a longitudinal analysis conducted by Robins et al. (47) also pinpoint an interaction between individual and group differences. These researchers found no significant differences between Jewish and Irish heavy drinkers in their rates of alcoholism. The same finding held true when comparing heavy drinking women with heavy drinking men. On the other hand, when other variables such as low family status, parental inadequacy and antisocial behavior in childhood were examined, differences did not disappear when the experience of heavy drinking was controlled. Robins et al. concluded that the high rate of alcoholism among those with the latter characteristics (low family status, etc.) reflects not only a greater exposure to heavy drinking but also a higher proportion of alcoholism-susceptible members. The higher rates of alcoholism among Irish vs Jews and men vs women, however, are explained not by a higher rate of Irish or men susceptible to alcoholism but rather by an exposure of a greater proportion of them to the experience of heavy drinking. Thus, the normative belief system concerning appropriate levels of drinking is an important variable explaining ethnoreligious and sex differences in rates of alcoholism. Stresses generated by poor family background, however, create more individuals susceptible to alcoholism.

Jellinek (48) made a similar observation: in cultures which prescribe a high daily intake of alcohol, even people with a small psychological vulnerability will become alcoholics. In cultures which expect low daily intakes of alcohol, it will take people with a high psychological vulnerability to become alcoholics. Thus the rate of alcoholism in a culture depends on the interaction of both psychological (vulnerability) and sociological (prescribed level of alcohol intake) variables. Lemert (2) agreed and claimed that psychiatric concepts of alcoholism may be a necessary if not sufficient factor in explaining more complex forms of drinking problems emerging in cultures with a low tolerance for heavy frequent drinking.

Other social scientists such as the Jessors (49) and Zucker (36) have also attempted to combine environmental and individual factors in describing the development of drinking behavior and problems. The Jessors' model, however, is not concerned with the etiology of alcoholism but rather is a general formulation of problem behavior. Zucker's model has been discussed earlier under the topic of socialization theories.

In the socioenvironmental model, psychological variables are tied very closely to the sociological variables. Brief reference is made to social learning theory which postulates that alcohol serves as a reinforcer. Eventually the drinking response becomes conditioned and dependence on alcohol develops. The reference to childhood traumas (e.g., family breakups, inadequate parents) are elements in the province of psychology, although a sociologist describing family life also finds them relevant. The point is that sociological and psychological variables interact too much in any discussion of the etiology of alcoholism or alcohol problems for either to be disregarded.

Likewise, I would maintain that a sociologist would profit from considering physiological and biochemical theories in the attempt to understand alcoholism. Perhaps the differential effect of alcohol is biologically or physiologically determined; some people may react differently to stress and tension. Perhaps some physiological or neurological attribute makes certain frustrations utterly unbearable for an alcoholism-prone individual and alcohol provides immediate relief. Also, sociologists should recognize the possibility of a genetic predisposition to alcoholism or to some types of drinking behavior (50). These ideas are just some of the many that come to mind when one thinks of all the disciplines that contribute to our understanding of alcoholism (51).

The etiology of alcoholism is far too complex to be examined solely from the perspective of one discipline. Its understanding requires the collaboration of scientists from the social, behavioral and natural sciences. In the past, researchers have paid lip service to the need for multidisciplinary projects, yet few have moved outside of their own discipline. There have been some interdisciplinary approaches which have linked sociological and psychological variables (e.g., 7, 46, 47). But there have been few which have combined social psychology with neurophysiology and biochemistry, etc. With the decade of the 1980s in front of us, it is time to ask all "alcohologists" to work together on truly multidisciplinary projects towards an understanding of alcoholism in terms of its causes, prevention, and treatment.

Concluding Remarks

Few sociologists were interested in alcohol problems at the time of the writing of *Alcohol, Science and Society*. Since then, a multitude of studies have been conducted. These studies range from surveys of alcohol-using behavior to ethnic–cross-cultural comparisons of attitudes and customs surrounding alcohol, to ethnographic studies of Skid Row, to studies of the impact of alcoholism, and alcohol problems on individuals and society, and finally, to studies of various prevention and intervention techniques. Sociological theorists have attempted to explain alcoholism applying deviance theories (such as labeling and anomie), socialization theories, and sociocultural theories. Yet, we are still far away from testing a model which incorporates all the social, cultural and environmental influences on alcohol problems. Thus the task for sociologists is to test existing models and theories and then to reformulate them.

The sociological literature has provided us with an understanding of cultural differences in rates of alcoholism and problem drinking. It has illuminated the social, cultural and environmental factors that are important in explaining the etiology of alcoholism. These studies have highlighted the basic importance of sociological research in the study of alcohol use and alcohol problems. When the original *Alcohol, Science and Society* was published it was generally accepted that the effects of alcohol were determined by chemical, physical and biological factors. Today it is realized that the social and cultural factors, as well as psychological factors, are equally important.

Not only is there a need for an integrated sociological model, there is a far greater need for an integrated universal model drawing together all the disciplines concerned with alcoholism. It is hoped that when *Alcohol, Science and Society Revisited* is revisited in 2015 we will have this interactive model and it will have been tested and have generated a theory on the etiology of alcoholism. With a grounded theory in mind, we will be able to develop prevention and intervention techniques that will be successful in alleviating one of the nation's most serious health and social problems.

REFERENCES

1. HORTON, D. The functions of alcohol in primitive societies. Pp. 153–177. In: Alcohol, science and society; twenty-nine lectures with discussions as given at the Yale Summer School of Alcohol Studies. New Haven; Quarterly Journal of Studies on Alcohol; 1945.

2. BACON, S. D. Sociology and the problems of alcohol; foundations for a sociological study of drinking behavior. Quart. J. Stud. Alc. 4: 402–445, 1943.

3. LEMERT, E. M. Sociocultural research on drinking. Pp. 56–64. In: Keller, M. and Coffey, T. G., eds. Proceedings of the 28th International Congress on Alcohol and Alcoholism. Vol. 2. Lectures in plenary sessions. Highland Park, N.J.; Hillhouse Press; 1969.

4. BLACKER, E. Sociocultural factors in alcoholism. Int. Psychiat. Clin., Boston 3 (No. 2): 51–80, 1966.

5. TARTER, R. E. and SCHNEIDER, D. U. Models and theories of alcoholism. Pp. 75–106. In: TARTER, R. E. and SUGERMAN, A. A., eds. Alcoholism; interdisciplinary approaches to an enduring problem. Reading, Mass.; Addison-Wesley; 1976.

6. ROEBUCK, J. B. and KESSLER, R. G. The etiology of alcoholism; constitutional, psychological and sociological approaches. Springfield, Ill.; Thomas; 1972.

7. CAHALAN, D. Problem drinkers; a national survey. San Francisco; Jossey-Bass; 1970.

8. JELLINEK, E. M. Cultural differences in the meaning of alcoholism. Pp. 382–94. In: PITTMAN, D. J. and SNYDER C. R. eds. Society, culture, and drinking patterns. New York; Wiley; 1962.

9. HORTON, D. The functions of alcohol in primitive societies; a cross-cultural study. Quart. J. Stud. Alc. 4: 199–320, 1943.

10. FIELD, P. B. A new cross-cultural study of drunkenness. Pp. 48–74. In: PITTMAN, D. J. and SNYDER, C. R., eds. Society, culture and drinking patterns. New York; Wiley; 1962.

11. BACON, M. K. BARRY, H., 3d and CHILD, I. L. A cross-cultural study of drinking. II. Relations to other features of culture. Quart. J. Stud. Alc., Suppl. No. 3, pp. 29–48, 1965.

12. MARSHALL, M., ed. Beliefs, behaviors and alcoholic beverages; a cross-cultural survey. Ann Arbor; University of Michigan Press; 1979.

13. EVERETT, M. W., WADDELL, J. O., HEATH, D. B., eds. Cross-cultural

approaches to the study of alcohol; an interdisciplinary perspective. Paris; Mouton; 1976.

14. BALES, R. F. Cultural differences in rates of alcoholism. Quart. J. Stud. Alc. **6**: 480–499, 1946.

15. FRANKEL, B. G. and WHITEHEAD, P. C. Drinking and damage; theoretical advances and implications for prevention, (Rutgers Center of Alcohol Studies, Monogr. No. 14.) New Brunswick, N.J.; 1981.

16. Room, R. Cultural contingencies of alcoholism; variations between and within nineteenth-century urban ethnic groups in alcohol-related death rates. J. Hlth soc. Behav. **9**: 99–133; 1968.

17. STRAUS, R. and BACON, S. D. Drinking in college. New Haven; Yale University Press; 1953.

18. SKOLNICK, J. H. Religious affiliation and drinking behavior. Quart. J. Stud. Alc. **19**: 452–470, 1958.

19. WILKINSON, R. The prevention of drinking problems; alcohol control and cultural influences. New York; Oxford University Press; 1970.

20. SNYDER, C. R. Alcohol and the Jews; a cultural study of drinking and sobriety. (Yale Center of Alcohol Studies, Mongr. No. 1.) New Haven; 1958.

21. GLASSNER, B. and BERG, B. How Jews avoid alcohol problems. Amer. sociol. Rev. **45**: 647–664, 1980.

22. HEATH, D. B. A critical review of the sociocultural model of alcohol use. Pp. 1–18. In: HARFORD, T. C. PARKER, D. A. and LIGHT, L., eds. Normative approaches to the prevention of alcohol abuse and alcoholism; proceedings of a symposium. . . . (NIAAA Research Monogr. No. 3; DHEW Publ. No. ADM-79-847.) Washington, D.C.; U.S. Govt Print. Off.; 1980.

23. PITTMAN, D. J. International overview; social and cultural factors in drinking patterns, pathological and nonpathological. Pp. 3–20. In: PITTMAN, D. J., ed. Alcoholism. New York; Harper & Row; 1967.

24. ULLMAN, A. D. Sociocultural backgrounds of alcoholism. Ann. Amer. Acad. polit. and social sci. **315**: 48–54, 1958.

25. MIZRUCHI, E. H. and PERRUCCI, R. Norm qualities and differential effects of deviant behavior; an exploratory analysis. Amer. sociol. Rev. **27**: 391–399, 1962.

26. MYERSON, A. Alcoholism; the role of social ambivalence. Pp. 306–312. In: MCCARTHY, R. G., ed. Drinking and intoxication. New Haven; Yale Center of Alcohol Studies; 1959.

27. ROOM, R. Ambivalence as a sociological explanation; the case of cultural explanations of alcohol problems. Amer. sociol. Rev. **41**: 1047–1065, 1976.

28. LEDERMAN, S. Alcohol, alcoolisme, alcoolisation. Vol. 1. Données scientifiques de caractére physiologique economique et social. (Institut National d'Études Démographiques; Travaux et Documents, Cah. No. 29.) Paris; Presses Universitaires de France; 1956.

29. LEDERMAN, S. Alcool, alcoolisme, alcoolisation, Vol. 2. Mortalité,

morbidité, accidents du travail. (Institut National d'Études Démographiques; Travaux et Documents, Cah. No. 29.) Paris; Presses Universitaires de France; 1964.

30. DE LINT, J. The epidemiology of alcoholism with specific reference to sociocultural factors. Pp. 323–339. In: EVERETT, M. W., WADDELL, J. O. and HEATH, D. B., eds. Cross-cultural approaches to the study of alcohol; an interdisciplinary perspective. Paris; Mouton; 1976.

31. DE LINT, J. and SCHMIDT, W. Consumption averages and alcoholism prevalence; a brief review of epidemiological investigation. Brit. J. Addict. **66:** 97–107, 1971.

32. HYMAN, M. M. The Ledermann Curve; comments on a symposium. J. Stud. Alc. **40:** 339–347, 1979.

33. PITTMAN, D. J. Primary prevention of alcohol abuse and alcoholism; an evaluation of the control of consumption policy. St. Louis; Washington University Social Science Institute; 1980.

34. WHITEHEAD, P. C. and HARVEY, C. Explaining alcoholism; an empirical test and reformulation. J. Hlth social Behav. **15:** 57–65; 1974.

35. MCCORD, W., MCCORD, J. and GUDEMAN, J. Origins of alcoholism. (Stanford Studies in Sociology, No. 1.) Stanford; Stanford University Press; 1960.

36. ZUCKER, R. A. Parental influences upon drinking patterns of their children. Pp. 211–238. In: GREENBLATT, M. and SCHUCKIT, M. A., eds. Alcoholism problems in women and children. New York; Grune & Stratton; 1976.

37. DURKHEIM, E. Suicide. (Spaulding, J. A. and Simpson, G., Trans.) New York; Free Press; 1951. (Orig., 1897.)

38. CHEINISSE, L. La race juive, jouit-elle d'une immunité à l'égard de l'acoolisme? Sem. medicale **28:** 613–615, 1908.

39. SNYDER, C. R. Inebriety, alcoholism, and anomie. Pp. 189–212. In: CLINARD, M. B., ed. Anomie and deviant behavior; a discussion and critique. New York; Free Press; 1964.

40. MERTON, R. K. Social theory and social structure. New York; Free Press; 1957.

41. LEMERT, E. M. Human deviance, social problems, and social control. Englewood Cliffs, N.J.; Prentice-Hall; 1967.

42. ROMAN, P. M. and TRICE, H. M. The sick role, labeling theory, and the deviant drinker. Int. J. social Psychiat. **14:** 245–251, 1968.

43. ROBINS, L. N. Alcoholism and labelling theory. Pp. 21–33. In: GOVE, W. R., ed. The labelling of deviance; evaluating a perspective. New York; Wiley; 1975.

44. CLINARD, M. B. and MEIER, R. F. Sociology of deviant behavior. New York; Holt, Rinehart & Winston; 1975.

45. SUTHERLAND, E. H. and CRESSEY, D. R. Criminology. Philadelphia; Lippincott; 1978.

46. TRICE, H. H. Alcoholism in America. New York; McGraw-Hill; 1966.

47. ROBINS, L. N., BATES, W. M. and O'NEAL, P. Adult drinking patterns

of former problem drinkers. Pp. 395–412. In: PITTMAN, D. J. and SNYDER, C. R., eds. Society, culture, and drinking patterns. New York; Wiley; 1962.

48. JELLINEK, E. M. The disease concept of alcoholism. Highland Park, N.J.; Hillhouse Press; 1960.

49. JESSOR, R. and JESSOR, S. L. Problem behavior and psychosocial development; a longitudinal study of youth. New York; Academic Press; 1977.

50. GOODWIN, D. W. Is alcoholism hereditary? New York; Oxford University Press; 1976.

51. COOPERATIVE COMMISSION ON THE STUDY OF ALCOHOLISM. Alcohol problems; a report to the nation. Prepared by T. F. A. PLAUT. New York; Oxford University Press; 1967.

14

Medical Issues: The Disease of Alcoholism

Charles S. Lieber, M.D.

In 1945, the medical issues related to the disease of alcoholism were not at the forefront because alcoholism was essentially considered to represent a social or behavioral problem and the intrinsic toxicity of alcohol was not fully appreciated. The prevalence of one medical problem alone, however, namely end-stage liver disease or cirrhosis of the liver, has reached such a magnitude that this complication of alcoholism now represents, by itself, a major public health problem: 75% of all medical deaths attributable to alcoholism are due to cirrhosis of the liver. In the U.S., cirrhosis has overtaken diabetes as the fifth cause of mortality; in large urban areas, it has become the third cause of all deaths among those aged 25 to 65. Although not all cirrhotic patients are alcoholics, it is now generally recognized that a majority do admit to excessive alcohol consumption. Until recently, however, alcohol intake was believed to be only indirectly responsible for the liver disease of alcoholics; liver disorders were considered to be due to the nutritional deficiencies so commonly associated with alcoholism. Given an adequate diet, alcohol was thought no more toxic to the liver than other calorie-rich substances such as fats or carbohydrates. But over the last two decades this view has undergone a major change, and the concept of a direct toxic effect of alcohol emerged, first for the liver, and then for the gut, pancreas and hematologic systems and, more recently, the central nervous system.

Alcohol and Diet in the Pathogenesis of Tissue Damage

Liver Injury

The following statement, quoted from an authoritative textbook, is fully representative of the dogma prevailing a few decades ago: it was "generally agreed that alcohol is not a hepatotoxin and that its

233

effects on the liver probably are secondary to an associated nutritional disturbance" (1, *p. 1500*). In fact, in the original *Alcohol, Science and Society*, the chapter discussing medical complications is called "Alcohol and Nutrition" and it was stated then that "it is certain that cirrhosis is not caused by the direct action of alcohol any more than beriberi" (2, *p. 78*).

The nutritional theory was plausible since alcoholics do commonly suffer from malnutrition. Long-term alcohol consumption interferes with food digestion and absorption because of its well-known effects on gut and pancreas. Furthermore, alcohol, high in caloric value, displaces other foods in the diet. Each gram of ethanol provides 7.1 Calories. Twelve ounces of an 86-proof beverage contains about 1200 Calories or about half of the recommended daily dietary allowance for energy. But, unlike regular food, alcoholic beverages contain few, if any, vitamins, minerals, proteins or other nutrients and, therefore, an alcoholic's intake of these nutrients may readily become insufficient. Economic factors may also reduce consumption of nutrient-rich foods, particularly those containing proteins. Indeed, in some epidemiological studies, alcoholics reported a history of poor protein intake (3). However, one cannot conclude from such an observation to what extent, if any, low protein intake contributed to liver injury or was a consequence of it. Interestingly, in other surveys, alcoholics developed cirrhosis despite an average daily dietary protein intake of 100 grams or more (4) which is about twice the daily recommended level (5).

The nutritional theory was also based on experimental studies. In rats, alcohol given in the drinking water was not capable of producing liver damage unless associated with a diet deficient in essential nturients, and it was concluded that "there is no more evidence of a specific toxic effect of pure ethyl alcohol upon liver cells than there is for one due to sugar" (6). Rats display a natural aversion for alcohol, however, and under the experimental conditions used intake was relatively low and blood alcohol concentrations negligible. When the aversion was counteracted by incorporation of alcohol in a totally liquid diet, and the alcohol consumption thereby increased to 36% of total calories (an intake still less than that of many alcoholics), fatty liver resulted despite a nutritious diet (7–9). The question was then raised whether lesions more severe than steatosis, particularly cirrhosis, can be produced by alcohol in the absence of dietary deficiencies. Studies in rodents had been unsatisfactory because, even with a liquid diet, alcohol intake does not reach the level of average consumption in alcoholics, namely 50% of total calories (3). Such a level of consumption was achieved in the baboon, again through incorporation of ethanol in totally liquid diets (10). The calories (exclusive of alcohol) were provided by protein (36%), fat (42%) and carbohydrate (22%); the diet was supplemented liberally with minerals and vitamins. All of the nutrients were calculated to exceed the normal requirements of the baboon, with choline being given at twice the recommended level. The fat and carbohydrate composition was calculated to mimic a situation in which an alcoholic may be trying to achieve a high protein diet with available natural foods while drinking. In fact, even if the alcoholic tried hard it would be difficult for him to consume a diet richer in protein than the one administered to the baboons. Nevertheless, the baboons developed not only fatty liver, but after 2 to 5 years one-third also had progression of liver damage to cirrhosis (11). It was, therefore, concluded that in addition to dietary factors, alcohol itself plays a key etiological role in the development of liver injury (Figure 1).

If we can extrapolate from these data to the human situation, it would seem that (over the long haul) heavy drinkers, even if they make an extra effort to maintain a nutritionally adequate high-protein diet, will not necessarily succeed in preventing the development of cirrhosis. Many questions, however, still remain unresolved, partic-

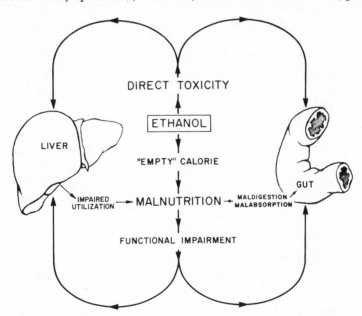

FIGURE 1.—*Interaction of Direct Toxicity of Ethanol on Liver and Gut with Malnutrition Secondary to Dietary Deficiencies, Maldigestion, and Malabsorption* From Lieber (12).

ularly the extent to which malnutrition promotes the toxic effect of alcohol. In primates, although some fibrosis occurred with protein deficiency alone, no actual cirrhosis developed. For cirrhosis to develop in primates, dietary conditions were required which are not normally achievable with natural foods (such as an association of low protein and a high cholesterol content). Thus, whether "across-the-board" nutritional deficiencies may potentiate alcohol-induced liver damage in primates has not been determined as yet. Selective deficiencies could play a contributory role. For instance, it is known that methionine deficiency may potentiate and methionine supplementation may partly alleviate alcohol-induced fatty liver (6, 13). Originally, this was attributed to choline sparing properties of methionine, but, as discussed subsequently, a more likely mechanism is the repletion of glutathione. Similarly, judicious correction of vitamin A deficiency may offset the potentiation of alcohol-induced liver damage caused by lack of vitamin A (14).

The Gastrointestional Tract, Bone, Muscles, Brain and Hematologic System; The Fetal Alcohol Syndrome

Alcohol is now believed to have direct toxic effects on the gut (15) (Figure 1), the heart (16), the testis (17) and the hematological system (18). Anemia, however, is not only due to the toxicity of alcohol, but seems to be a major consequence of associated nutritional deficiencies, primarily folate deficiency. Similarly, it is not clear to what extent the abnormalities in the metabolism of bone (19), calcium and other minerals, now more distinctly recognized in alcoholics, are due to nutritional factors and to what extent direct effects of alcohol are involved.

    The mechanisms whereby excessive alcohol use leads to necrosis in pancreas, muscle, and other tissues have not been elucidated. Pancreatitis, though not as common as cirrhosis, can evolve into a very dramatic lethal complication of alcohol misuse. It occurs at an earlier age than cirrhosis. In addition to its acute manifestations, it may also engender a chronic incapacitating condition. Theories have been proposed to explain the development of pancreatitis on the basis of blocked pancreatic secretion, either at the level of the Sphincter of Oddi or through precipitation of protein plugs in the ductal system (20, 21). Virtually nothing, however, is known about the possible effects of ethanol or its metabolites on the pancreatic tissue itself. Similarly, although direct effects of ethanol on muscle have been described (22), the mechanism involved and how this effect relates to the myopathy which develops in some alcoholics is unknown. Cardiovascular diseases are associated with alcohol consumption in varying degrees. The relationship of consumption to a specific cardiomyopathy, or disease of the heart muscle, is well established. Some studies suggest that modest levels of alcohol consumption result in decreased risk for coronary heart disease, including myocardial infarction, compared with abstaining and alcoholic populations; however, coronary heart disease mortality rates among alcoholics are as much as 400% greater than among nonalcoholics.

Birth anomalies, ranging from decreased birth weight to the fetal alcohol syndrome, may affect thousands of babies each year. Current conservative estimates suggest an incidence of fetal alcohol syndrome of 1 per 2000 live births, or about 1650 cases for the 3.3 million live births in 1977 in the U.S. Although malnutrition may play a role, alcohol appears to exert direct teratogenic effects.

Mental disorders also are associated with alcohol consumption. Alcoholic psychoses, manifested in delirium tremens, hallucinosis, and paranoia, exceed 100,000 cases per year in the U.S. Affective disorders (depressive or manic-depressive psychoses) and depression are related to alcohol use in up to 42% of all cases, and at that level account for 35,000 hospitalizations each year in the U.S. Motor and cognitive impairments also are closely associated with excessive alcohol use.

Though the specific mechanisms of alcohol's action on the brain are still uncertain, it appears that energy metabolism and derangements of the neurotransmitters are not primarily affected. The main effect of alcohol seems to be exerted by its interaction with lipid or protein components of the neuronal membrane or both, increasing membrane fluidity. In this respect alcohol mimics the effects exerted by other cerebral depressants. Some changes in the brain membranes may be related to the observed decrease in the normal Na^+ influx into the neuronal membrane and may, in turn, induce a decrease in the action current. Organic brain diseases associated with alcoholism have been attributed primarily to vitamin deficiency, although the direct toxic effects of alcohol on the peripheral and central nervous systems are now being recognized (23). It seems fairly clear that such entities exist independent of nutritional deficits, but may at times be aggravated by them. There is also good evidence that cerebral protein synthesis is impaired with prolonged alcohol intake.

Thus, there is hardly any tissue that is spared from the damaging effects of excessive alcohol use. Alcoholism has now replaced syphilis as the "great imitator" of medical disease; in most instances, however, our knowledge of the pathogenesis involved is scant.

Role of Dietary Therapy

At present, the optimal diet for alcoholics is not established. Obviously, specific nutritional deficiencies, such as lack of thiamine and folate, should be avoided and when present corrected. These deficiencies, though decreasing in incidence, are still common and cause hematological and neurological complications. Concerning proteins, the situation is more complex. In rats, carbon tetrachloride-induced cirrhosis can be prevented by a low protein diet. The situation in alcoholic liver injury has not been clarified. Early studies relating beneficial effects of high protein diets were uncontrolled. Subsequently, the risk of dietary-induced encephalopathy in cirrhotic patients has become more apparent. Therefore, in the absence of experimental data to the contrary, high protein diets do not seem indicated at the present time. Until the issue is resolved it may be prudent to settle for an intake of protein that does not exceed the recommended amount (5) or the individual protein tolerance of the cirrhotic patient, whichever is lower.

In any event, the concept of the toxicity of alcohol has had a major impact on the management of alcoholics. When confronted with a patient reluctant to stop drinking, it was common practice for the physician to advise his patient that he could preserve normal liver function provided an adequate diet was maintained. It is now clear that such advice could be considered unjustified or even misleading and that the emphasis must shift from dietary control to control of alcohol intake. The demonstration of the direct toxicity of alcohol generated a number of studies of the mechanisms involved since better understanding of the mode of action of ethanol may provide more effective management of medical complications and ultimately perhaps even of the problem of alcoholism itself.

Mechanisms of the Toxicity of Alcohol

Metabolism

The demonstration of the toxicity of alcohol raised the question of the mechanisms of this action. The large caloric load that alcohol represents has been alluded to before, but alcohol has little in common with other energy-rich compounds. Carbohydrates and fats can be synthesized in the body as well as ingested in the diet, but alcohol is essentially foreign to the body. As with carbohydrates and fats, alcohol has a high caloric value and is readily absorbed from the gastrointestinal tract; unlike them, however, it is not effectively stored in the tissues. Moreover, very little alcohol can be disposed of through the lungs or the kidneys, so that essentially the body can get rid of alcohol only by oxidizing it. Again, unlike fats and carbohydrates, which can be oxidized in most tissues, alcohol is burned primarily in the liver, the organ that contains the bulk of the enzymes necessary for initiating the process. The organ-specificity of alcohol explains to a large extent the concentration of so many of alcohol's deleterious effects in the liver, which is the body's main chemical plant and the primary site of metabolic processes ranging from the synthesis of proteins to the detoxification of drugs and other foreign substances.

The first step in alcohol's primary metabolic pathway is catalyzed by the enzyme alcohol dehydrogenase (Figure 2). (In spite of its name, alcohol dehydrogenase serves as a generalized remover of hydrogen atoms from various compounds including some steroids;

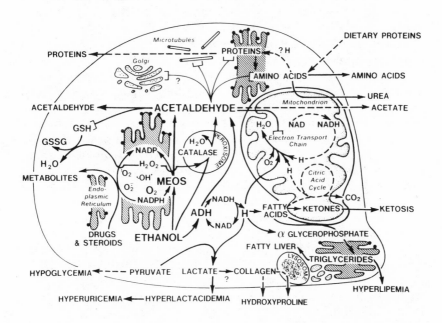

FIGURE 2.—*Oxidation of Ethanol in the Hepatocyte and Link of the Two Metabolites (Acetaldehyde and H) to disturbances in Intermediary Metabolism, including Abnormalities of Amino Acid and Protein Metabolism.* NAD denotes nicotinamide adenine dinucleotide, NADH, reduced NAD, NADP, nicotinamide adenine dinucleotide phosphate; NADPH, reduced NADP; MEOS, the microsomal ethanol-oxidizing system; and ADH, alcohol dehydrogenase. The broken lines indicate pathways that are depressed by ethanol. The symbol —⊏ denotes interference or binding by the metabolite. From Lieber (24).

it was therefore probably present in the liver of the prehistoric men who first sampled alcohol, available to take on what has since become a major function.) Alcohol dehydrogenase catalyzes the transfer of hydrogen atoms from ethanol to a coenzyme, nicotinamide adenine dinucleotide (NAD), converting the ethanol into acetaldehyde. The acetaldehyde is then oxidized, primarily in the liver, to form acetate, which is eventually converted into carbon dioxide and water. A number of the metabolic effects of alcohol are directly linked to the two first products of its oxidation: hydrogen and acetaldehyde.

Effects of Excess Hydrogen Generation, Including Hypoglycemia, Acidosis, Hyperuricemia and Lipid Abnormalities

The excess hydrogen from alcohol unbalances the liver cell's chemistry (25). In order to live, the cell must get rid of the hydrogen, and it does so by shunting hydrogen ions into one or more of several pathways dependent on them, sometimes with deleterious effects. One such pathway is the process whereby amino acids (derived from protein breakdown in the liver) are converted, with pyruvate as an intermediate, into glucose. In the presence of excess hydrogen ions the process is turned in a different direction; the pyruvate is reduced to lactate instead of being converted into glucose. Blood sugar is derived primarily from three sources: gluconeogenesis, or synthesis from amino acids in the liver, breakdown of glycogen stored in the tissues and conversion of carbohydrates in the diet. If alcoholics have been drinking and not eating, there are no dietary carbohydrates and glycogen may be used up; if gluconeogenesis is then blocked by the diversion of pyruvate to lactate, the level of sugar in the blood will be lowered (26). Low blood sugar, or hypoglycemia, is known to be a complication of acute alcoholic states, but it is often overlooked. When an intoxicated person is brought into a hospital emergency room, it is important to test for hypoglycemia, since crucial organs, including the brain, can be critically affected by a lack of sugar; some deaths of "unknown origin" in alcoholics may be attributable to the condition.

The increase in lactate from excess hydrogen has other effects. The lactate moves into the blood, bringing on lactic acidosis. In the kidney it interferes with the excretion of uric acid (27). A high uric acid level in the blood (hyperuricemia) exacerbates gout. This process may explain the ancient clinical observation that excessive drinking can trigger or aggravate gout attacks.

There are other ways in which the liver cell can rid itself of the excess hydrogen, several of which involve the formation of lipids, or fat. The hydrogen can be shunted directly into the synthesis of α-glycerophosphate and fatty acids. Those are the two precursors of the triglycerides, and triglycerides are the lipids that accumulate in alcoholic fatty liver. The main mechanism for disposing of hydrogen is more indirect, but it has a similar result. The hydrogen is transferred to the mitochondria, the cell organelles that produce the energy for liver functions. Normally it is fat that is oxidized—in effect burned—in the mitochondrial citric acid cycle to produce usable energy in the form of phosphate ions in an energized state. The plentiful hydrogen from alcohol, however, provides an alternate fuel that is oxidized instead of the hydrogen from fat. In promoting the oxidation of alcohol the substitution exacts a price: the lipids accumulate, again leading to fatty liver. If alcohol is ingested along with a diet containing fat, the fat of dietary origin accumulates in the liver; even when alcohol is taken with a low-fat diet, fat made in the liver itself is deposited there. In addition, when alcohol is ingested in very large quantities, it can trigger hormonal discharges that mobilize fat from stores of adipose tissue and move it toward the liver (Figure 3).

Proliferation of the Endoplasmic Reticulum and Possible
Consequences in Metabolism of Lipids, Drugs, Hormones, other
Hepatotoxic Agents and Carcinogens; Role of the Microsomal
Ethanol Oxidizing System

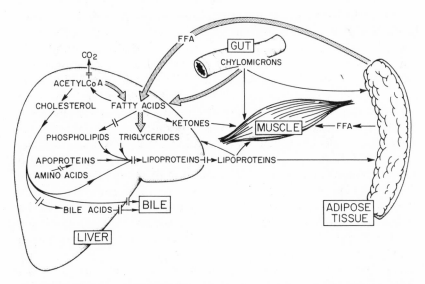

FIGURE 3.—*Possible Mechanisms of Fatty Liver Production through either Increase* (—►) *or Decrease* (—⊣ —►) *of Lipid Transport and Metabolism.*

What can the liver do with the accumulating fat? One possibility is for it to be secreted into the bloodstream, which delivers blood lipids to provide fuel for peripheral tissues such as muscle and deposits extra supplies for storage in adipose tissue. The secretion is complicated by the fact that lipids have to be made water-soluble by being wrapped in a thin coat of protein to form lipoproteins. The assembly of lipoproteins is carried out in the liver in the smooth membranes of the endoplasmic reticulum. We observed proliferation of the smooth endoplasmic reticulum after heavy alcohol intake in both rats and humans (28, 29). The proliferation is reflected in the increased activity of certain enzymes in the smooth reticulum, which enlarges the liver's capacity for secreting lipoproteins. A liver that has thus adapted to alcohol after being conditioned by a period of heavy alcohol intake will respond with exaggerated secretion of lipoproteins even after a normal meal is eaten, producing hyperlipemia, or an abnormally high level of fat in the blood (30, 31). The effect is of particular significance in people who have underlying abnormalities of either lipid or carbohydrate metabolism and therefore a propensity for elevated concentrations of lipids in the blood. Some of the lipids represent a major predisposing factor in heart attacks whereas others (high-density lipoproteins) may be protective. There is still debate on which effect is prevailing.

Still another way for the liver to dispose of excess fat is to convert some of it to water-soluble ketone bodies and secrete them into the bloodstream. In some people the response may be exaggerated, resulting in an elevation of ketone bodies in the blood that mimics the condition known as ketoacidosis in diabetic patients (32).

The conversion of lipids into lipoproteins is just one of a number of functions of the endoplasmic reticulum of liver cells, which also inactivates a wide variety of drugs and other foreign substances and converts them into water-soluble products that can be excreted. I therefore wondered whether the proliferation of smooth reticulum after lengthy alcohol consumption would be reflected in an increased capacity of the liver to metabolize various drugs. That turned out to be the case. After repeated administration of alcohol (but before the evolution of severe liver injury) the enzymes of the smooth reticulum that inactivate tranquilizers, anticoagulants and other drugs and that detoxify certain food additives, cancer-causing substances and insecticides do increase their activity, enhancing the body's capacity to rid itself of these compounds. For example, Misra et al. (33) gave the tranquilizer meprobamate to volunteers and measured its rate of disappearance in the bloodstream. The time for the blood concentration of the drug to fall to half its original value decreased from 16 hr during a control period to 8 hr after a month of alcohol consumption.

Anesthesiologists had known for many years that larger doses of sedatives are required to achieve a given effect in alcoholics than in other people. That drug tolerance was attributed to adaptation in the central nervous system (increased resistance to sedatives by the brain). Our findings pointed to a metabolic adaptation as well: increased capacity by the liver to inactivate and excrete sedatives and other compounds detoxified in the endoplasmic reticulum. At the adaptive stage of their disease (which, as we shall see, does not continue indefinitely), alcoholics therefore require larger doses of many drugs. That is true, however, only when the individual is sober. When alcoholics have been drinking, the effect is quite the opposite. The reason, we found, is that one of the drugs the smooth reticulum metabolizes is alcohol itself, by way of an accessory pathway that supplements the basic alcohol dehydrogenase system.

We demonstrated the accessory pathway by spinning liver tissue in the ultra-centrifuge and isolating the endoplasmic reticulum as what is called the microsomal fraction. We then found that a preparation of the microsomal fraction would oxidize alcohol. We called the accessory pathway the microsomal ethanol-oxidizing system (34, 35). We were able to obtain it in a semipurified form (36). We have found that this microsomal pathway comes into operation after the blood alcohol reaches a certain level. The alcohol then enters into competition with other drugs whose metabolism shares some elements of the microsomal system, thereby slowing their metabolism and enhancing their effects. That is why simultaneous drinking and taking of tranquilizers is particularly dangerous: the alcohol can accentuate the action of the drug not only because the effect of the two drugs on the brain may be additive, but also because the presence of alcohol can interfere with the liver's capacity to inactivate the drug, so that a given dose remains active for a longer time.

Along with the rest of the microsomal enzyme systems, the ethanol-oxidizing system adapts to heavy alcohol consumption by increasing its activity, thus contributing to tolerance. Alcoholics' ability to drink more than most nonalcoholics is due primarily to central nervous system tolerance, a progressive decrease in the effectiveness of alcohol's action on the brain. Alcoholics also develop an increased capacity to metabolize alcohol, not only through the microsomal system but also through the alcohol dehydrogenase pathway and perhaps a third pathway that depends on the enzyme catalase. After heavy or prolonged drinking, however, adaptation can be offset by progressive liver injury, so that the liver's over-all ability to handle alcohol remains about the same or even decreases.

The microsomal changes are beneficial in that they help to

rid the liver of fat and speed the detoxification of many drugs, food additives and other foreign compounds. There are undesirable concomitants of the adaptive process, however. Some foreign substances are activated rather than inactivated by conversion in the endoplasmic reticulum. Certain potentially cancer-causing substances become carcinogenic only after activation by the microsomes (37) and other substances become toxic to the liver after such activation. For example, exposure to carbon tetrachloride is known to cause liver damage, but the familiar dry-cleaning agent is harmless to the liver until it is activated by the endoplasmic reticulum. The increase in microsomal activity induced by alcohol enhances the toxicity of carbon tetrachloride: rats chronically treated with alcohol were much more susceptible to the toxic effects of carbon tetrachloride than matched control animals (38). This effect presumably explains the clinical observation that alcoholics are particularly susceptible to carbon tetrachloride poisoning in dry-cleaning plants that still use the compound. The enhanced susceptibility of alcoholics very probably extends to a number of other foreign compounds that are harmless to most people but that may become toxic in alcoholics as an undesirable side effect of increased microsomal activity (39). This is the case for commonly used drugs (e.g., isoniazid) and analgesics (e.g., acetaminophen) when taken in excessive amounts (40).

Another side effect is energy waste. Microsomal activity requires energy. Moreoever, it is a peculiarity of the various microsomal oxidations that they produce heat without conserving chemical energy. This could conceivably contribute to the impaired growth observed in animals that are fed alcohol, since the production of heat beyond what is required to maintain body temperature represents a waste of energy. Such waste may explain at least in part the observation that addition of alcohol to the diet results in less weight gain than the addition of the same number of calories from other sources (41). Thus, with respect to body weight, calories from alcohol apparently do not fully "count"—at least in alcoholics and when the alcohol intake is large.

Endogenous substrates, such as steroid hormones, also serve as microsomal substrates. As for consequences, the ethanol-induced microsomal changes have also some endocrine repercussions. The major perturbation in the endocrine system is due to effects of alcohol at the hypothalamic–pituitary and glandular loci (42). However, significant effects of alcohol on the peripheral events concerned with the transport and clearance of the hormones occurs. Indeed, Gordon et al. (43) reported that alcohol increased the activity of the hepatic testosterone 5-A-Ring reductase in the rat and man after 1 month of alcohol exposure. We have recently demonstrated that the increase in enzyme activity extends to other nonhepatic tissues such as the prostate and gingiva (44). These alterations in enzyme levels may play a role, along with hemodynamic changes and alterations in binding parameters, in the increased metabolic clearance rate of testosterone noted during short-term alcohol use (4 weeks). This effect, in addition to the direct toxic action of ethanol on the testes, contributes to the decrease in plasma testosterone after alcohol consumption (43).

Dependence and Tolerance

It is a reasonable assumption that not only exogenous, but also endogenous, substances undergo increased microsomal metabolism. Lack of these compounds, in turn, could have adverse effects. In the presence of ethanol, however, metabolism of endogenous microsomal substrates is probably inhibited. Therefore, physical dependence on alcohol could be linked tentatively to the sequence of events shown in Figure 4: prolonged alcohol consumption induces

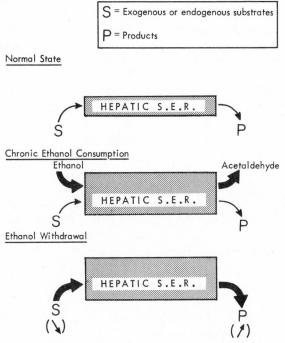

FIGURE 4.—*Increase in Hepatic Smooth Endoplasmic Reticulum (S.E.R.) after Chronic Ethanol consumption; Its Possible Role in the Development of Ethanol Tolerance and Dependence and in the Interaction of Ethanol with Drug Metabolism* From Lieber (45).

the activity of microsomal enzymes that metabolize exogenous as well as endogenous compounds. As a result, alcoholics, on sobering, experience accelerated metabolism of the endogenous compounds with untoward effects; these effects can be alleviated by further alcohol ingestion, the metabolism of which spares that of the postulated endogenous substrates, thereby promoting the craving for alcohol. It is not implied that these changes necessarily represent the sole mechanism responsible for the addictive process. Alterations of the internal milieu secondary to the described microsomal changes may, however, play a contributory role but, obviously, various neuroendocrine changes and psychological and socioeconomic factors are involved in the misuse of alcohol (46, 47) and physical dependence may result primarily from brain changes. Neurochemical–neurophysiological adaptation of the central nervous system adequately explains most aspects of tissue (cerebral) tolerance. In the case of withdrawal, rapid decrease or removal of the depressive effects of alcohol without concomitant timely readjustment of the adaptation results in temporary cerebral overstimulation until normal equilibrium is regained. Thus, it seems reasonable to consider cerebral tolerance and withdrawal from alcohol as two expressions of the same basic phenomenon. This does not mean, of course, that they may not be dissociable temporarily or for some specific manifestations, perhaps because of differential brain sensitivity in the syndromes.

The precise nature of adaptation is still unclear. The best current evidence favors a change in neuronal membranes, with decrease in fluidity after lengthy exposure to alcohol. Such a physical change in turn may express itself neurophysiologically and neurochemically in altered sodium-ion flux across the membrane, perhaps affecting the action current changes in the membrane or in synaptosomal calcium-ion binding, altering neurotransmitter release.

Changes in cerebral energy metabolism may occur in withdrawal, but there are no compelling data to conclude that these are a causal phenomena. Likewise, the available information on most neurotransmitter concentrations and turnover, particularly those of serotonin and catecholamines, is conflicting and confusing. This does not mean that changes in serotonin concentration or flux in brain may not have a role in preference for alcohol or the expression of some forms of tolerance or perhaps withdrawal. Furthermore, there is the possibility that condensation products of acetaldehyde with certain biogenic amines (broadly called the TIQ hypothesis), rather than alcohol per se, may be involved (see the chapter by Roach in this volume).

Toxic Role of Acetaldehyde

Apart from the metabolic complications of excess hydrogen and of the adaptive changes in microsomal activity, heavy alcohol consumption has direct toxic effects caused by acetaldehyde. This product of alcohol metabolism is extremely reactive and affects most tissues in the body. Most of the acetaldehyde is converted to acetate by the liver mitochondria, but some of it escapes into the bloodstream. Both the alcohol dehydrogenase and the microsomal ethanol-oxidizing pathways become saturated when the liver is loaded with a large amount of alcohol, so that the acetaldehyde concentration in the blood reaches a plateau and stays there until the alcohol concentration drops. Korsten et al. (48) found that the acetaldehyde plateau is significantly higher in alcoholics than it is in nonalcoholics, even when the same amount of alcohol has been given to both groups and the same blood alcohol concentration has been attained. This abnormally high acetaldehyde concentration in alcoholics could result from faster metabolism of alcohol to form acetaldehyde (because of the adaptive increase in microsomal activity) or from slower disposition of the acetaldehyde (because of impaired acetaldehyde metabolism).

Probably both factors are involved, at least in the beginning. Indeed, striking alterations in the mitochondria were revealed by the electron microscope even in the early stages of heavy alcohol consumption. Hasumura et al. (49) isolated the damaged mitochondria and found that their capacity for metabolizing acetaldehyde to acetate is reduced. Acetaldehyde itself may be responsible for part of the decrease in mitochondrial function. Alcoholics may therefore be victims of a vicious circle: a high acetaldehyde level impairs mitochondrial function in the liver, acetaldehyde metabolism is decreased, more acetaldehyde accumulates and causes further liver damage.

Another possible mode of toxicity of acetaldehyde is its interaction with amino acids. Aldehydes react quite readily with mercaptans. Thus L-cysteine could react with acetaldehyde forming a hemiacetal. Furthermore, cysteine is one of the three amino acids that constitute glutathione (GSH). Binding of acetaldehyde with cysteine or GSH or both may contribute to a depression of liver GSH (50). GSH represents one of the mechanisms for the scavenging of toxic free radicals; a reduction in GSH favors peroxidative damage of membranes, and the damage may possibly be compounded by the increased generation of active radicals by the "induced" microsomes following repeated alcohol consumption (Figure 5).

The acetaldehyde affects other tissues as well. For example,

248   Charles S. Lieber

Schreiber et al. (51) reported that acetaldehyde concentrations no higher than those we have measured in the blood of alcoholics can inhibit the synthesis of proteins in heart muscle. Such an effect could in part explain the impaired cardiac function that is common in alcoholics. Acetaldehyde may affect the functioning of other muscles as well.

Acetaldehyde has striking effects on the brain. Several investigators have suggested that acetaldehyde, rather than alcohol itself, may be responsible for the development of the dependence that, along with tolerance, characterizes alcohol addiction. Dependence manifests itself by a state of extreme discomfort, often accompanied by physiological disturbances such as tremors and seizures, produced by withdrawal of a drug. Several mechanisms mediated by acetaldehyde have been proposed to explain dependence but none have yet been confirmed. One proposal begins with the fact that certain amine neurotransmitters, which transmit nerve impulses from one cell to another in the brain, are inactivated by the enzyme monoamine oxidase to form an aldehyde that is then converted into an acid. The conversion to acid requires an enzyme that is also active in the metabolism of acetaldehyde. Davis and Walsh (52) suggested that if acetaldehyde is present in the brain, it may compete for the enzyme; unmetabolized neurotransmitter aldehydes may therefore accumulate; the aldehydes may then combine with the neurotransmitter, forming compounds that are startlingly similar to certain morphine derivatives known for their ability to promote dependence.

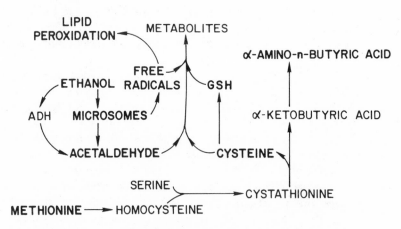

FIGURE 5.—*Possible Link between Microsomal Induction, Enhanced Acetaldehyde Production, Lipid Peroxidation and Alpha Amino-N-Butyric Acid Generation after Alcohol Consumption*

Another possibility pointed out by Cohen and Collins (53) is that the acetaldehyde combines directly with amines to form iso-quinoline derivatives, potent psychoactive compounds that could play a role in developing dependence. Alcohol dependence is probably determined by a number of factors acting in concert. It is at least possible, however, that acetaldehyde is involved in the predisposition to alcoholism: a predisposed person may have deficiencies of acetaldehyde metabolism that make for a higher blood acetaldehyde concentration and a propensity to dependence because of the effect of the higher concentrations on the brain. It is noteworthy that relatives of alcoholics may also have higher blood acetaldehyde concentrations than controls after drinking (54).

Progression of Tissue Injury Including Development of Cirrhosis and Its Complications

As discussed before, if heavy drinking continues, reversible fatty liver progresses, in most people, toward more severe and irreversible liver disease: hepatitis and then cirrhosis. Just why and how the liver may lose its ability to adapt to alcohol is not clearly established. Even in the fatty-liver stage there are some indications that more severe lesions are developing. There may be a ballooning of the liver cells, which is usually attributed to fat accumulation. Baraona et al. (55) noted that proteins synthesized by the liver also accumulate in the cells because their export is somehow depressed by large amounts of alcohol (perhaps by interference with microtubular functions and structure) (56). The engorgement of liver cells with fat and protein interferes with normal functioning. So does the previously mentioned reduced energy production. For all these reasons some liver cells may die, and the necrosis can trigger an inflammatory process that is characteristic of alcoholic hepatitis. The acute reduction in liver cell function at this stage is enough to cause death in some patients.

Cell necrosis and inflammation in turn promote the next stage: fibrosis, or the development of scar tissue, the hallmark of cirrhosis. It is also possible that alcohol may directly promote the production of collagen (57). The fibrous connective-tissue barriers between groups of liver cells interfere with the flow of blood to and from the cells, further decreasing liver function. Partly obstructed in the liver, blood backs up, increasing the pressure in the portal system, which brings blood to the liver from the intestines. Abnormal channels may therefore develop in the venous system so that some of the blood can bypass the circulatory block established by the cirrhotic liver.

Some newly overloaded veins (notably in the esophagus) may become distended, and such veins, called varices, can rupture and hemorrhage. Bleeding varices are a major cause of death in cirrhosis. Moreover, under high pressure, plasma leaks out of the portal-system blood vessels. Extra lymph is also formed, and it leaks out of the lymphatic vessels. Both sources contribute to ascites, the accumulation of fluid in the abdominal cavity. A final complication stems from the liver's inability to clear from the blood ammonia and other nitrogenous compounds produced by bacteria in the intestines. As such compounds accumulate they act on the brain and may cause functional disturbances, hepatic coma and death. Death may also result from complications in other tissues, such as acute pancreatitis, whereas damage to the brain results in more chronic incapacitating conditions, although the acute phase of the Wernicke–Korsakoff syndrome, if not treated with vitamin B, may also lead to death.

A Public Health Strategy Against Alcoholism and Its Complications

Alcoholism obviously encompasses a variety of forms with complex interactions of psychosocial, behavioral, biochemical and neuropharmacological factors. The traditional approach to prevention has been to determine psychological causes for the development of dependence and to detect individuals prone to develop such a disorder. Until such an approach is successful, however, there is a real need for a public health approach to the problem with a special effort at controlling the medical diseases associated with alcoholism. Indeed, it is now becoming increasingly apparent that a problem of such magnitude is not likely to be brought under control merely by an individualized psychiatric or psychological approach. The traditional psychiatric treatment can be very helpful, but because of its cost in manpower it cannot be expected to meet all the needs of millions of alcoholics and, more importantly, of those many millions of heavy drinkers in whom severe complications eventually develop. For these reasons, I wish to outline how a public health strategy against alcoholism might be developed.

Medical complications of alcoholism have now become a major public health problem, particularly liver disease. At the present time, medical complications usually are detected only at a late, irreversible stage. Patients afflicted with irreversible and often chronic complications suffer not only from a great personal tragedy, but they constitute an inordinate load on our medical system: half of the beds in most hospitals are occupied by patients with some type of complication related to excessive drinking. There is an obvious need to recognize these complications at an early, still reversible stage. For. instance, only about 25 to 50% of heavy drinkers eventually develop cirrhosis, depending on the dose and duration of intake (58). Obviously, it would be extremely useful to detect at an early pre-cirrhotic stage individuals prone to progress to cirrhosis if they continue to drink. Our studies of alcoholic cirrhosis in baboons have provided an initial tool towards such a goal. As mentioned before, collagen is deposited in the liver at the early stages in the so-called pericentral or perivenular areas of the lobule. In those baboons which progressed to cirrhosis, pericentral sclerosis invariably occurred in the fatty-liver stage; by contrast, animals that did not show the lesion did not progress beyond the stage of fatty liver. The data suggest that, at least in the baboon, pericentral sclerosis is a common and early warning sign of impending cirrhosis if drinking continues (59). Preliminary studies have indicated that in man too the occurrence of these lesions can be used to detect individuals prone to develop cirrhosis rapidly if they continue to drink (60).

At present the lesion of perivenular sclerosis can be detected only by liver biopsy, obviously not a practical tool for mass screening purposes. What is now needed are some more accessible methods to detect some of the medical complications of alcoholism early. Here again some progress has been made from the application of our basic knowledge acquired in the studies of pathogenesis of alcohol-induced liver disease.

Spill-over in the blood of liver enzymes, especially of trans-aminases, is commonly used as a marker of liver cell damage. Blood transaminase values, however, are a poor reflection of liver cell necrosis as revealed on biopsy, especially in alcohol-induced liver cell injury. In alcoholic hepatitis, for example, levels of transaminases are only moderately elevated and normal values can occasionally be found (61). Gamma-glutamyl transpeptidase (γ-glutamyltransferase) content (GGTP) is not more reliable, although some correlation with liver cell necrosis exists. Elevation from nonhepatic origin is common and in some alcoholics elevated levels may only reflect microsomal induction (62). In contrast with transaminases and GGTP, which show a considerable overlap between patients with and without significant cell necrosis on biopsy, glutamate dehydrogenase (GDH) more accurately reflects the degree of underlying cell damage (63). The high liver content of this enzyme, its mitochondrial origin and its predominantly centrolobular localization (the area which suffers the major impact of alcoholic liver injury) could explain the advantage of the enzyme as an index of liver cell damage in alcoholic patients. To help define the severity and progressive nature of liver injury, however, GDH determination must be carried out within 1 to 2 days after the drinking episode; thereafter, the values drop rapidly, whereas liver necrosis persists. Thus, except for the early stage, GDH offers no advantage over other enzyme tests (64).

There is obviously a need to continue the development of better and more practical tools for early assessment of medical complications. Further studies of the early tissue changes produced by ethanol may yield some more specific and sensitive markers which may also be used in the detection of heavy drinking. Ideally one would wish to have a marker which reflects alcoholism and which may detect such propensity in heavy drinkers prior to the alcoholism stage. Such a marker may be difficult to find. Much more readily available are early consequences of heavy alcohol use which can be used to detect and assess drinking. Even relatively nonspecific biochemical changes can be extremely useful as has been illustrated in the case of diabetes by blood glucose. A rise in blood sugar is neither specific nor pathognomonic for diabetes; it can be influenced by the nutritional state, as well as by diseases of various organs including liver and pancreas. Yet the recognition that an elevation in blood sugar is commonly associated with uncontrolled diabetes has been extremely useful in its management. Similarly, the availability of a marker for heavy drinking can be expected to be extremely useful in at least three situations: (1) to detect relapse in patients who have been rehabilitated; (2) to provide an objective means of assessing and comparing various treatment modalities; and (3) to screen for heavy drinking in populations at risk.

A beginning has been made in the development and use of such markers. As mentioned before, α-amino-n-butyric acid (AANB) increases after alcohol consumption, both in experimental animals and in volunteers, possibly as a by-product of enhanced methionine degradation in response to glutathione depletion caused by acetaldehyde and the active radicals produced by the "induced" microsomes (Figure 5). Investigations of plasma AANB in alcoholics revealed an increase related to alcohol consumption (66). However, the level in the plasma is also decreased by protein deficiency. Indeed, clinical protein malnutrition is associated with characteristic plasma amino acid abnormalities, including depression of branched-chain amino acids and AANB. Thus, in alcoholics the level of AANB may reflect at least two factors: prolonged alcohol consumption, which tends to increase the amino acid, and dietary protein deficiency, which tends to decrease it. To use the level of plasma AANB as a reflection of alcohol consumption Shaw et al. (67) found it necessary to control for nutritional factors. Since AANB and leucine are depressed to a similar degree by dietary protein restriction, the level of AANB was expressed relative to the branched-chain amino acids of which leucine was selected for convenience. The plasma AANB concentration relative to leucine was increased in a large number of alcoholics regardless of dietary protein deficiency or the presence of moderate liver injury. The increase was reversible on cessation of drinking, persisted for days to weeks, did not require the presence of alcohol in the blood and was not present in nonalcoholics with moderate liver injury of viral etiology. Further studies have revealed that the concentration of AANB relative to leucine is not linear over the range of leucine values and, therefore, replacement of the simple ratio by a normal curve was recommended (68). In addition, determination of GGTP was found to enhance the specificity and sensitivity of the test. In alcoholics with severe liver injury, however, the AANB : leucine ratio may be increased because of markedly depressed values of plasma leucine and result in false positive determinations. This group of patients does not usually present a problem with respect to differential diagnosis. Similarly, in patients with extremely severe metabolic abnormalities, such as diabetic ketoacidosis, massive obesity undergoing starvation and subjects consuming an experimental diet with very low carbohydrate and high fat content, AANB may increase relative to leucine. To exclude such subjects, the simple use of a "dip Stix" test was recommended to detect acetone in the urine (68).

Although AANB may not offer a practical tool in screening for alcoholism in populations heterogenous in nutrition and degree of liver disease, there is now good evidence that, in rehabilitation programs in which the patient can serve as his own control, measurement of AANB can be useful in documenting treatment success as well as relapse (69). To that effect, it was more useful than measurement of mean corpuscular volume of red blood cells and more accurate than GGTP. For such purposes, the patient's AANB concentration (at discharge) is used for comparison; therefore, no correction of AANB with leucine is needed and the absolute values of AANB can be usefully interpreted.

These studies represent only a beginning. The development of markers both more practical and more specific will provide us with tools needed for the early detection of heavy drinking prior to medical and social disintegration, thereby allowing a truly preventive approach to the medical complications of alcoholism.

Conclusions

The last three decades have seen significant advances in our understanding of the medical complications of alcoholism. Major advances resulted from our recognition of the toxicity of alcohol in various tissues, particularly the liver, and some of the mechanisms involved. The knowledge and experimental tools acquired, especially the experimental models, allow us now to contemplate a truly public health strategy in prevention. Determining what makes people drink excessively or predisposes them to loss of control might enable us to begin to solve the problem of alcoholism itself. Such difficult questions, however, may not be resolved in the immediate future. It is of utmost importance that the lack of resolution of these basic questions does not deter us from other, less fundamental, approaches to the problems of alcoholism which may alleviate much of the suffering involved. Much can be learned from successful approaches to other major public health problems.

There are, in fact, many similarities between alcoholism and some other major public health problems which have confronted our society over the last century. One example is tuberculosis. Of course, there are many dissimilarities between tuberculosis and alcoholism, but some of the common aspects might suggest possible avenues for intervention in alcoholism. In both cases, the etiologic agent is known—Koch bacillus and alcohol. Both agents can affect most tissues of the body, but the most severe adverse effects usually occur in one organ, namely, the lungs in tuberculosis and the liver in alcohol. In both instances, exposure to the agent was or is widespread: a majority of the population had a positive reaction to tuberculin indicating contact with the Koch bacillus; similarly, the majority of the population consumes alcohol—but only a minority of those exposed develop complications.

One of the successful approaches to the problem of tuberculosis has been the early detection of the affected subject through appropriate laboratory tests; alcoholism could lend itself to a similar approach. It is obvious that among the alcohol users there is a subpopulation of very heavy drinkers who are particularly at risk for the development of alcoholism and its complications. One goal should be early detection prior to social or medical disintegration. This might be achieved through some chemical marker of heavy drinking, such as a blood test that could be performed by automated procedures on a large scale. There is reasonable hope that ongoing studies might, in the near future, provide us with tools for the early detection of heavy drinking.

Another cornerstone of the public health approach to tuberculosis has been the early detection of its major complication through mass screening for pulmonary lesions. A similar approach might also be feasible in alcoholism through improved means to detect liver lesions at an early stage. Such tests should now be incorporated in the routine work-up of heavy drinkers to detect those with significant liver complications.

Because cirrhosis does not develop in all heavy drinkers, one major task in an effort at prevention is to define the susceptible population. Success will also derive from determination of environmental factors which favor or oppose the development of the disease. Correction of social problems and nutritional imbalance has helped in the fight against tuberculosis; such avenues, particularly the nutritional approach, might also be used in alcoholism.

Since alcoholism and its complications, particularly cirrhosis, do not develop in all heavy drinkers, it is reasonable to speculate that some factor of heredity or individual susceptibility might play an important role. Obviously, research should be encouraged; however, study should not be limited to such an approach because the question is an extremely difficult one, and we may find an answer in the foreseeable future. For instance, we still do not know why pulmonary cavities develop in only a minority of subjects exposed to the Koch bacillus. We suspect that some genetic factor is important and that individual susceptibility plays a role, but despite a century of research these questions have not been answered. Yet, for all practical purposes, we have arrested the ravages of tuberculosis not only because of improved social and nutritional conditions, but also primarily by interfering with the etiologic agent, namely the Koch bacillus, through the development of antimicrobials.

Similarly, in the alcoholism field, although we may not understand why alcoholism develops in only some drinkers and why liver disease develops in only a minority of heavy drinkers, we might nevertheless be able to intervene sucessfully if we could acquire a better understanding of how alcohol affects the chemistry of the body. It is not unreasonable to hope that better explanations of the mechanisms whereby alcohol affects various tissues might give us some means with which to interfere with the actions of ethanol, perhaps altering the development of alcoholism and its major complications. Currently, there are relatively few studies on the basic effects of alcohol. Actually, we sometimes hear that such fundamental studies might be somewhat irrelevant to alcoholism because they do not appear to deal directly with the problem of compulsive drinking and associated psychiatric pathology. Such a narrow view may be stifling. Indeed, in the long run, understanding the biochemical effects of alcohol might provide a reasonable approach to alcoholism by allowing us to intervene through interference with the action of ethanol. Studies in the last decade have shown significant biochemical differences between alcoholics and nonalcoholics; further refinement may be extremely useful as a basis for the biological approach to alcoholism. Eventually, extrapolation from these studies may provide some tools to detect individuals prone to develop alcoholism prior to their exposure to alcohol. A possible example of such an evolution is provided by acetaldehyde: at the same dose of alcohol, resulting blood levels of acetaldehyde have been found to be higher in alcoholics than in nonalcoholics, and recently in relatives of nonalcoholics as well (54). The latter observation is still embroiled in methodological controversies; nevertheless, it is of interest since it illustrates how studies of consequences of excessive drinking may eventually lead to better insight in detecting individuals prone to develop alcoholism and what determines such behavior.

Of course, alcoholism is multifaceted. Behavioral, social and psychological problems play a major role and should not be neglected. However, we cannot escape the fact that conventional approaches based on traditional concepts have not succeeded in stemming the tide of alcoholism. Moreover, attempts at elimination of the etiologic agent (through prohibition) have failed. One virtue of the biologic outlook is to make possible the public health strategy which has been briefly outlined here. Public health intervention, in conjunction with the more traditional treatment and grass-root movements (such as Alcoholics Anonymous), might enable us to get a grasp on the problem of alcoholism.

Of course, not all aspects of alcoholism can be solved by a single approach. But if greater reliance on a public health strategy were to result in only moderate reduction in the incidence of alcoholism, or even if it were only to diminish some of its severe complications (e.g., cirrhosis), such a partial success could be expected to have a significant impact on morbidity and mortality.

REFERENCES

Because of space limitations, the references include primarily those of the author's unit, and a few other key citations. References to the numerous contributions of other centers can be found in Lieber (70).

1. HARRISON, T. R., ed. Principles of internal medicine. New York; McGraw-Hill; 1958.

2. JOLLIFFE, N. Alcohol and nutrition; the diseases of chronic alcoholism. Pp. 73–82. In: Alcohol, science and society; twenty-nine lectures with discussions as given at the Yale Summer School of Alcohol Studies. New Haven; Quarterly Journal of Studies on Alcohol; 1945.

3. PATEK, A. J., Jr., TOTH, I. G., SAUNDERS, M. G., CASTRO, G. A. M. and ENGEL, J. J. Alcohol and dietary factors in cirrhosis; an epidemiological study of 304 alcoholic patients. Arch. intern. Med. **135:** 1053–1057, 1975.

4. PEQUIGNOT, G. Die Rolle des Alkohols bei der Ätiologie von Leberzirrhosen in Frankreich. Münch. med. Wschr. **103:** 1464–1468, 1962.

5. NATIONAL ACADEMY OF SCIENCES. Recommended dietary allowances. 9th ed. Washington, D.C.; 1980.

6. BEST, C. H., HARTROFT, W. S., LUCAS, C. C. and RIDOUT, J. H. Liver damage produced by feeding alcohol or sugar and its prevention by choline. Brit. Med. J. **2:** 1001–1006, 1949.

7. LIEBER, C. S., JONES, D. P., MENDELSON, J. H. and DeCARLI, L. M. Fatty liver, hyperlipemia and hyperuricemia produced by prolonged alcohol consumption, despite adequate dietary intake. Trans. Ass. Amer. Physicians **76:** 289–300, 1963.

8. LIEBER, C. S., JONES, D. P. and DeCARLI, L. M. Effects of prolonged ethanol intake; production of fatty liver despite adequate diets. J. clin. Invest. **44:** 1009–1021, 1965.

9. LIEBER, C. S. and DeCARLI, L. M. Quantitative relationship between amount of dietary fat and the severity of alcoholic fatty liver. Amer. J. clin. Nutr. **23:** 474–478, 1970.

10. LIEBER, C. S. and DeCARLI, L. M. An experimental model of alcohol feeding and liver injury in the baboon. J. med. Primatol. **3:** 153–163, 1974.

11. POPPER, H. and LIEBER, C. S. Histogenesis of alcoholic fibrosis and cirrhosis in the baboon. Amer. J. Pathol. **98:** 695–716, 1980.

12. LIEBER, C. S., ed. Medical disorders of alcoholism; pathogenesis and treatment. Philadelphia; Saunders. [In press.]

13. LIEBER, C. S. and DeCARLI, L. M. Study of agents for the prevention of the fatty liver produced by prolonged alcohol intake. Gastroenterology **50:** 316–322, 1966.

14. LEO, M. A., SATO, M., ARAI, M. and LIEBER, C. S. Liver injury after vitamin A deficiency in excess; potentiation by ethanol. Gastroenterology **79:** 1034, 1980.

15. BARAONA, E., PIROLA, R. C. and LIEBER, C. S. Small intestinal damage and changes in cell population produced by ethanol ingestion in the rat. Gastroenterology **66:** 226–234, 1974.

16. FRIEDMAN, H. S. and LIEBER, C. S. Cardiotoxicity of alcohol. Cardiovas. Med. **2:** 111–112, 1977.

17. GORDON, G. G., SOUTHREN, A. L. and LIEBER, C. S. Hypogonadism and feminization in the male; a triple effect of alcohol. Alcsm clin. exp. Res. **3:** 210–212, 1979.

18. LINDENBAUM, J. and LIEBER, C. S. Hematologic effects of alcohol in man in the absence of nutritional deficiency. New Engl. J. Med. **281**: 333–338, 1969.

19. SAVILLE, P. D. and LIEBER, C. S. Effect of alcohol on growth, bone density and muscle magnesium in the rat. J. Nutr. **87**: 477–484, 1965.

20. PIROLA, R. C. and DAVIS, A. E. Effects of ethyl alcohol on sphincteric resistance at the choledocho-duodenal junction in man. Gut, Lond. **9**: 577–560, 1968.

21. SARLES, H. Chronic calcifying pancreatitis—chronic alcoholic pancreatitis. Gastroenterology **66**: 604–616, 1974.

22. RUBIN, E., KATZ, A. M., LIEBER, C. S., STEIN, C. S., STEIN, E. P. and PUSZKIN, S. Muscle damage produced by chronic alcohol consumption. Amer. J. Path. **83**: 499–516, 1976.

23. NOBLE, E. P. and TEWARI, S. Metabolic aspects of alcoholism in the brain. Pp. 149–185. In: LIEBER, C. S., ed. Metabolic aspects of alcoholism. Baltimore; University Park Press; 1977.

24. LIEBER, C. S. Metabolism and metabolic effects of alcohol. Seminars Hematol. **17**: 85–99, 1980.

25. LIEBER, C. S. and DAVIDSON, C. S. Some metabolic effects of ethyl alcohol. Amer. J. Med. **33**: 319–327, 1962.

26. FREINKEL, N. and ARKEY, R. A. Effects of alcohol on carbohydrate metabolism in man. Psychosom. Med. **28**: 551–563, 1966.

27. LIEBER, C. S., JONES, D. P., LOSOWSKY, M. S. and DAVIDSON, C. S. Interrelation of uric acid and ethanol metabolism in man. J. clin. Invest. **41**: 1863–1870, 1962.

28. ISERI, O. A., LIEBER, C. S. and GOTTLIEB, L. S. The ultrastructure of fatty liver induced by prolonged ethanol ingestion. Amer. J. Pathol. **48**: 535–555, 1966.

29. LANE, B. P. and LIEBER, C. S. Ultrastructural alterations in human hepatocytes following ingestion of ethanol with adequate diets. Amer. J. Pathol. **49**: 593–603, 1966.

30. BARAONA, E. and LIEBER, C. S. Effects of chronic ethanol feeding on serum lipoprotein metabolism in the rat. J. clin. Invest. **49**: 769–778, 1970.

31. BOROWSKY, S. A., PERLOW, W., BARAONA, E. and LIEBER, C. S. Relationship of alcoholic hypertriglyceridemia to stage of liver disease and dietary lipid. Dig. Dis. Sci. **25**: 22–27, 1980.

32. LEFÈVRE, A., ADLER, H. and LIEBER, C. S. Effect of ethanol on ketone metabolism. J. clin. Invest. **49**: 1775–1782, 1970.

33. MISRA, P. S., LEFÈVRE, A., ISHII, H., RUBIN, E. and LIEBER, C. S. Increase of ethanol, meprobamate and pentobarbital metabolism after chronic ethanol administration in man and in rats. Amer. J. Med. **51**: 346–351, 1971.

34. LIEBER, C. S. and DECARLI, L. M. Ethanol oxidation by hepatic microsomes: adaptive increase after ethanol feeding. Science **162**: 917–918, 1968.

35. LIEBER, C. S. and DECARLI, L. M. Hepatic microsomal ethanol-oxidizing system; in vitro characteristics and adaptive properties in vivo. J. biol. Chem. **245:** 2505–2512, 1970.

36. OHNISHI, K. and LIEBER, C. S. Reconstitution of the microsomal ethanol-oxidizing system; qualitative and quantitative changes of cytochrome P-450 after chronic ethanol consumption. J. biol. Chem. **252:** 7124–7131, 1977.

37. LIEBER, C. S., SEITZ, H. K., GARRO, A. J. and WORNER, T. M. Alcohol-related diseases and carcinogenesis. Cancer Res. **39:** 2863–2886, 1979.

38. HASUMURA, Y., TESCHKE, R. and LIEBER, C. S. Increased carbon tetrachloride hepatotoxicity, and its mechanism, after chronic ethanol consumption. Gastroenterology **66:** 415–422, 1974.

39. LIEBER, C. S., SEITZ, H. K., GARRO, A. J. and WORNER, T. M. Alcohol as a cocarcinogen. Pp. 320–335. In: BERK, P. D. and CHALMERS, T. C., eds. Frontiers in liver disease. New York; Thieme-Stratton; 1981.

40. SATO, C., MATSUDA, Y. and LIEBER, C. S. Increased hepatotoxicity of acetaminophen after chronic ethanol consumption in the rat. Gastroenterology **80:** 140–148, 1981.

41. PIROLA, R. C. and LIEBER, C. S. Energy wastage in alcoholism and drug abuse; possible role of hepatic microsomal enzymes. Amer. J. clin. Nutr. **29:** 90–93, 1976.

42. GORDON, G. G., VITTEK, J., WEINSTEIN, B., SOUTHREN, A. L. and LIEBER, C. S. Acute and chronic effects of alcohol on steroid hormones with emphasis on the metabolism of androgens and estrogens. Pp. 89–102. In: AVOGARO, P., SIRTORI, C. R. and TREMOLI, E., eds. Metabolic effects of alcohol. New York; Elsevier; 1979.

43. GORDON, G. G., ALTMAN, K., SOUTHREN, A. L., RUBIN, E. and LIEBER, C. S. Effect of alcohol (ethanol) administration on sex-hormone metabolism in normal men. New Engl. J. Med. **295:** 793–797, 1976.

44. VITTEK, J., GORDON, G. G., SOUTHREN, A. L., RAPPAPORT, S. C., MUNNANGI, P. R. and LIEBER, C. S. Effect of ethanol intake on the cellular regulation of testosterone-5α-reductase in rat oral tissue. J. pharmacol. exp. Ther. **217:** 411–415, 1981.

45. LIEBER, C. S. Possible role of microsomal changes induced by ethanol intake in the development of tolerance and dependence. Pp. 59–78. In: GROSS, M. ed. Alcohol intoxication and withdrawal; experimental studies. (Advances in Experimental Medicine and Biology, Vol. 35.) New York; Plenum; 1973.

46. MENDELSON, J. H. Biologic concomitants of alcoholism. New Engl. J. Med. **283:** 24–32, 71–81, 1970.

47. LIEBER, C. S. Ethanol metabolism and biochemical aspects of alcohol tolerance and dependence. Pp. 135–161. In: MULE, S.J. and BRILL, H., eds. Chemical and biological aspects of drug dependence. Cleveland; Chemical Rubber Company Press; 1972.

48. KORSTEN, M. A., MATSUZAKI, S., FEINMAN, L. and LIEBER, C. S. High blood acetaldehyde levels after ethanol administration; differences

between alcoholic and nonalcoholic subjects. New Engl. J. Med. **292:** 386–389, 1975.

49. HASUMURA, Y., TESCHKE, R. and LIEBER, C. S. Characteristics of acetaldehyde oxidation in rat liver mitochondria. J. biol. Chem. **251:** 4908–4913, 1976.

50. SHAW, S., JAYATILLEKE, E., ROSS, W. A. and LIEBER, C. S. Ethanol induced lipid peroxidation; potentiation by chronic alcohol feeding and attenuation by methionine. J. lab. clin. Med. **98:** 417–425, 1981.

51. SCHREIBER, S. S., ORATZ, M., ROTHSCHILD, M. A., REFF, F. and EVANS, C. Alcoholic cardiomyopathy. II. The inhibition of cardiac microsomal protein synthesis by acetaldehyde. J. molec. cell Cardiol. **6:** 207–213, 1974.

52. DAVIS, V. E. and WALSH, M. J. Alcohol, amines and alkaloids; possible biochemical basis for alcohol addiction. Science **167:** 1005–1007, 1970.

53. COHEN, G. and COLLINS, M. Alkaloids from catecholamines in adrenal tissue; possible role in alcoholism. Science **167:** 1749–1751, 1970.

54. SCHUCKIT, M. A. and RAYSES, V. Ethanol ingestion; differences in blood acetaldehyde concentrations in relatives of alcoholics and controls. Science **203:** 54–55, 1979.

55. BARAONA, E., LEO, M. A., BOROWSKY, S. A. and LIEBER, C. S. Alcoholic hepatomegaly; accumulation of protein in the liver. Science **190:** 794–795, 1975.

56. MATSUDA, Y., BARAONA, E., SALASPURO, M. and LIEBER, C. S. Effects of ethanol on liver microtubules and Golgi apparatus; possible role in altered hepatic secretion of plasma proteins. Lab. Invest. **41:** 455–463, 1979.

57. FEINMAN, L. and LIEBER, C. S. Hepatic collagen metabolism; effect of alcohol consumption in rats and baboons. Science **176:** 795, 1972.

58. LELBACH, W. K. Cirrhosis in the alcoholic and its relation to the volume of alcohol abuse. Ann. N.Y. Acad. Sci. **252:** 85–105, 1975.

59. VAN WAES, L. and LIEBER, C. S. Early perivenular sclerosis in alcoholic fatty liver; an index of progressive liver injury. Gastroenterology **73:** 646–650, 1977.

60. WORNER, T. M. and LIEBER, C. S. Prognostic value of perivenular sclerosis in the progression of alcoholic liver injury. Gastroenterology **75:** 995, 1978.

61. BECKETT, A. G., LIVINGSTONE, A. V. and HILL, K. R. Acute alcoholic hepatitis. Brit. med. J. **2:** 1113–1119, 1961.

62. TESCHKE, R., BRAND, A. and STROHMEYER, G. Induction of hepatic microsomal gamma-glutamyltransferase activity following chronic alcohol consumption. Biochem. biophys. Res. Commun. **75:** 718–724, 1977.

63. VAN WAES, L. and LIEBER, C. S. Glutamate dehydrogenase; a reliable marker of liver cell necrosis in the alcoholic. Brit. med. J. **2:** 1508–1510, 1977.

64. WORNER, T. M. and LIEBER, C. S. Plasma glutamate dehydrogenase

(GDH) as a marker of alcoholic liver injury. Pharmacol. Biochem. Behav. **13** (Suppl. No. 1): 107–110, 1980.

65. LIEBER, C. S. Alcohol, liver injury and protein metabolism. Pharmacol. Biochem. Behav. **13:** (Suppl. No. 1): 17–30, 1980.

66. SHAW, S. and LIEBER, C. S. Plasma amino acids in alcoholic liver injury; contrast with protein malnutrition. Clin. Res. **23:** 459A, 1975.

67. SHAW, S., STIMMEL, B. and LIEBER, C. S. Plasma alpha amino-*n*-butyric acid to leucine; an empirical biochemical marker of alcoholism. Science **194:** 1057–1058, 1976.

68. SHAW, S., LUE, S.-L. and LIEBER, C. S. Biochemical tests for the detection of alcoholism; comparison of plasma alpha amino-*n*-butyric acid with other available tests. Alcsm clin. exp. Res. **2:** 3–7, 1978.

69. SHAW, S., WORNER, T. M., BORYSOW, M. F., SCHMITZ, R. E. and LIEBER, C. S. Detection of alcoholism relapse; comparative diagnostic value of MCV, GGTP, and AANB. Alcsm clin. exp. Res. **4:** 297–301, 1979.

70. LIEBER, C. S., ed. Metabolic aspects of alcoholism. Baltimore; University Park Press; 1977.

15

Alcoholism: An Overview of Treatment Models and Methods

A. Arthur Sugerman, M.D.

When *Alcohol, Science and Society* first appeared in 1945, treatment was discussed in several separate chapters. These included chapters on medical treatment, social case work and pastoral counseling together with a chapter on the role of religious bodies in the treatment of inebriety and also one on the fellowship of Alcoholics Anonymous. When Mark Keller wrote the preface to the sixth printing in 1954 he added an update on treatment, mentioning psychotherapy, sensitization therapy by disulfiram (Antabuse), conditioned-reflex or aversion therapies, other adjunctive therapies such as steroid hormones and massive doses of vitamins, and the alcoholism clinic approach.

Although terminology has changed somewhat since *Alcohol, Science and Society* was published, and we now speak of alcoholism and alcoholics rather than inebriety and inebriates, and although there have been improvements in the medications available for detoxification, very little in the original work is obsolete, and much of it is basic to contemporary treatment of sufferers from alcoholism.

Dr. Robert Fleming's chapter (1) on the "Medical Treatment of the Inebriate" describes how excessive drinking may be a form of self-medication; the patient drinks to obtain relief from symptoms of underlying physical, psychological or social pathology. Such symptomatic drinking can be cured if the underlying condition can be relieved or removed. On the other hand, if this type of drinking continues, sooner or later new pathology appears and the sufferer will seek relief by recourse to more alcohol. When the vicious circle has been established true addiction is present. The essential first step in the treatment of alcohol addiction is breaking the circle; according to Dr. Fleming, after this has been accomplished the original basis for the early symptomatic drinking, "often heavily overlaid and encrusted by years of dissipation," must be dealt with. Few therapists today would feel capable of dealing with such underlying causes, but all would agree that the vicious circle of drinking to relieve symptoms caused by previous drinking must be interrupted. When the addictive cycle has been broken, special medical and social problems must be treated, e.g., acute intoxication or withdrawal symptoms, unemployment, homelessness or poverty. The role of the physician and the social worker are just as important as they were in 1945. Only after the crisis has passed, when the individual is medically and socially stable, can the core problem be attacked.

There are many ways of developing alcoholism and many ways of initiating treatment, but the essential first step is helping the patient to recognize that he (or she) has a problem with alcohol. The patient may realize that he has lost control of his drinking and that this has affected his health, job, family or other aspects of his life. More often the problem is brought to his attention by his physician, his family or his employer. Some degree of coercion by the patient's spouse or employer may be required so that he agrees to accept treatment. Treatment approaches may be medical or nonmedical; medical treatment may be in- or outpatient. After the acute episode is past rehabilitation may proceed in a residential setting, away from the patient's home and former drinking situations; or it may take place while he remains at home and returns to work and receives care from physician or psychotherapist, or attends a self-help program, e.g., Alcoholics Anonymous.

There are several models of treatment in alcoholism rehabilitation, including the medical model, the behavior modification model, the psychological model, the social model and the Alcoholics Anonymous model. Modern approaches to treatment are multimodal and multidisciplinary, so that different therapies are available to different kinds of patients, and different programs are available at different stages of the illness. Treatment is available from the private medical sector, from general, state and private psychiatric hospitals, nonmedical residential centers, specialized alcoholism treatment programs funded by federal, state and city agencies, and from A.A. What treatment a patient obtains should be based on his special needs, but often it depends on what is available at the time. The increasing availability of insurance coverage for alcoholism treatment, however, and the provision of specialized alcoholism treatment programs under legislative mandate, have greatly increased the prospects of obtaining treatment.

Acute Alcoholic States

The treatment of an acute episode is least controversial; it is generally agreed that this is a proper area for the application of the medical model and the use of medications. The treatment of acute intoxication is the same in alcoholics and nonalcoholics, and although much research has been directed at finding a treatment which might alter the metabolism of alcohol to produce more rapid sobering, hormones, vitamins or stimulants have not been effective and intravenous fructose, which increases alcohol metabolism, may have undesirable side effects. A patient in coma after an overdose of alcohol must be treated as in serious danger of death from respiratory depression. Cardiopulmonary resuscitation may be required; the airway must be kept clear and ventilation supported if necessary. Intravenous fluids should be started and blood samples obtained to check levels of alcohol, electrolytes, blood gases and glucose, and hepatic and renal function tested.

The presence of coma is of course an indication for hospital admission; there are many complications of alcoholism which also justify admission. These include malnutrition, hepatic cirrhosis, polyneuropathy, chronic obstructive pulmonary disease associated with excessive smoking, pulmonary tuberculosis, diabetes, heart disease, gastritis and others. However, the most frequent reason for admission is the alcohol withdrawal syndrome (sometimes called abstinence syndrome). This occurs when the ingestion of alcohol is stopped or diminished. It is an expression of release from the depressant effect of alcohol on the central nervous system, and so the symptoms of withdrawal are in many ways opposite to those of intoxication. Instead of diminished psychomotor activity with relaxation, depressed respirations, increased delta and decreased rapid eye movement sleep, and an increased seizure threshold, the patient suffering from alcohol withdrawal shows increased psychomotor activity, with anxiety, tremors, sensory hyperacuity, hyperventilation, increased rapid eye movement and decreased delta sleep, hallucinations, delirium and convulsions.

Withdrawal symptoms may be very mild and require no treatment, or so severe as to threaten life. If the history and physical and mental examinations reveal disorientation, marked psychomotor agitation, hallucinations, withdrawal seizures or severe hypertension, or coexisting illnesses, the patient should be hospitalized.

Outpatient Detoxification

Detoxification on an outpatient basis should be carried out only after thorough evaluation and only when the patient can be very carefully observed by daily visits. The patient treated without admission to hospital should have only mild to moderate symptoms such as tremulousness, insomnia and apprehension. A history of delirium tremens or a history of consumption of one pint or more of liquor daily may indicate the need for immediate and appropriate anti-anxiety medication to prevent more serious complications. The medications most commonly used are diazepam (Valium) or chlordiazepoxide (Librium) although oxazepam (Serax) and lorazepam (Ativan) are at least as effective and have certain advantages with regard to predictability and duration of their effects because of their lack of active metabolites. Patients who have had frequent episodes of withdrawal may know which drug works best for them.

A patient undergoing withdrawal must be supervised carefully by a friend or relative or by a staff member in a halfway house. He should not medicate himself but should be given medication by the "significant other" who is instructed about the usual course and possible complications of withdrawal. Daily vitamins should also be given. Brief, daily follow-up appointments are required for the next few days to evaluate the clinical course and adjust the medication. After withdrawal symptoms have been suppressed for 24 to 36 hr the dosage may be reduced over the following week.

Even though the treatment of the withdrawal syndrome by medical means is relatively uncontroversial it must be recognized that halfway houses and some outpatient facilities which emphasize psychosocial intervention are commonly prejudiced against the use of any psychotropic medication. It is important to enlist the support and cooperation of nonmedical staff so that prescribed medications are given properly.

Inpatient Detoxification

If patients cannot be thoroughly screened for outpatient detoxification, if adequate supervision is not available, if withdrawal symptoms are severe, or if any of the physical complications or concomitant illnesses noted above are present, treatment should be carried out in a hospital.

The essential aspects of adequate detoxification are sufficient sedation, good nursing care and strong psychological support. Sedation is accomplished through the use of long-acting drugs crosstolerant with alcohol, and at the present time in the U.S. the benzodiazepines are the drugs of choice, and, of these, chlordiazepoxide is most commonly used. The dose should be enough to control symptoms, but not sever the link with reality, and the medication should be given continuously, not as needed when symptoms appear. Other benzodiazepines such as diazepam, lorazepam and oxazepam may also be used and older drugs such as barbiturates, chloral hydrate and paraldehyde are used effectively in some hospitals. In Great Britain and Europe chlormethiazol (Heminevrin) is the drug of choice because of its sedative hypnotic and anticonvulsant properties but it is not available in the U.S.

The dose of sedative medication must be progressively decreased over the first week, by at least 25% each day, and the drug should be completely discontinued in about seven days.

Vitamins are given routinely. The most important is thiamine. If the patient's history suggests a risk of multiple vitamin deficiencies or if examination of the blood shows macrocytosis it is advisable to assess the folic acid and vitamin B_{12} levels and prescribe these if needed.

When hallucinations are present antipsychotic medication should be administered. Several researchers have found haloperidol more effective than other drugs in relieving the severe symptoms of alcohol withdrawal and delirium tremens. It has a low incidence of side effects and is non-addicting. If hallucinations and delusions persist electroconvulsive therapy should be considered. Convulsions can occur as a complication of alcohol withdrawal, usually during the first 48 hr. If convulsions occur after the first few days the possibility of preexisting epilepsy, other neurological disorder, or multiple drug withdrawal should be investigated. If a patient has had seizures in the past anticonvulsant medication should be given from the start of treatment. Serum electrolytes should be measured at the time of admission; if serum potassium is low it may be replaced; oral fluids containing potassium such as fruit juices should be given. Propranolol (Inderal) has been tried with and without chlordiazepoxide. It is effective in reducing tremor, blood pressure and heart rate, but it does not have anti-seizure activity and should be used with caution in patients with tendencies to hypoglycemia, bronchospasm or cardiac depression. It cannot substitute for chlordiazepoxide in the treatment of withdrawal.

Nursing care and psychological support are tremendously important during detoxification. The environment should be such that the patient can relax and feel comfortable. The staff should provide support and reassurance; restraints should be avoided. During the early stages it is important to maintain the patient's links with reality. He should be kept in a lighted environment, other people should be present or within call, familiar objects—a clock, radio, or television—should be close by. The patient should be encouraged to carry out simple tasks, to walk rather than stay in bed and to participate in daily activities. As soon as possible the patient should meet with a qualified alcoholism counselor who will orient him to ongoing programs.

The Post-Withdrawal Phase

The post-withdrawal or subacute phase starts at the end of the detox-ification period and lasts about six weeks. Treatment in this phase lays the groundwork for long-term therapy.

Anxiety and depression are frequently present and if not treated may lead to premature termination of treatment. Every effort should be made to avoid the use of psychotropic medication during this period, as the symptoms are often alleviated by individual and group psychotherapy, involvement in the educational, occupational and recreational programs of the hospital and the gradual influence of time. If a patient requires medication it should be prescribed in the lowest dose possible to relieve symptoms. Continuation of chlordi-azepoxide in gradually decreasing doses may be very effective; other physicians recommend the use of phenothiazines such as mesori-dazine (Serentil), or antidepressants such as amitriptyline (Elavil).

As the patient begins to feel well, perhaps for the first time in years, he will be ready to learn about his illness and what he must do to avoid further problems with alcohol. An inpatient unit can provide informational lectures, films and discussions with a trained counselor, as well as the traditional individual and group psychoth-erapy, psychodrama, classes in relaxation techniques and autohyp-nosis. This is also the time to introduce the patient to Alcoholics Anonymous and his family to Al-Anon and Alateen. Some or all of these treatments may be available in outpatient clinics, halfway houses, rehabilitation centers or recovery houses.

When the patient is in full contact with reality an evaluation should be made of his physical, psychological, family, social and occupational problems and needs.

The hepatic, cardiac, pulmonary and neurological disorders commonly found in older alcoholics may need intensive and pro-longed care. Psychiatric and psychological assessment may show endogenous affective disorder or schizophrenia requiring chemo-therapy and psychotherapy; lithium may be needed in manic-depressive disorders or recurrent depression. Neuroses or character disorder may be revealed and may require psychotherapy. A history of hyperactivity in childhood with psychological evidence of atten-tion disorder may suggest a trial of pemoline (Cylert).

Marital or family problems may need to be addressed. Changes in family attitudes are very important if the alcoholic's change in life-style is to be maintained. If the patient is employed his job may be protected by informing his employer of his progress in treatment (this of course should only be done with the patient's permission). If he is unemployed he may require educational and vocational training and assistance in finding work. If he is homeless or without income he should be helped to find shelter and welfare benefits. He may require legal or other kinds of professional assistance to help him return to adequate socioeconomic functioning. These are some of the many kinds of special problems which may require adjunctive help before long-term treatment can proceed.

Long-Term Treatment: An Overview of Models and Methods

In the acute stage alcoholism appears at its most homogeneous. The causes and treatment of acute episodes are quite similar in every patient, as are the causes and treatment of withdrawal symptoms. In the post-withdrawal stage many different problems may appear, requiring a variety of different treatments. In long-term treatment a host of differences emerge in personality and constitutional factors, childhood upbringing, social and familial factors, precipitating causes, reaction to alcoholism and motivation for recovery which indicate the uniqueness of the individual and his illness. Many different kinds of therapy are employed in the treatment of alcoholism in an effort to deal with the core problem of the psychological dependence on alcohol. Different aspects of the core problem are stressed by therapists of different orientations: some stress the biological predisposition, some the psychological or psychodynamic and others the social and familial. Many different models of treatment are offered in attempts to help alcoholics to cope without depending on alcohol. Various classifications of the models have been suggested; the following is one proposed by Benjamin Kissin (2).

The Medical Model

The medical model views alcoholism as a disease, which is therefore best treated by a physician. It stresses the biological mechanisms of alcoholism with much less attention to the psychological and social. It emphasizes the use of medications such as antianxiety or antipsychotic agents and disulfiram. In a 1972 survey, over 65% of 15,000 physicians who responded to a mailed questionnaire prescribed one or more tranquilizers as well as other medication in the treatment of alcoholism. In this model the patient is typically detoxified in the manner described and continued on benzodiazepine antianxiety medication for weeks to months afterwards. The physician sees the patient for brief supportive interviews at weekly intervals at first, focusing largely on the medical symptoms, with emphasis on gradual recovery and occasional discreet reminders of the damage which drinking again would cause. One study of alcoholics in outpatient clinics found that internists and general practitioners were the most successful of all primary therapists, with social workers rather less successful and psychiatrists much less so. Although firm conclusions cannot be drawn from the study we have no reason to believe that this form of treatment is less successful than any other.

The Behavior Modification Model

This model focuses on attempts to alter behavior without reference to psychodynamic factors which assume individual susceptibility to develop alcoholism. In the positive reward operant conditioning paradigm, the susceptible individual ingesting alcohol is rewarded either by reduction of tension or induction of a sense of well-being, or both. The original experience of gratification which many alcoholics remember as their first significant exposure to alcohol is reinforced by subsequent experience, resulting in the development of primary psychological dependence. This is a powerful psychological drive for alcohol-seeking behavior. Techniques of behavior modification therapy include aversive conditioning and more recently developed positive reinforcement methods.

Aversive Conditioning. Aversive conditioning involves pairing the sight, smell and taste of alcoholic beverages with an unpleasant stimulus such as electric shock or a chemical compound which induces nausea, vomiting or other distressing effects. Electric shock was used as an aversive stimulus 50 years ago in the U.S.S.R. with apparently good results. Recent studies in Great Britain and the U.S. do not support the value of this treatment, although it may be useful when combined with other behavioral methods.

Chemical aversion therapy uses chemicals such as emetine or apomorphine to produce nausea and vomiting. The chemical is paired repeatedly with the sight, taste, smell or touch of alcohol. The technique was used extensively at the Shadel Sanatorium in Seattle and is described in Fleming's chapter (1) in *Alcohol, Science and Society*. It is still used in the same hospital (now called Schick's Shadel Hospital) and in some other private hospitals mostly in the western United States. The aversive chemical is given shortly before the patient's favorite alcoholic beverage so that the beverage is taken at the time when the effect of the chemical is most prominent. Two or three trials are given in a typical 45-min treatment session. Four to six sessions given every other day constitute the standard regimen.

Although the results reported from Shadel and from another hospital in Oregon are impressive (over 50% abstinent by self-report after 4 years in one study, 63% abstinent for 1 year in another), they have been criticized on the grounds that most of the patients were better motivated and of higher educational and socioeconomic level than those in other studies. However, even taking into account that the patients were a selected group, and that other therapies were used at the same time (such as group therapy and individual and family counseling), the results appear to be better than most others and so deserving of further research.

Covert aversion or covert sensitization is a method which employs aversive images instead of shock or chemicals. It has theoretical and practical advantages over other aversive methods but its place in therapy has not yet been validated by adequate outcome studies. Cautela (3), its creator, has advocated its use together with relaxation training and systematic desensitization to reduce anxiety.

Disulfiram therapy was originally an aversive technique. After the patient had been taking disulfiram for several days, challenge doses of alcohol were given to demonstrate the unpleasant effects, which usually start 15 to 30 min afterwards. They include flushing, throbbing headache, changes in blood pressure, palpitations, weakness, nausea and vomiting, cold sweats and perhaps unconsciousness; the severity and duration of the symptoms is proportional to the amount of alcohol taken. At the present time the effects are well enough known so that the challenge doses of alcohol generally are not given, and disulfiram is considered to act as a deterrent rather than as an aversive stimulus.

Although disulfiram, like other medications, is found unacceptable by some A.A. members, others have found it of great value and advocate its use. Marty Mann (4) has said that it is "immensely helpful to someone who really doesn't want to drink but has trouble staying away from it." The individual has to make only one decision a day (to take the disulfiram) rather than many (to drink or not). She viewed the medication as most useful in the early stages of therapy, helping alcoholics to stay sober long enough to benefit from treatment.

Operant Methods of Modifying the Drinking Response. In contrast to the aversive techniques which aim to make drinking itself aversive, operant approaches aim to reduce drinking by manipulating its consequences. Laboratory experiments have aimed to reduce drinking through reward of moderate drinking by providing an enriched living environment while pairing excessive drinking with an impoverished environment. Following drinking responses by 10 min of isolation in a small booth has also effectively led to reduced drinking. Other social reinforcers such as a weekend pass to leave the hospital have been used to reinforce moderate drinking. Negative reinforcement by electric shock, either self-administered or subject-administered, may also be effective. Many techniques have been devised to teach more moderate drinking in the laboratory, but it is too early to be sure of their usefulness in natural settings, where so many uncontrolled events may effect outcome.

Overview of Behavioral Treatment

At the present time aversive conditioning alone has not been shown to be an effective treatment for excessive drinking. Of the various available techniques, chemical aversion therapy may be the most useful because the nausea and vomiting produced chemically relates much more directly to the natural consequences of alcohol excess. Operant conditioning works well in the laboratory but may not be so effective outside. The use of a variety of behavioral and other techniques simultaneously such as the "broad spectrum" approach of Lazarus (5) appears to be more effective; such an approach might include medical care for physical disabilities, aversive conditioning to modify drinking behavior, construction of anxiety hierarchies for systematic desensitization based on tests and interviews, assertive training, standard relaxation techniques, hypnosis and marital therapy.

A great deal of controversy has arisen around the use of techniques to teach moderate or "controlled" drinking. Many studies have found that some alcoholics can learn to drink socially, or more usually in a carefully controlled way. This outcome usually appears spontaneously and was discovered serendipitously in examining the results of outcome studies and was not due to deliberate attempts to induce controlled drinking. The findings that some alcoholics could somehow resume nonproblem drinking, however, led to reconsideration of the generally accepted belief that alcoholism is a progressive disease rather than a relapsing and remitting one. Although several elaborate studies have used behavioral therapy to teach controlled drinking to selected patients, we do not yet have enough information to enable us to select patients rationally for a program aimed at control rather than abstinence. Alcoholics who have been able to achieve abstinence are not suitable subjects for such a program; they should not return to any form of drinking. Research will undoubtedly continue and in time show the proper place of behavior therapy aimed at controlled drinking in the total approach to the treatment of alcoholism.

The Psychological Model

The psychotherapeutic approach to alcoholism implies the use of psychotherapy to modify psychological conditions which predispose to alcoholism; these may be either proximal or distal, i.e., either current events which precipitate a drinking episode, or early developmental factors. Psychotherapy may also be directed towards conditions secondary to alcoholism and conditions brought about by the cessation of drinking; psychotherapeutic intervention may thus be required to treat the complications of problem drinking and to prevent relapse.

Many techniques of psychotherapy are used with alcoholic patients; a definition of psychotherapy should allow for a variety of approaches and practitioners. Blane (6) suggests: "Psychotherapy is a structured emotional experience occurring in a close relationship between two persons, in which a trained individual helps another to achieve greater self-understanding, objectivity and personal growth through a series of contacts in which relevant inner experiences and life situations of the latter are discussed."

This definition can encompass the modifications of psychoanalytic therapy as well as transactional analysis, gestalt therapy and counseling by former alcoholics. These and many other forms of psychotherapy have been used in the treatment of alcoholism; all may be useful for specific individuals.

Zimberg (7) believes that many alcoholics share a common conflict, which must be recognized in psychotherapy, consisting of a lack of self-esteem along with feelings of worthlessness and inadequacy. The effects of alcohol are to subdue anxiety and create feelings of power and invulnerability. When the alcoholic recovers from drinking, however, problems still exist, intensifying the basic conflict. The use of alcohol to produce an artificial feeling of power and control feeds the alcoholic's need for grandiosity. Zimberg calls this intense need "reactive grandiosity." The grandiosity leads to trying harder, with inevitable failure, and so to more anxiety, depression, anger and guilt. The conflict alone is not enough to produce alcoholism. It is one predisposing force, which acts together with genetic, familial and social forces.

The traditional psychoanalytic approach involves working backwards from the defenses—repression, denial and reactive grandiosity—to the underlying psychological conflicts. The process of therapy must uncover the unconscious conflict; it must be fully experienced in conscious awareness. The approach has not been very successful; the pharmacological effects of alcohol are too strong to be altered through insight alone, and the technique of uncovering produces anxiety which often leads to the craving to drink.

The therapist should work actively to establish a positive relationship in the opening phase of treatment by being warm and open and fulfilling appropriate requests. In later phases of therapy positive transference is encouraged and other modifications are made which are usual for nonintensive short-term or goal-limited forms of individual therapy.

One of the most important principles in psychotherapy with alcoholics is that first the drinking must be terminated. Although "slips" may occur, they should be dealt with individually as violations of the general rule that drinking should not continue during treatment and abstinence should be an ultimate goal. The transference relationship is intensely ambivalent because of the alcoholic's dependence and grandiosity; he may show hostile, manipulative and testing behavior, and in the euphoria of the early stages of abstinence he may believe that he can control his drinking and his life. The patient's provocative behavior may lead to a characteristic countertransference. The therapist feels frustrated and angry at his inability to control the drinking. The therapist, however, must realize no one can stop the patient from drinking if he is determined to do so. The therapist must set limits on the patient's behavior and set conditions for continuing treatment. If the conditions cannot be met therapy should be terminated, leaving open the option of resuming the process in the future.

As in the case of behavior therapy individual psychotherapy is best used as one of a number of modalities brought into play concurrently or consecutively. Zimberg, one of the leading advocates of individual psychotherapy in alcoholism, divides the progress of treatment into three stages.

In the first stage the alcoholic "cannot drink" because of external pressures from employer or spouse threatening his job or marriage, or other external controls such as the use of disulfiram. Entering an inpatient program which also enforces abstinence will serve the same purpose, which is to stop the individual's drinking temporarily while attitudes towards drinking and denial of alcoholism have not yet changed. In this stage the individual must be helped in ways of facing problems without alcohol. Family therapy and referral of the family to Al-Anon are helpful in changing the family's attitudes and giving them greater understanding.

When an alcoholic has stopped drinking at this early stage

he may feel over-confident about his sobriety; Zimberg interprets this as a reaction to the lack of control over drinking which is now experienced as a certainty of control over drinking and other aspects of life. The situation may be unstable enough to lead to a return to drinking, or, through counseling and further A.A. involvement, to the second stage where the alcoholic "won't drink."

Controls have now become internalized and the individual's attitudes towards drinking have changed considerably. The conflict about drinking is now unconscious and may show itself in fantasies and dreams. At this stage disulfiram may be discontinued or used only in stressful situations. Directive and supportive psychotherapy should be continued to develop other means of coping with stress and unpleasant feelings. This is a reasonably stable stage which may lead only occasionally to "slips" after years of sobriety. Reactive grandiosity is redirected towards control over a previously uncontrollable problem, with ego-enhancing feelings of success. Active A.A. members may find an outlet for their need for grandiosity in their work with other alcoholics ("twelfth-stepping").

In the third stage the alcoholic "does not have to drink." This stage is achieved only through insight and resolution of personality problems and conflicts, and requires several years of psychotherapy and self-understanding. It is stable, provided the individual refrains from drinking and it is relatively easy to maintain.

The Social Model

In this model the major factors in the development of alcoholism are seen as social factors which result in psychological dependence. Although the social model has proved more directly useful in drug addiction it has wide application in alcoholism, where such factors as socioeconomic status, ethnic origins and cultural mores are important in determining susceptibility to alcoholism. Psychological dependence may develop as an adaptive skill to aid survival in adverse social circumstances. This, then, is best extinguished through a variety of social techniques, e.g., through social rehabilitation, education and vocational training or through finding a place to live, a job, etc. Behavior modification of a social kind may be used to extinguish the learned response constituting psychological dependence; peers' social approval may act as a powerful reward for non-drinking and social disapproval as a powerful but constructive punishment for drinking. The treatment system includes establishment of facilities and services to meet alcoholics' varied socioeconomic needs. The value of social rehabilitation may be visibly demonstrated by the use of ex-alcoholic counselors; the success of A.A. and its very widespread incorporation into the treatment system may be largely explained by the social model. Recovering alcoholic counselors and A.A. members may be seen as best equipped to deal with some of the problems of alcoholics because they are successful and can serve as examples to reinforce socially acceptable behavior. Used by itself, however, the social model of treatment suffers from lack of adequate involvement in psychological origins of increased dependence as well as ignorance of the facts of physical dependence.

The Alcoholics Anonymous Model

The organization and history of Alcoholics Anonymous is discussed by Milton Maxwell elsewhere in this volume; however, its unique approach to treatment merits a brief description here.

The last chapter in *Alcohol, Science and Society* is "The Fellowship of Alcoholics Anonymous" by W. W., better known as Bill W., one of the two founders of A.A. (8). The historic meeting with Dr. Bob had taken place only nine years before the lecture, but by 1945 the main elements were well developed. The concept of alcoholism as a disease, different from that of the medical model, considered alcoholism as "a sort of allergy of the body that guarantees that we shall die if we drink, an obsession of the mind which guarantees that we shall go on drinking." Many of the "Twelve Steps" of A.A. which Bill W. based on the Oxford Groups' emphasis on the principles of self-survey, confession, restitution and the giving of oneself in service to others had been enunciated earlier in the basic textbook, *Alcoholics Anonymous* (9). From the beginning the A.A. fellowship required its members to admit their powerlessness over alcohol, vow abstinence, and commit their lives to helping other alcoholics. A.A. insisted that alcoholics must abstain from alcohol, but did not prohibit alcohol to the majority who did not suffer from the disease of alcoholism.

The desire to stop drinking is the only qualification for A.A. membership. Would-be members may attend open A.A. meetings or contact A.A. in an emergency, when two members are sent as a crisis team to establish rapport, provide emotional support, help the problem drinker define his problem and clarify alternative courses of action. As in formal A.A. meetings the crisis team members may talk about their own personal experiences to illustrate their doubts at the beginning of their membership and how they were resolved. The potential member may identify with his rescuers and feel less unique and hopeless. He is given information on how to contact A.A. again.

Meeting A.A. members who are "recovered" provides the opportunity for identifying with others who have successfully dealt with similar problems and, as well as providing role models to be emulated, the new interpersonal relationships provide a gratification to replace that of alcohol. When a problem drinker joins A.A. he chooses a sponsor who undertakes voluntarily to work with him; the sponsor believes that this effort furthers his own sobriety. A.A. groups seem to be most successful when they are socially and culturally homogeneous. A free and open exchange of confidential material is easiest in such a setting, and group meetings permitting and encouraging such interchange are the primary therapeutic tools of A.A. A.A. provides fellowship wherever members may be; members must keep in touch with the organization when they travel. The program avoids feelings of guilt over personal dependence by arranging that whenever a member calls for help a different person is likely to answer. Help is always available, no matter how harshly the alcoholic judges himself or is judged by others. A.A. emphasizes that an alcoholic is always recovering but never recovered. The A.A. member is always potentially a drinker, however long he has been sober.

The Multivariant Model: Selecting the Appropriate Treatment

It is apparent that there are many models of treatment of the alcoholic, and a treatment facility may provide modalities of treatment belonging to different models, simultaneously or consecutively. Moreover, there is little evidence that any one approach is more effective than any other when applied globally to the heterogeneous population of alcoholics as they appear after recovery from withdrawal. The multivariant model rejects the traditional concept of a unitary syndrome of alcoholism: that all persons afflicted are substantially the same, experience a similar progressive deterioration, and respond to a singular treatment which results in one specific outcome, abstinence. Pattison (10), a leading exponent of the multivariant model, conceptualizes it thus: "What alcoholism syndromes at which stage of their development and in what kinds of patients respond under what conditions in what short—and long—range ways to what measures administered by whom?" The model implies that alcohol dependence subsumes a variety of syndromes, that an individual's use of alcohol can be considered on a continuum from nonuse, to nonproblem drinking, to various degrees of harmful drinking; that the development of alcohol problems follows various patterns over time; that abstinence bears no necessary relation to rehabilitation; and that psychological and physical dependence are separate and not necessarily related. Continued drinking of large doses of alcohol over an extended period of time is likely to initiate a process of physical dependence; i.e., no specific or unique metabolic characteristic is necessary (11). The state of physical dependence may change from time to time. Since the population of individuals with alcohol problems is multivariant, treatment intervention must also be multivariant, addressing the severity of alcohol use, the particular problems and consequences of drinking, and the patient's ability to achieve specific treatment goals. The model recognizes that alcohol problems are usually related to other life problems, so that rehabilitation should aim at specific changes in drinking behavior suitable for each individual, and also at specific changes in problem areas of life function. Emphasis should be placed on treatment procedures that relate to the drinking environment; these may include temporary or permanent changes of environment and study of the individual's interactions with his environment leading to planned interventions with family, friends and others in his social network.

Essential elements in treatment and rehabilitation services providing continuity of care over an extended period are: effective identification of alcoholics, referral mechanisms in the early phases, immediate and long-term treatment services and follow-up aftercare. Evaluation of treatment should take into account initial disability, potential for change, and individual dysfunction in various life areas. Pattison suggests dividing the individual's total adaptation into five parts: drinking health, emotional health, interpersonal health, vocational health and physical health.

At the beginning of this chapter I noted that, broadly speaking, the treatments available today have not made the treatments of 1945 obsolete, and in fact our modern treatments existed either in practice or in concept 35 years ago. The major change in treatment emphasis is that we have moved toward an attempt to match the patient with specific and appropriate treatment facilities, personnel and methods.

At the present time the multivariant model appears to provide the most adequate conceptual approach to alcoholism treatment, and it is likely that research based on this model will provide the framework for selecting "the appropriate treatment for the appropriate patient."

REFERENCES

1. FLEMING, R. Medical treatment of the inebriate. Pp. 387–401. In: Alcohol, science and society; twenty-nine lectures with discussions as given at the Yale Summer School of Alcohol Studies. New Haven; Quarterly Journal of Studies on Alcohol; 1945.

278   A. Arthur Sugerman

2. KISSIN, B. Theory and practice in the treatment of alcoholism. Pp. 1–
    51. In: KISSIN, B. and BEGLEITER, H., eds. The biology of alcohol-
    ism. Vol. 5. Treatment and rehabilitation of the chronic alcoholic.
    New York; Plenum; 1977.
3. CAUTELA, J. R. The treatment of alcoholism by covert sensitization.
    Psychotherapy, Chicago 7: 86–90, 1970.
4. MANN, M. Marty Mann answers your questions about drinking and
    alcoholism. New York; Holt, Rinehart & Winston; 1970.
5. LAZARUS, A. A. Towards the understanding and effective treatment
    of alcoholism. South Afr. med. J. 39: 736–741, 1965.
6. BLANE, H. T. Psychotherapeutic approach. Pp. 105–160. In: KISSIN,
    B. and BEGLEITER, H. The biology of alcoholism. Vol. 5. Treatment
    and rehabilitation of the chronic alcoholic. New York; Plenum; 1977.
7. ZIMBERG, S. Principles of alcoholism psychotherapy. Pp. 3–18. In:
    ZIMBERG, S., WALLACE, J. and BLUME, S. B., eds. Practical ap-
    proaches to alcoholism psychotherapy. New York; Plenum; 1978.
8. W., W. The fellowship of Alcoholics Anonymous. Pp. 461–473. In:
    Alcohol, science and society; twenty-nine lectures with discussions
    as given at the Yale Summer School of Alcohol Studies. New Haven;
    Quarterly Journal of Studies on Alcohol; 1945.
9. Alcoholics Anonymous; the story of how more than one hundred men
    and women have recovered from alcoholism. New York; Works; 1939.
10. PATTISON, E. M. The selection of treatment modalities for the alco-
    holic patient. Pp. 125–227. In: MENDELSON, J. H. and MELLO, N. K.,
    eds. The diagnosis and treatment of alcoholism. New York; McGraw-
    Hill; 1979.
11. PATTISON, E. M., SOBELL, M. B. and SOBELL, L. C. Emerging con-
    cepts of alcohol dependence. New York; Springer; 1977.

16

Human Behavioral Research on Alcoholism, with Special Emphasis on the Decade of the 1970s

Peter E. Nathan

This review of behavioral research on alcoholism between 1945 and 1980 puts most emphasis on the decade of the 1970s, when behavioral approaches reached their peak of innovation and promise. And because the pace and significance of research and clinical reports differed markedly during the first and second halves of the decade, the work during these two periods of time will be contrasted. The comparison is interesting because many of the directions that behavioral research on alcoholism took in the 1970s paralleled the investigations on other behavioral disorders by behavioral researchers.

To set the stage for this review of the veritable explosion of behavioral research and treatment of alcoholism that developed shortly after the change of decades from the 1960s to the 1970s, I will review very briefly both behavioral and nonbehavioral views on, and studies of, alcohol, alcoholics, and alcoholism before 1970.

Behavioral Research: 1945–1970

Antecedents of the 1970s' developments in behavioral assessment extended back one decade, and more than four in the treatment of alcoholism. With some rare exceptions (e.g., 1) it was not until the 1960s that controlled studies of drinking by alcoholics and nonalcoholics generally became possible. Before then the idea of giving alcohol to alcoholics, even for research purposes, met with general disfavor. Understandably, the first studies involving the administration of alcohol to alcoholics and nonalcoholics tended to focus on

This chapter is based in part on an address given at the annual meeting of the American Psychological Association in September, 1979. Support for its preparation was provided by NIAAA grant No. AA00259-10.

the effects of measured doses of ethanol on physiological, metabolic, and behavioral functioning. For example, Docter and Bernal (2), reported that alcohol led to increased heart rate, rapid eye movements during sleep, and brain alpha wave activity in alcoholics. The same investigators reported that performance on a task demanding continuous attention to an auditory signal did not worsen after the alcoholics had begun to drink. This report, like that of Talland and his co-workers published the same year (3), suggested that even relatively large amounts of alcohol do not interfere with some perceptual and motor performances by alcoholics. The early studies led to studies of prolonged drinking, first reported by Mendelson and his colleagues in 1964 (4).

The first reported behavioral treatment for alcoholism, electrical aversion conditioning, was introduced more than 50 years ago, in 1928, by Soviet physician N. V. Kantorovich (5). Kantorovich treated 40 Russian alcoholics by pairing the sight, taste, and smell of a variety of alcoholic beverages with repeated, painful, fingertip shocks. Follow-up of the treatment lasted from 3 weeks to 20 months and indicated that 70% of the patients remained abstinent. But despite the encouraging nature of the findings, it was not until the 1960s that electrical aversion was reintroduced as a prime treatment for alcoholism. During the middle portion of that decade, positive reports from several laboratories (6–9) encouraged wider use of electrical aversion in alcoholism treatment. Subsequent analyses of those reports by others, however, as well as independent disconfirmation of the efficacy of electrical aversion in inducing conditioned aversion to alcohol (e.g., 10, 11), led to a dramatic decline in its use by the end of the 1960s. Electrical aversion is now infrequently used as a treatment of alcoholism.

Chemical aversion, which employs the drug emetine or apomorphine to induce nausea and vomiting to condition aversion to the sight, taste and smell of alcoholic beverages, has achieved a more prominent place in the armamentarium of behavioral treatments of alcoholism. Inspired by Kantorovich's early success within the Pavlovian conditioning treatment paradigm, physicians Voegtlin and Lemere reported during the 1940s and 1950s (12–14) the productive use of this procedure at the Shadel Hospital in Seattle. Essentially the same treatment approach is still employed at the same Seattle hospital, as well as at 17 other alcoholism treatment hospitals scattered throughout the west (15). This expansion in use of chemical aversion accurately indicates that this behavioral treatment, in conjunction with other supportive and reeducative treatment methods, some of which are behavioral, is a treatment of choice for well-motivated patients prepared for its rigors. A more detailed analysis of this approach to treatment is provided elsewhere (16).

Thus, prior to the 1970s, behavioral assessment efforts, including early studies of drinking by alcoholics and nonalcoholics, had just begun. In treatment, however, studies of electrical and chemical aversion encouraged researchers and clinicians to believe that both were efficacious. More recent research suggests, however, that chemical aversion methods are more promising. Neither "behavioral" treatment method, however, had much effect on the behavioral treatment research undertaken during the 1970s.

View on Alcohol and Alcoholism: 1970

Alcoholics Anonymous was clearly the most popular approach to treatment in 1970. Accompanying A.A.'s dominance in treatment was the prevailing disease orientation to alcoholism which it espoused: once a person had become an alcoholic, he or she would remain an alcoholic because his or her body would always respond to alcohol as a poison. This orientation, in turn, heightened the importance of psychological constructs such as dependence, tolerance, loss of control and craving, mechanisms which automatically assume responsibility for the addictive process independent of personality traits, cognitive functioning or environment. As a consequence, the treatment goal for alcoholism could only be abstinence since, if alcoholism is a disease, one cannot presume to treat it with any other goal than abstinence from the disease-causing pathogen, ethanol.

In 1970, there had not been that much research on the genetics of alcoholism; in fact, very little was known about any causative factors. About the only alternative to the A.A. position was the psychoanalytic view of alcoholism that prevailed among many professionals (reviewed by Gomberg in this volume). Interestingly, since the psychoanalytic view of the etiology of alcoholism is also based on a disease model, it was broadly consonant with the A.A. view.

What had taken place before 1970 so far as behavioral research on alcoholism was concerned can also be summarized very briefly. Jack Mendelson and Nancy Mello and their colleagues reported in 1964 on initial studies of the effects, in alcoholics, of the prolonged administration of ethanol (4). Mendelson and Mello broke important new ground by having the persistence and vision to give alcohol to alcoholics on a continuing basis to examine the behavioral and physiological consequences. In no other way could many of the data be gathered. A few years later, a behavioral laboratory with similar goals (17) was developed so that, by the late 1960s, there were two laboratories studying the effect of alcohol on alcoholics' behavior.

The focus of my research was on the relative reinforcement value of alcohol and social interaction to alcoholics, on the quantification of drinking behavior which Mello and Mendelson had begun, and on the development and evaluation of behavioral treatment methods. Other behavioral researchers interested in applied issues, including Australians Syd Lovibond and Glenn Caddy, California researchers Mark and Linda Sobell and Roger Vogler, Peter Miller, then working at the Jackson, Mississippi, Veterans Administration Hospital, the group at Baltimore City Hospitals (Ira Liebson, Louis Faillace, Miriam Cohen, George Bigelow, and Roland Griffiths), Alan Marlatt, who first worked at the University of Wisconsin and then the University of Washington, began at the same time. So, by 1970, alcoholism research in general stood at the threshold of a new era. The new combination of augmented public interest and public funding offered enormous potential in behavioral research and in research on the genetics of alcoholism, the pathophysiology of alcohol ingestion, the mechanisms of ethanol metabolism, and the epidemiology of the disorder. Given this picture, let us now ask, How much of the behavioral research potential was realized?

Behavioral Research: 1970–1975

A great deal of the behavioral research conducted during the years 1970–1975 centered on assessment of drinking and of alcoholics' associated behavioral deficits and excesses. These areas were a natural focus because little was known about alcoholics' behavior relevant to their alcoholism.

Drinking behavior was studied in three different settings. The first was the operant ward setting. The laboratory in Boston was one of the first such laboratories (17, 18). It contained a "mini-environment," a closed, controlled living situation. Alcoholics brought into the laboratory each had their own bedroom, shared a large living area, and took all their meals in a separate dining area. The subjects had to work to earn points which could be used to purchase alcohol or, if desired, other reinforcers including the opportunity to leave individual living spaces to interact with other patients and ward staff. The work involved repetitive small motor behaviors—pushing a button, pulling a knob, or passing a finger through an electric eye-beam. The design of the research permitted quantification of the relative reinforcement value, at any given moment, of alcohol, people, and other available reinforcers. We found that shortly after the alcoholics began to drink, alcohol was immensely reinforcing; nothing else compared to the reinforcement value of ethanol at that time (17, 18). But when the alcoholics were not drinking, the opportunity to be with other people was quite reinforcing. Alcoholism, as we observed it, entailed an isolation paralleled only by the isolation of the schizophrenic.

We also found a biphasic pattern of alcohol consumption. During an initial week or two of drinking, the blood alcohol concentration (BAC) averaged 0.30–0.40% (a *very* high level of consumption). Then our subjects, quite spontaneously, would reduce their consumption markedly so that BACs would return to and remain at around 0.10–0.15% for weeks on end. This pattern of consumption, observed as well by Mello (19), appeared to recreate the pattern of binge drinking seen outside the laboratory.

At about the same time another research group at Baltimore City Hospitals contrasted the relative reinforcement value of alcohol and people in a different way (20). They reported that alcoholics would choose to drink moderately to live in a socially-enriched environment rather than to drink immoderately and live in an environmentally-impoverished setting. The results of this research also questioned the traditional view that alcoholics were incapable of moderating their drinking.

Concurrently, another group of researchers established an experimental bar in a hospital ward setting to examine styles of drinking (21). Schaefer and his colleagues at the Patton (California) State Hospital observed that alcoholics tended to gulp their drinks, avoid using ice or mixers, and leave very little or nothing in their bottles or glass when they were done. In contrast, social drinkers used ice and mixers more often, sipped their drinks, and left something in the glass or bottle. While the results surprised no one, the findings did confirm tangible differences in drinking per se by alcoholics and nonalcoholics, differences substantial enough to lead some observers to suggest that the most direct way to cure alcoholism was to induce alcoholics to add ice to mixed drinks! While this early "radical-behavioral" view of treatment has now been largely discounted by more sophisticated views, the assessment on which it was based nonetheless represents a milestone in behavioral research on alcoholism.

A third approach to the assessment of drinking was by the way of "taste tests" by Peter Miller (22) at the Jackson, Mississippi, Veterans Hospital and Alan Marlatt (23, 24) at the University of Washington. These studies imposed a believable deception on the alcoholics and nonalcoholics who came to a psychology laboratory "to rate alcoholic beverages along several taste dimensions." Actually, the researchers were measuring how much the subjects drank. This type of research permitted study of variables which may affect alcohol consumption, such as stress, failure, anger and frustration. The taste test appeared to be a reasonably valid way to present these antecedents of drinking in a controlled manner, and to study their effects on drinking itself.

These efforts to assess alcoholics' behavior and drinking were of value, first, because many of the results suggested convincingly that alcoholics can moderate their drinking if the environment supports moderation.

Almost as important was the conclusion, based on empirical observation, that the drinking of alcoholics differs on both macro and micro dimensions from that of nonalcoholics. The importance of this finding lies in its potential to provide empirically-derived bases for distinguishing between alcoholics and nonalcoholics as well as between alcoholics whose dependencies differ in severity and chronicity. For example, comparisons made around this time in our laboratory between men and women drinkers and alcoholics (25) showed the women alcoholics to resemble men heavy social drinkers more than they resembled men alcoholics, both in their drinking and in their finding social interaction as reinforcing as alcohol, even during periods of heavy drinking. We concluded that the drinking of women alcoholics appears to be under better stimulus control, even during its most intense stages, than that of men.

These techniques may have a potential for identifying alcoholics and problem drinkers who are most likely to benefit from behavioral treatment and to differentiate them from those who probably will not. Data from one taste test study (22) indicated that alcoholics in treatment who drank most or all of the beverage alcohol available to them in the taste test situation were less likely to benefit from behavioral treatment than alcoholics who moderated their taste test drinking.

Another interesting research venture of this era involved assessment of consumption in the natural environment. Researchers started going into bars and taverns equipped with increasingly sophisticated coding techniques to record the amount of drinking by patrons, the circumstances of their drinking, what they were drinking, and who they were drinking with, all in an effort to determine whether the drinking behavior observed in the laboratory and in the natural environment were comparable (26, 27). We have recently conducted research of this kind at a local university drinking establishment (28), the results of which remind us again of the power of environmental stimuli in determining drinking behavior. When a male college student drinks with a woman, he drinks less than when he drinks with a man. Conversely, when a female student drinks with a man, she drinks more than when she is with a woman. When men drink in groups, they drink more than when they are in dyads; when women drink in dyads, they drink more than when they are in groups. This research emphasizes again the powerful role of social determinants of drinking.

Still another contribution of assessment studies was a gradual increase in the sophistication of behavioral views of the etiology of alcoholism. Until the early 1970s the most widely-accepted behavioral view of the etiology of alcoholism was a simple one. Derived from Conger's shock avoidance research with rats (29), it assumed that alcoholismic drinking is a learned means to reduce conditioned anxiety; necessary assumptions were that alcoholics are more anxious than nonalcoholics and that they have learned that alcohol reduces anxiety. Behavioral research in the early 1970s created problems for this theory, i.e., alcoholics actually become more, rather than less, anxious when they drink (4). Several investigators pointed out that men and women, unlike the rats that Conger studied, think and remember and imagine.

An appropriately complex behavioral theory of alcoholism etiology has replaced this theory. The theory attributes partial responsibility for the development of alcoholism to Pavlovian mechanisms, the same learning elements central to the anxiety-reduction model. But additional components of the theory point to operant mechanisms as well as to modeling and vicarious reinforcement to account for development of alcoholismic behavior; these mechanisms attribute responsibility to peer interaction during teen-age years as well as to adult social pressures as factors in etiology.

Our etiologic research, begun in 1972, suggests that physiological and behavioral factors may well interact to cause alcoholism. We have found alcoholics to be much less able than nonalcoholics to judge level of intoxication on the basis of internal cues (30). We hypothesize that this relative inability stems from a genetic predisposition to rapid development of tolerance to small amounts of alcohol; we have also reported the observation that tolerance directly influences alcoholics' estimation of blood alcohol concentrations. Inability to detect relative changes in intoxication, of course, exerts a direct and immediate consequence on drinking: If you don't know how drunk you are, if you don't suffer the untoward consequences of drinking, or both, you are likely to drink more than you would if drinking produces negative effects. We conclude that the most accurate view of the etiology of alcoholism combines environmental and behavioral factors with genetic–physiologic ones.

Another important contribution during the early 1970s was the empirical testing of "craving" and "loss of control," the two bulwarks of the disease model of alcoholism. According to that model, all nondrinking alcoholics crave ethanol. Furthermore, when alcoholics do drink, they quickly and irresistably lose control over their drinking with the result that they are likely to drink as much as they can for as long as they can. Many investigators during the early 1970s—one of the first was Henry Cutter (31)—reported that alcoholics who were unaware that they had consumed alcohol showed no change in self-reported levels of craving and no increase in tendency towards loss of control, despite their belief in both phenomena. Viewed another way, believing in the inevitability of craving and loss of control may play a greater role than consuming alcohol and experiencing its effects in accounting for self-reports of those behaviors, believed previously to be involuntary and drug-induced.

Research by Terry Wilson through the 1970s (32) has expanded on these data to suggest that expectancy and attitudinal effects play important roles in determining alcohol's influence on a wide range of behaviors. Research begun in the early 1970s has convinced many scientists that expectancy bulks large in explaining how people act after drinking.

Let us turn to consideration of a different research enterprise, that of treatment outcome. Without question, the most controversial research was that which led to questioning of traditional goals of treatment for alcoholism. During the 1970s, publications by Lovibond and Caddy (33), the Sobells (34, 35) and Vogler and colleagues (36) questioned whether abstinence ought to be the only goal of treatment for alcoholism. While an extended discussion of the merits of this question is inappropriate here, it is appropriate to note that (1) the question was raised, and (2) the data justified asking the question. The findings, in essence, suggested that significant numbers of problem drinkers and a lesser number of alcoholics, either following treatment or spontaneously, developed and maintained patterns of nonproblem drinking for varying lengths of time. (I do not agree to treat alcoholics who wish nonproblem drinking to be a treatment goal. In the final analysis, a nonabstinent treatment goal is probably not cost-effective with persons whose drinking has never been under effective personal control.)

My own views on the matter notwithstanding, the process by which the question of alternative treatment goals was raised was, on the whole, salutory. It established the right of scientists and professionals to question, and it generated efforts, many of them behavioral, to design innovative treatment methods that could support a nonproblem drinking outcome. Although some critics have claimed that many alcoholics have suffered because the question was raised, I have seen no data which support such claims.

Among the new treatment approaches developed during the early 1970s were some which focused on the associated behavioral deficits and excesses that typically accompany the development of alcoholism. Programs were developed to modify stress and teach stress management, to teach assertiveness in situations in which the patient might wish to tell others he or she did not wish to drink, and to alter vocational ineptitude and peer, social, and familial incompetence. The assumption was that helping alcoholics develop skills to deal more effectively with a spouse, co-workers, or employer, would reduce the chances of a return to excessive drinking.

At the same time there was a general movement by the treatment community away from a focus on the needs of Skid Row-type alcoholics towards those of problem drinkers and middle-class alcoholics, said to represent the bulk of this country's alcoholics. To this end, clinicians like Roger Vogler (36), Ovide Pomerleau (37) and Arthur Wiens (15) and their co-workers began to study the efficacy of behavioral treatment packages developed specifically for individuals who possessed more effective social and vocational skills and retained more ties to the community than the individuals on whom treatment had been focused previously.

In response to the new thoughts on alcohol and alcoholism which characterized the 1970s rising from behavioral research, some social scientists, life scientists, the public and other adherents of traditional views maintained strong opinions about what the behavioral researchers and clinicians were reporting. Some strongly condemned behavioral methods because they were said to remove responsibility for change from the patient to the therapist, whose treatment methods appeared to be so potent that they worked without the patient's cooperation. Others attacked all behavioral clinicians because some espoused controlled drinking treatment goals.

Behavioral Research: 1975–1980

Unhappily, the promise, excitement and controversy has given way to disappointment, to a few elaborations of earlier findings, to the fulfillment of only a small portion of early promise, and to continued simmering of the controlled drinking controversy.

What has been fulfilled during this time? To begin with, the work on expectancy effects begun in the early 1970s has blossomed. Reflecting behavioral psychologists' newly-developed recognition of the influence of cognitive factors on behavior, the research now reveals that alcohol's impact on mood, aggression, sexual arousal, tension, and pain perception is subject to robust and predictable expectancy effects (32). At moderate BACs (up to 0.10% or so), the pharmacological effects of alcohol are relatively inconsequential while the expectancy effects are quite powerful. Facilitating a clear view of alcohol's expectancy effects was the application, in the middle of the decade, of a research design that permits direct comparison of the drug and expectancy effects of alcohol. The "balanced placebo" design requires half the members of a group of subjects to consume alcohol and the other half to ingest a placebo. Half of those in each group are told that they will be given alcohol to drink while the other half are informed that they are going to receive placebo. The design permits an unencumbered view of the relative effect of drug and expectancy on behavior. If subjects believe the deception (one must do a careful post-experimental inquiry to determine whether they have), their behavior will reflect how they believe alcohol affects them even if they did not consume alcohol but were told that they would.

One of the few exciting developments in the assessment area during the past five years has been the publication and subsequent validation of comprehensive rating scales designed to capture treatment climate at out- and inpatient treatment facilities. Rudolph Moos at Stanford Medical School is most closely identified with this innovative research. Unlike most other instruments designed to predict treatment outcome from therapist's or patient's personality patterns, these scales focus on more enduring and robust factors which change when important elements of the environment change, and appear to differentiate among treatment facilities on the basis of interpersonal climate, a variable predictive of treatment outcome (38).

The past five years have also seen a heightening of awareness of the impact of modeling processes on drinking behavior. Modeling exerts its influence at two points—early in an individual's drinking history, when modeling of peer or parental drinking may play a role in the development of alcoholism, and later, when a specific drinking episode may be influenced by a heavy- or light-drinking companion. Caudill and Marlatt (39) observed that persons drinking with a heavy-drinking confederate were apt to drink more than when they were with a light-drinking confederate. Other research confirming the impact of modeling on drinking has been reported more recently.

In the treatment arena, Marlatt and his colleagues have developed an interesting approach which anticipates patients' discouragement at the difficulties involved in maintaining total abstinence (40). According to Marlatt, patients taking a drink following a period of abstinence may lose heart altogether and immediately resume a pattern of excessive drinking. In anticipation of this common phenomenon, Marlatt et al. developed a method which "programs," either in the patient's imagination or in fact, a "slip" from which the patient can quickly recover to resume a pattern of abstinence. Though still awaiting empirical confirmation, this approach, by anticipating a major pitfall of treatment, may prevent disruption of otherwise successful treatment for many patients.

Unhappily, beyond the above research, the last five years have seen little of the innovation in theory and practice that characterized the first five years of the decade. Instead, accompanying these few original contributions have been a series of elaborations on earlier themes. Continued research on craving and loss of control confirmed that they relate in large part to expectancy. There has been an elaboration in the design of multifaceted treatment programs containing a variety of different treatment techniques; one introduced social skills assessment and training to existing treatment programs (41). Other treatment evaluations have weighed the differential impact of one or another sets of behavioral treatment on drinking, research not undertaken during the early years of the decade because all energies were directed toward producing an over-all effect on drinking via massive behavioral intervention.

Still awaiting development are studies on the differential outcome of abstinence-oriented and controlled-drinking-oriented treatments, research to identify predictors of successful therapeutic outcome following treatments with differing treatment goals, studies of A.A. designed to focus on the active ingredients of this successful treatment mode, and inquiry into the respective contributions in- and out-patient treatment make to successful outcome.

Why have none of these studies been undertaken? One reason was that federal support for behavioral research lessened during the latter half of the decade. Another reason, related to the first, was the severe criticism behavioral researchers had received for espousing nonproblem drinking goals of treatment even though, in fact, many researchers had not joined their colleagues in this espousal. Finally, both researchers and funding agencies, observing the failure of behavioral theory and practice to "deliver" on its high promise of the early 1970s, became disenchanted.

The most notable—and controversial—psychosocial research

event of the 1975–1980 era was the study by the Rand Corporation (42, 43). An investigation of the fate of patients seen at federally funded alcoholism treatment centers during the early 1970s, the Rand report on 6- and 18-month treatment outcome (42) was both interesting and surprising. To my way of thinking, one of the most surprising findings was that 70% of patients followed up at 18 months (unfortunately, quite a small number) had shown improvement in drinking behavior, in life circumstance, or in both. Another surprising finding not well-received by many treatment personnel was that, at both the 6- and 18-month follow-ups, as long as patients had remained in treatment, they were apt to get better, regardless of the kind of treatment they received. Depending on one's point of view, that finding was either good or bad. Those strongly committed to a particular approach to treatment, for example, probably did not savor the finding very much.

The aspect of the 1976 Rand report receiving the greatest attention was the observation, made as well in the 1980 4-year follow-up (43), that a significant number of alcoholics had developed and maintained a so-called nonproblem drinking pattern. The 1980 report also cited data demonstrating that nonproblem drinking was not associated with either more psychopathology or a greater risk of return to problem drinking than abstinence.

Both reports were roundly condemned by those who reject the possibility of nonproblem drinking by alcoholics and by those who hold to the primacy of a particular mode of treatment for alcoholism. Both studies were applauded by persons who saw in them the chance to use empirical data in the process of formulating policies on alcoholism control and treatment.

Another notable event of the last half of the decade was a report by Griffith Edwards and Jim Orford (44) of their treatment of 100 men alcoholics. The study, categorized by some as a "tongue-in-cheek" effort, contrasted the effects of a very intensive treatment program lasting several months with a single counseling session attended by a patient, his wife, and the research social worker, psychologist and psychiatrist. At this session, a series of structured questions were asked, after which the psychiatrist told the patient that he was suffering from alcoholism, that his treatment goal should be total abstinence, that he should continue or return to work, and that he and his wife should work to improve their marriage. They were also told that they held responsibility for the husband's drinking in their own hands. No difference in treatment efficacy was revealed, which disturbed many clinicians. The study was, however, sufficiently idiosyncratic to the English treatment scene that generalization of its results is problematic.

Concluding Remarks

Will behavioral approaches to alcoholism experience a renaissance in this new decade? Did the interregnum of 1975–1980 prepare behavioral clinicians and researchers for new, innovative attacks on the problems of alcoholism? Or have the behavioral methods had their try at solving the age-old problem of alcoholism and, like those which came before them, faded from the treatment scene? My own admittedly wishful response to these questions is a prediction that behavioral approaches to both the assessment and treatment of alcoholism will continue to have an important place in the armamentarium of efforts to confront alcohol problems.

REFERENCES

1. ISBELL, H., FRASER, H. F., WIKLER, A., BELLEVILLE, R. E. and EISEN-MAN, A. J. An experimental study of the etiology of "rum fits" and delirium tremens. Quart. J. Stud. Alc. **16:** 1–33, 1955.
2. DOCTER, R. F. and BERNAL, M. E. Immediate and prolonged psychophysiological effects of sustained alcohol intake in alcoholics. Quart. J. Stud. Alc. **25:** 438–450, 1964.
3. TALLAND, G. A., MENDELSON, J. H. and RYACK, P. Experimentally induced chronic intoxication and withdrawal in alcoholics. Pt. 4. Tests of motor skills. Quart. J. Stud. Alc., Suppl. No. 2, pp. 53–73, 1964.
4. MENDELSON, J. H., ed. Experimentally induced chronic intoxication and withdrawal in alcoholics. Quart. J. Stud. Alc., Suppl. No. 2, 1964.
5. KANTOROVICH, N. V. (An attempt at associative-reflex therapy in alcoholism.) [Russian text.] Nov. Refleksol. Fiz. nerv. Sist. **3:** 436–447, 1929.
6. BLAKE, B. G. The application of behavior therapy to the treatment of alcoholism. Behav. Res. Ther., Oxford **3:** 75–85, 1965.
7. BLAKE, B. G. A follow-up of alcoholics treated by behavior therapy. Behav. Res. Ther., Oxford **5:** 89–94, 1967.
8. MacCULLOCH, M. J., FELDMAN, M. P., ORFORD, J. F. and MacCUL-LOCH, M. L. Anticipatory avoidance learning in the treatment of alcoholism; a record of therapeutic failure. Behav. Res. Ther., Oxford **4:** 187–196, 1966.
9. McGUIRE, R. J. and VALLANCE, M. Aversion therapy by electric shock; a simple technique. Brit. med. J. **1:** 151–152, 1964.
10. NATHAN, P. E. Alcoholism. Pp. 3–44. In: LEITENBERG, H., ed. Behavior modification and behavior therapy. Englewood Cliffs, N.J.; Prentice-Hall; 1976.
11. WILSON, G. T., LEAF, R. and NATHAN, P. E. The aversive control of excessive drinking by chronic alcoholics in the laboratory setting. J. appl. Behav. Anal. **8:** 13–26, 1975.

12. LEMERE, F. and VOEGTLIN, W. L. An evaluation of the aversion treatment of alcoholism. Quart. J. Stud. Alc. **11**: 199–204, 1950.

13. VOEGTLIN, W. L. The treatment of alcoholism by establishing a conditioned reflex. Amer. J. med. Sci. **199**: 802–810, 1940.

14. VOEGTLIN, W. L. Conditioned reflex therapy of chronic alcoholism; ten years' experience with the method. Rocky Mtn med. J. **44**: 807–812, 1947.

15. WIENS, A. N., MONTAGUE, J. R., MANAUGH, T. S. and ENGLISH, C. J. Pharmacological aversive counterconditioning to alcohol in a private hospital; one year follow-up. J. Stud. Alc. **37**: 1320–1324, 1976.

16. NATHAN, P. E. and LIPSCOMB, T. R. Behavior therapy and behavior modification in the treatment of alcoholism. Pp. 305–357. In: MENDELSON, J. H. and MELLO, N. K., eds. The diagnosis and treatment of alcoholism. New York; McGraw-Hill; 1979.

17. NATHAN, P. E., TITLER, N. A., LOWENSTEIN, L. M., SOLOMON, P. and ROSSI, A. M. Behavioral analysis of chronic alcoholism; interaction of alcohol and human contact. Arch gen. Psychiat. **22**: 419–430, 1970.

18. NATHAN, P. E. and O'BRIEN, J. S. An experimental analysis of the behavior of alcoholics and nonalcoholics during prolonged experimental drinking: a necessary precursor of behavior therapy? Behav. Ther., N.Y. **2**: 455–476, 1971.

19. MELLO, N. K. Behavioral studies of alcoholism. Pp. 219–291. In: KISSIN, B. and BEGLEITER, H., eds. The biology of alcoholism. Vol. 2. Physiology and behavior. New York; Plenum; 1972.

20. COHEN, M., LIEBSON, I. A., FAILLACE, L. A. and ALLEN, R. P. Moderate drinking by chronic alcoholics; a schedule-dependent phenomenon. J. nerv. ment. Dis. **153**: 434–444, 1971.

21. SCHAEFER, H. H., SOBELL, M. B. and MILLS, K. C. Baseline drinking behavior in alcoholics and social drinkers; kinds of drinks and sip magnitude. Behav. Res. Ther., Oxford **9**: 23–27, 1971.

22. MILLER, P. M., HERSEN, M., EISLER, R. M. and ELKIN, T. E. A retrospective analysis of alcohol consumption on laboratory tasks as related to therapeutic outcome. Behav. Res. Ther., Oxford **12**: 73–76, 1974.

23. MARLATT, G. A., DEMMING, B. and REID, J. B. Loss of control drinking in alcoholics; an experimental analogue. J. abnorm. Psychol. **81**: 233–241, 1973.

24. MARLATT, G. A., KOSTURN, C. F. and LANG, A. R. Provocation to anger and opportunity for retaliation as determinants of alcohol consumption in social drinkers. J. abnorm. Psychol. **84**: 652–659, 1975.

25. TRACEY, D. A. and NATHAN, P. E. Behavioral analysis of chronic alcoholism in four women. J. cons. clin. Psychol. **44**: 832–842, 1976.

26. CUTLER, R. E. and STORM, T. Observational study of alcohol consumption in natural settings; the Vancouver beer parlor. J. Stud. Alc. **36**: 1173–1183, 1975.

27. KESSLER, M. and GOMBERG, C. Observations of barroom drinking; methodology and preliminary results. Quart. J. Stud. Alc. **35**: 1392–1396, 1974.

28. ROSENBLUTH, J., NATHAN, P. E. and LAWSON, D. M. Environmental influences on drinking by college students in a college pub; behavioral observation in the natural environment. Addict. Behav., Oxford **3:** 117–121, 1978.

29. CONGER, J. J. The effects of alcohol on conflict behavior in the albino rat. Quart. J. Stud. Alc. **12:** 1–29, 1951.

30. NATHAN, P. E. Studies in blood alcohol level discrimination. Pp. 161–175. In: NATHAN, P. E., MARLATT, G. A. and LØBERG, T., eds. Alcoholism; new directions in behavioral research and treatment. (NATO Conference Series: III, Human factors; v. 7.) New York; Plenum; 1978.

31. CUTTER, H. S. G., SCHWAAB, E. L., JR. and NATHAN, P. E. Effects of alcohol on its utility for alcoholics and nonalcoholics. Quart. J. Stud. Alc. **31:** 369–378, 1970.

32. WILSON, G. T. Booze, beliefs, and behavior; cognitive processes in alcohol use and abuse. Pp. 315–339. In: NATHAN, P. E., MARLATT, G. A. and LØBERG, T., eds. Alcoholism; new directions in behavioral research and treatment. (NATO Conference Series: III, Human factors, v. 7.) New York; Plenum; 1978.

33. LOVIBOND, S. H. and CADDY, G. Discriminated aversive control in the moderation of alcoholics' drinking behavior. Behav. Ther., N.Y. **1:** 437–444, 1970.

34. SOBELL, M. B. and SOBELL, L. C. Alcoholics treated by individualized behavior therapy; one year treatment outcome. Behav. Res. Ther., Oxford **11:** 599–618, 1973.

35. SOBELL, M. B. and SOBELL, L. C. Individualized behavior therapy for alcoholics. Behav. Ther., N.Y. **4:** 49–72, 1973.

36. VOGLER, R. E., COMPTON, J. V. and WEISSBACH, T. A. Integrated behavior change techniques for alcoholics. J. cons. clin. Psychol. **43:** 233–243, 1975.

37. POMERLEAU, O., PERTSCHUK, M. and STINNETT, J. A critical examination of some current assumptions in the treatment of alcoholism. J. Stud. Alc. **37:** 849–867, 1976.

38. BROMET, E. J., MOOS, R. H. and BLISS, F. The social climate of alcoholism treatment programs. Arch. gen. Psychiat. **33:** 910–916, 1976.

39. CAUDILL, B. D. and MARLATT, G. A. Modeling influences in social drinking; an experimental analogue. J. cons. clin. Psychol. **43:** 405–415, 1975.

40. MARLATT, G. A. Craving for alcohol, loss of control, and relapse; a cognitive-behavioral analysis. Pp. 271–314. In: NATHAN, P. E., MARLATT, G. A. and LØBERG, T., eds. Alcoholism; new directions in behavioral research and treatment. (NATO Conference Series: III, Human factors, v. 7.) New York; Plenum; 1978.

41. EISLER, R. M. Assessment of social skills deficits. In: HERSEN, M. and BELLACK, A. S., eds. Behavioral assessment; a practical handbook. New York; Pergamon; 1976.

42. ARMOR, D. J., POLICH, J. M. and STAMBUL, H. B. Alcoholism and treatment. Prepared for the National Institute on Alcohol Abuse and Alcoholism. Santa Monica; Rand Corporation; 1976.

43. POLICH, J. M., ARMOR, D. J. and BRAIKER, H. B. The course of alcoholism; four years after treatment. Prepared for the National Institute on Alcohol Abuse and Alcoholism. Santa Monica; Rand Corporation; 1980.

44. EDWARDS, G., ORFORD, J., EGERT, S., GUTHRIE, S., HAWKER, A., HENSMAN, C., MITCHESON, M., OPPENHEIMER, E. and TAYLOR, C. Alcoholism; a controlled trial of "treatment" and "advice." J. Stud. Alc. 38: 1004–1031, 1977.

17

Alcoholics Anonymous

Milton A. Maxwell

Alcoholics Anonymous was still very young when cofounder Bill (William G. Wilson) gave his talk on A.A. at the 1944 Summer School of Alcohol Studies. It had been only five years before, when they could barely count 100 sober members, that they had ambitiously published a book (1) which pulled together their pioneering recovery experience. Included also were the alcoholism and recovery stories of Bill and Dr. Bob (Robert L. Smith, M.D.), the Akron cofounder, plus those of 27 others. The title selected, *Alcoholics Anonymous*, finally gave the infant fellowship a name.

Slowly, word about the book—and the reality of individual recoveries from alcoholism—began getting around. Eighteen months later, there were 22 groups in 14 states and the District of Columbia. The public, however, remained unaware of the fledgling fellowship until March 1941 when the story of A.A. was told to the nation in the *Saturday Evening Post*. A surge of growth followed so that by the time of Bill's Summer School talk, he estimated a membership approaching 15,000, with groups in over 360 places, including some in Canada.

That was Alcoholics Anonymous in 1944. What is A.A. like as we enter the 1980s?

Continued growth is the most obvious fact. A.A. does not keep individual membership records, but it does keep records of the groups which register (not all do) with the General Service Office in New York. These show that by the beginning of 1980 there were about 26,000 groups in the U.S. and Canada, including over 1200 groups in treatment facilities and over 1100 in correctional facilities. The total membership in the two countries was estimated to be about 800,000. (A.A. in the U.S. and Canada is considered, in every respect, one organizational entity.)

At the same time, there were at least an additional 14,000 groups in 90 other countries. While some of the 90 had only 1 or 2 groups, A.A. was well enough established and organized in 17 of them to send delegates to the 1980 biennial World Service Meeting.

The growth of A.A. is well illustrated by the sale of A.A.'s basic text (1) in English. It took 34 years (from 1939 to 1973) to sell the first million copies, but only 4½ to sell the second million. In 1979, 310,000 copies were sold. Altogether, some part of A.A. literature is printed in 13 languages, including Icelandic and Japanese.

Al-Anon Family Groups, a very similar but totally separate fellowship for spouses, family members or friends of alcoholics (whether in A.A. or not) has also experienced rapid growth. In early 1980, there were 16,500 groups world-wide, including 2300 Alateen groups of children of alcoholics.

Not only has A.A. kept growing, but it has also shown a considerable capacity to adapt and change without losing its integrity or effectiveness. Today, we have almost forgotten that A.A. began as an essentially all-male society of so-called "low bottom" (late-stage) alcoholics. Of the first 100 members, only 3 were women; and there were more than a few men who doubted whether A.A. could survive an influx of women. Yet, by 1968, when A.A. undertook its first survey of its U.S.–Canadian membership, women constituted 22% of the total sample. In the 1977 survey the proportion increased to 30%— and 32% of the members who had come into A.A. during the preceding 3 years were women.

When earlier-stage ("high-bottom") alcoholics first began to attend A.A. meetings, many old-timers resisted, believing that the new people were not "real" alcoholics and therefore did not belong in A.A. Today, middle and early-stage alcoholics are probably the majority in most A.A. groups.

There were similar feelings toward newcomers who were "too young." How could anyone under 35 be ready for A.A.? They had not yet suffered enough, had not yet been disabused of all their rationalizations. I recall that many professional workers were also skeptical. Yet A.A.'s 1977 survey showed the proportion of 35-and-under men and women to be 22%, and those 30-and-under to be 11.3%.

How much A.A. membership has changed in terms of socio-economic status it is difficult to say, for A.A. has always cast a wide net. In the 1977 survey, 40% classified themselves as business and professional persons; 25% as skilled or semi-skilled labor. But, judging by members I have met or know about, A.A. has members from every economic and status level—from the wealthy and famous to those just emerging from Skid Row. About 1500 men and women are affiliated with International Doctors in A.A. Not infrequently, professionals and executives will quietly hold meetings of their own.

As for minorities like Blacks, Indians and Hispanics, they still remain underrepresented in A.A., but in each case membership is definitely increasing. For the most part, Blacks belong to predominantly White groups. But, there are also all-Black or predominantly Black groups in every major American city, in the north and the south.

Indians were long considered to be hard to reach, not only by A.A. but by other alcoholism programs as well. Nevertheless, in early 1980, there were 130 Indian groups listed in the U.S. and Canada. Once an Indian group is formed in a given region, Indian members will help start A.A. groups in other Indian communities.

A.A. was also slow in getting underway among Spanish-speaking alcoholics. During the 1970s, however, rapid growth occurred. By 1980, there were over 300 Spanish-speaking groups in the U.S. alone and they had set up their own intergroup offices in Los Angeles, San Francisco, Chicago, Newark, New York and San Juan. Since 1972, they have also held annual Spanish-speaking national conferences.

It should also be noted that French-speaking Canadians organized A.A. groups at a much earlier date; and that 1980 began with about 900 French-speaking groups in Canada and the U.S.

Probably the most noteworthy change among A.A. members is the modification of attitudes toward members of the helping professions. A.A. as such has never taken anything but a friendly, cooperative stance with regard to the professional community. But it is true that there have been many individual members, over the years, who have held negative attitudes toward professionals in general; people who "haven't been there," who "don't understand," and who "don't know how to help." Such attitudes often stemmed from their own experience with individual professionals; but frequently the atitutdes were rooted in lay pride and over-possessive feelings about "their own" A.A. Such attitudes are now much less prevalent. A substantial change has been occurring, as documented by the same 1977 membership survey.

One section of the questionnaire dealt with "treatment or counseling *other than A.A.*: i.e., medical, psychological, spiritual, etc." which the responding members had received for their alcoholism before coming to A.A., and, also, after being in A.A.

About half (48%) of the respondents stated that they had received such help before coming to A.A., and, of these, almost two-thirds (65%) indicated that this help had "played an important part in directing them to A.A."

As for using professional help after being in A.A., 46% had done so. Moreover, of these, more than four out of five (84%) stated that the professional help had "played an important part in their recovery from alcoholism." This means that nearly 40% of the entire sample had received some important, supplementary help from some non-A.A. professional source after coming into A.A.

That this many A.A. members had received important professional help in addition to A.A. (the survey sample was composed entirely of meeting-attending members) should dispel the stereotype that decidedly negative attitudes toward the professional community are typical of A.A. members. Furthermore, a comparison of the findings with A.A.'s survey taken 3 years previously (the first in which this cluster of questions was asked) reinforces my own observation that a greater openness to professional help is definitely a trend among A.A. members. This trend reflects, I believe, better experience with professionals in treatment centers and elsewhere. The trend also reflects a recognition, on the part of a growing number of professionals, of the desirability of having their patients or clients become involved, on a continuing basis, in the kind of social environment which A.A. provides.

However, to complete our perspective, we should also note that greater openness to professional help is not evenly distributed throughout A.A. This is illustrated by a small survey I made in 1975 of 43 groups in a two-county area. I found that just over 40% of the 1115 members in these groups had been in a residential treatment center (usually 4 weeks' duration) just before or shortly after coming into A.A. This was the over-all trend, but the 43 groups varied greatly in the proportions of members who had experienced this particular type of professional exposure. In 10 groups, the proportions were 60% or higher, half of them ranging from 80 to 100%. On the other hand, 14 groups had proportions below 20%, 6 groups below 10%, including 1 group in which no member had been to a residential treatment center.

The findings portray a wide variation in such treatment center experience, even among groups within a fairly small geographic area. The findings also remind us that considerable variability is generally found at the local group level, not only with regard to such matters as treatment center exposure but also with regard to ever so many other aspects of group life. It is easy for those of us not in A.A. to be misled by A.A. group phenomena which we may observe in certain local groups or hear about from certain individual members, unless we can also become aware of the fellowship-wide patterns and trends.

In addition to more interaction between professionals and A.A. members, there are a number of other developments, entirely outside the alcoholism field, which are stimulating more professional and scientific interest in A.A., and which provide perspective for a more insightful appraisal of the dynamics in the A.A. program and social environment.

The first of these developments, in recent decades, has been the remarkable proliferation of "self-help" groups, or, as I prefer to call them, "mutual-help" groups, with programs for doing something about a wide range of personal problems: not only alcoholism but also drug addiction, compulsive gambling, mental illness, over-eating, bereavement, physical and emotional disablements of one kind or another. Instead of feeling threatened, or disdainful, more professionals are learning complementary ways of working with these groups.

A growing appreciation for supportive and therapeutic dynamics of a nonprofessional nature is found in another quarter, that of "community psychiatry" (2). Because there can never be enough direct professional help for the millions upon millions who need some degree of help with personal problems, the community psychiatry approach focuses on a greater use of the kind of support to be found in a community. This includes both the unorganized "natural support systems" of family, neighborhood and friendship circles and the organized support systems of not merely the traditional community institutions and organizations, but also the newer mutual-help groups which are primarily or entirely in the hands of nonprofessional persons. Systematic attention is being given to ways in which the helping professions can relate supportively to mutual-help support systems, such as A.A., without disturbing the lay character and control considered basic to their effectiveness.

Another development, the growing recognition of the basic factors which the various schools of psychotherapy have in common (3), is enabling more practitioners to recognize many of the same dynamic factors in nonprofessional settings such as A.A. The metaphors differ, but more professionals are coming to see that A.A. makes a lot of sense in psychotherapeutic terms and I, for one, have written along these lines (4).

Additionally, however, I have come to the view that we miss something very important about A.A. if we do not also look on it as a new society and culture. A.A. is more readily and completely understood when we perceive so-called "recovery" in A.A. to be, in large part, a socialization and enculturation process. It helps if we see A.A. as a specialized society which not only possesses a "way of life" (a culture) but also (and this is crucially important) provides an unusually favorable social environment in which to absorb the new cultural ways.

Let us recall what happens to a person in the process of

becoming alcoholic. No matter what preceded or what else happens during the "becoming" process, the person undergoes increasing changes in his social and cultural worlds. To some degree, often to a substantial degree, his attitudes and values undergo change. So do the facts and ideas held to be true. The meaning of drinking itself changes as it becomes more central to life. There are increasing changes in perception of self, of others and relationships to them, and in generalized outlook on life. In a very real sense, the "world" changes. To some degree, the person comes to live in a different social and cultural world.

Unless such changes in an alcoholic's world are recognized, and their implications grasped, words like "treatment" and even "recovery" have a limiting logic. What is needed, above all, is a new orientation: a changed set of facts, ideas and values, and changed perceptions of one's "self," of others, and of life in general.

A.A. may accordingly be perceived as a society in which such changed perceptions constitute the norms of its way of life. Furthermore, one of A.A.'s great strengths lies in the quality of its social environment: the empathic understanding, the acceptance and concern which alcoholics experience there which, along with other qualities, make it easier to interiorize the new ways of thinking, feeling and doing.

When professional helpers can view A.A. not as a single therapeutic modality but as a many-faceted, ongoing society, they can also distinguish more clearly between what they can do for an alcoholic at some point or points in his recovery–reorientation career and what A.A. can do. The various professions do have contributions to make. As one of my physician friends in A.A. emphasizes, however, what no single professional, nor all of them combined, can do is to provide their patients or clients with such a potent, continuing social environment. It appears to me that more and more professional persons are viewing the A.A. program and society in this way, not as competitive but as complementary, and frequently indispensable.

Now, what about the future of A.A.? Perhaps the best clues, in addition to the trends already noted, are to be found in internal factors which have enabled A.A. to endure so well, to date. We know that many social phenomena flourish for a time and then fade away. Few of them survive their founders. Certainly, such a loose-knit fellowship of alcoholics would seem to have been a most unlikely candidate for surviving at all.

At the time of Bill's talk, the basic ingredients of A.A.'s recovery program had been worked out and A.A. had begun to grow rapidly. This very growth, however, spawned new problems. Earlier, members of the oldest groups, often traveling business men, had helped form new groups in other cities. By word of mouth, they also transmitted some of their experiences about the principles and practices which fostered a favorable group climate for personal recovery and which also helped them to "hang together."

But with expansion, there were many new groups whose members were unaware of this collective experience. Struggling along by themselves, improvising their own group patterns, it is not surprising that many of the earlier mistakes were frequently repeated. Some power-driven or prestige-seeking "leaders" were too domineering or exploited their A.A. affiliation. (All for the good of A.A., of course). Some went ego-tripping in headlines. Borrowing from organizations they knew, some tried patterns which were not appropriate for A.A. purposes. Some set up exclusive membership rules. Sometimes money, property, and organizational difficulties disrupted groups. So did questions of leadership, authority, and power. Personal and jurisdictional rivalries were not unknown. In short, there were all kinds of deviation from the most functional patterns, thus not only diluting the recovery atmosphere within groups, but also threatening the welfare of A.A. as a whole.

The problem was not merely one of pulling together what they had already learned but also one of communicating it in an acceptable manner, as among equals. The eventual solution was a distillation of experience into a clearly formulated set of "traditions," first outlined in a series of articles by Bill in 1947 and 1948 issues of the fellowship's monthly magazine, the *A.A. Grapevine*. Citing abundant examples from their experience, Bill made a pragmatic case for the principles and practices by which A.A. members could best conduct their own group affairs and best relate to the surrounding world.

There is not space here for an adequate presentation and discussion of the traditions. Their full significance can be gained only by a study of both the short and long forms of the *Twelve Traditions*, to be found in the appendix of *Alcoholics Anonymous* (1), and in *Twelve Steps and Twelve Traditions* (5) and the experiences out of which they evolved, as detailed in *Alcoholics Anonymous Comes of Age* (6). But, they boil down to singleminded attention to A.A.'s primary purpose of recovery from alcoholism, and keeping considerations of money, property, authority, power and prestige-seeking from blurring the focus.

As for relations with the surrounding world, the A.A. name is not to be drawn into public controversy. Nor is A.A. to be affiliated with any outside entity, although A.A. may cooperate where appropriate. Individual members are not to use their A.A. identity at the level of the public media. A.A.'s public relations policy is to be based on attraction rather than promotion. There is to be no solicitation or acceptance of outside contributions; A.A. is to be fully self-support-ing. Not only does this self-support policy contribute to self-respect, but in today's alcoholism world it gives groups and the total fellow-ship an enviable freedom and independence.

The Twelve Traditions, born out of A.A.'s early collective experience, are as much a part of A.A. as the Twelve Steps. Even though individuals and groups will at times deviate from the Tra-ditions and get into trouble of one kind or another (and they do), they can always find their way again by consulting the Traditions and the experiences out of which these guiding practices and prin-ciples emerged.

A.A.'s third major asset consists of the structures through which its collective work is done and its collective decisions are made. In urban areas, the local groups will themselves operate and support a central or intergroup office to provide various services such as pub-lishing a listing of local meeting times and places; carrying a stock of A.A. literature for the groups and the public; filling requests for speakers to non-A.A. groups; and, sometimes, publishing a newslet-ter. There may be other services, but foremost is the handling of calls for help from those in trouble with drinking, their own or that of a family member or close friend. When an alcoholic is the caller, and is willing, the office arranges to have a member or two from a nearby group visit the caller and personally follow through. In smaller communities, the group or groups will at least have an A.A. number listed in the telephone directory.

A.A. in the U.S. and Canada also has a central office, called the General Service Office (GSO), located in New York (P.O. Box 459, Grand Central Station, New York, NY 10163). This is the major communications and service center, not only serving the groups in North America, but also acting as the senior service center to A.A. around the world. It serves groups directly in countries which do not yet have a working structure, but it is policy to encourage A.A. groups in other countries to develop their own national service offices and structures. Another major function is the publishing and distrib-uting of A.A.'s own books and abundant literature.

The General Service Office and the publishing activities are both managed by a nonprofit corporation, A.A. World Services, which has its own working directors but is wholly owned by the top legal body, the General Service Board. This board of trustees also owns the separate nonprofit corporation which publishes A.A.'s official magazine, the *A.A. Grapevine.*

The final organizational structure by which all U.S. and Canadian groups are tied in with the policy-making process and the conduct of fellowship-wide affairs is the unique "Conference" structure.

During A.A.'s early years, the cofounders were the chief link between the groups and the operation of so called "Headquarters." But, when Dr. Bob lay seriously ill (he died in late 1950) and Bill was reminded of his own mortality, serious thought was given to creating a structure by means of which the fellowship itself could take over from the founders and successfully guide its collective affairs. The result is a General Service Conference, meeting each April, which is composed not only of delegates from the 91 geographic "delegate areas" but also includes the trustees of the General Service Board, the directors of its two corporate entities, and the A.A. staff members of the General Service Office and the Grapevine.

The Conference is not a governing body: its purpose is the developing of consensus; its products are *advisory actions* based on informed sharing. Furthermore, the Conference operates on the general principle that no policy or position is to become an advisory action until there is "substantial unanimity." Lacking that, the issue will receive further consideration or postponement until greater agreement emerges. Once arrived at, however, advisory actions are taken seriously. They are thought of as representing the "group conscience" of the fellowship as a whole. They are not set in concrete, however, for experience may lead a subsequent Conference to modify or even nulify an earlier advisory action.

Begun on a trial basis in 1951, the Conference plan was approved by the 20th Anniversary International Convention in 1955. It has been working well ever since, and has become the model for A.A. in other countries.

In effect, the establishment of the General Service Conference completed the creation of A.A.'s three major components. Called "The Three Legacies," they are the recovery program, the guiding Traditions, and the General Service Conference with its related service structure in each of the delegate areas.

With this accomplished, Bill put into writing the story of the 20-year development in an aptly titled book: *Alcoholics Anonymous Comes of Age*. (6) This book provides a fascinating review of the historical experiences and pragmatic reasons for the development of each of the three components. This book does much to explain why A.A. has taken the shape it has.

The conduct of its own affairs was now in A.A.'s own hands. Bill continued, however, to play a useful senior statesman role. He helped in refining the service structure. He repeatedly reminded one and all of the experiences which had led to the fashioning of every aspect of A.A. He also put into writing a further exposition of principles and philosophy to provide additional clarification, emphasis or guidance. Called *The Concepts*, (7) these were adopted by the 1962 General Service Conference and are highly valued by an increasing number of thoughtful A.A. members.

When I interviewed Bill in his home, 15 months before his death in January, 1971, he struck me as a man who had not only had the satisfaction of guiding the development of a pioneering and unique mutual-help movement, but also had had the rare privilege of being able to complete his creative contributions. He had done all he could to prepare the fellowship for a future of continued strength.

What Bill, Dr. Bob and some key others helped to design *is* different in many respects. Nevertheless, I suggest that the keys to A.A.'s effectiveness and survival are to be found, in large part, in these very differences. In a larger society characterized by competitive striving for status, recognition, power, and their material symbols, Alcoholics Anonymous has a recovery program based upon opposite values—on moving toward a nonegocentric way of life.

Furthermore, A.A. has a collective life—traditions and structure—which is remarkably consonant with, and supportive of, the basic recovery program. There is no confusion of ends and means. There is singleness of purpose. There is an internal harmony of program, principles and practices which stands in striking contrast to the operations of most organizations and agencies in our society. While it is true that individuals and groups in A.A. often fall quite short of the mark, and equally true that no social organization is immune to drift and foundering, it appears to me that Alcoholics Anonymous has some unusual assets for keeping itself on course in the foreseeable future.

REFERENCES

1. ALCOHOLICS ANONYMOUS WORLD SERVICES. Alcoholics Anonymous, the story of how many thousands of men and women have recovered from alcoholism. 3d ed. New York; 1976.
2. CAPLAN, G. Support systems and community mental health. New York; Human Sciences Press; 1974.
3. YALOM, I. D. The theory and practice of group psychotherapy. 2d ed. New York; Basic; 1975.
4. MAXWELL, M. A. Alcoholics Anonymous; an interpretation. Pp. 577–585. In: PITTMAN, D. J. and SNYDER, C. R., eds. Society, culture, and drinking patterns. New York; Wiley; 1962.

5. ALCOHOLICS ANONYMOUS WORLD SERVICES. Twelve steps and twelve traditions. New York; 1953.
6. ALCOHOLICS ANONYMOUS WORLD SERVICES. Alcoholics Anonymous comes of age, by a co-founder. New York; Harper; 1957.
7. ALCOHOLICS ANONYMOUS WORLD SERVICES. The concepts. New York; 1962.

18

Alcoholism and the Family

Peter Steinglass, M.D.

Thirty-five years ago the notion that the family might play a significant role in the development and course of alcoholism generated little attention in the biomedical and psychosocial scientific communities. Activity was devoted instead to the documenting of alcoholism as a major public health problem. Early formulations about stages in alcoholism and the proposal that a set of signs and symptoms could be predictably associated with excessive alcohol use, i.e., that alcoholism was a disease (1), raised the enticing possibility that alcoholism could be treated as a biomedical condition.

Although a few argued that illnesses involving complicated behavior as well as physiological properties could not be adequately understood at a biomedical level alone—note especially the early struggles with psychosomatic concepts of disease (2)—the prevailing atmosphere was one of faith in the power of biological research, a "scientific optimism" that is still very much alive today.

It is hardly surprising that we are attracted to ideas which promise to reduce conditions such as alcoholism to a series of critical biochemical or physiological properties of the "disease-susceptible" individual. More dispassionate voices have cautioned against the naiveté associated with this approach, proposing instead that the "answer" will be as complex as the subject under study. George Engel, a pioneer of behavioral medicine, argued convincingly that a bio-psycho-social model of disease is the most appropriate one in studying complex conditions such as alcoholism (3). John Cassel alerted us to the importance of social environment in explaining the host's relative resistence to illness (4). The interest in "life events" research in large part has been in response to a desire to tease out these fascinating relationships (5).

The critical conclusion emerging from this work is that the actual development of a "disease," that is, the expression of a characteristic set of signs and symptoms, is the result of a complex series of interactions between etiological agents, host factors, and environmental factors. Simple cause and effect hypotheses no longer appear to suffice. They leave too much unexplained. It is only recently, however, with the growing interest in family medicine, and the contributions of the behavioral science disciplines of medical sociology and anthropology, that these views have been receiving serious attention.

For psychiatrists, psychologists, biomedical researchers, and even sociologists, therefore, emphasis 35 years ago was placed on the ramifications of viewing alcoholism as a disease of the individual. The conception of this individual, jointly held, by the way, by both the general public and the scientific community, was of a lower social-class male, physiologically addicted to a drug, and secondarily undergoing deterioration of psychological and social aspects of his life. Alcoholism most often came to the attention of the physician only at its "terminal" phase. Evidence of physiological addiction in such individuals was usually a prominent presenting complaint. Withdrawal symptoms were often life-threatening, and the physical consequences resulting from severe nutritional abnormalities were often devastating.

Treatment of such debilitated individuals was difficult, to say the least. It is not surprising that in such a setting a popular conception arose, especially in the medical community, that the treatment of alcoholism centered around stabilization of the patient's physiological environment, but considerable skepticism existed about the ability to meaningfully affect the patient's psychosocial environment. The phenomenal rates of recidivism associated with alcoholism tended to support this pessimistic point of view. This therapeutic nihilism reinforced the opinion that the only answer was the pursuit of the biological etiology of alcoholism.

Juxtaposed against this picture of life in the world of alcoholism 35 years ago, our current conceptions of the condition have undergone dramatic changes. Starting in the 1950s, a series of reports directed our attention to the marital relationship as a critical factor not only for an understanding of certain behavioral aspects of alcoholism, but also to alert us to the importance of including family members in the treatment process (6, 7). Survey data emerging in the 1960s and 1970s indicated that we had underestimated the extent of alcoholism as a public health problem (8, 9). Furthermore, it turned out that only a small percentage of alcoholics were homeless individuals living in the Skid Row settings of our large cities (10). The majority were living in more or less intact nuclear family environments.

Our earlier view has proven, therefore, to have been a limited one, and the family could no longer be ignored by the alcoholism treatment specialist. Speculation quickly arose that the psychological and socioeconomic impact of alcoholism on spouses and children of alcoholics was perhaps the equal to or greater than the impact on the alcoholic himself. Hence the entry into the field of both family-oriented research and family treatment techniques.

Research into Alcoholism and the Family: 1945–1980

The family is an extremely difficult entity to study. Although it is the single most influential group in a person's life, its sources of strength remains elusive. It is even difficult to identify accurately all the individuals that can reasonably be included as members of a particular family. A dizzying array of family types and family life styles have emerged over the past two decades. Previously aberrant styles of family life have joined the single-marriage, nuclear family as the norm of American middle-class life. Much has been made of the inherently stressful nature of growing up and living as an adult in such unsettled family circumstances. Efforts to isolate alcoholism as a special variable and pin down its particular contribution to the stress of modern family life is therefore a formidable task.

Even a cursory review of the field makes this point clear. A typical study has been designed around the collection of data from a group of families whose sole reason for inclusion in the study is the presence of an alcoholic in their midst. Social class, racial, cultural and ethnic variables were almost never dealt with. It is not merely that these variables were rarely systematically controlled in the design of the study: after all, it is not easy to entice families to participate as research subjects in studies, no matter what the inherent social value of the project. But investigators often seemed unaware that such variables might be worth considering. Add variables related to family composition, ages of various family members, the family as the product of a first, second, or blended marital arrangement, and the picture begins to appear almost totally hopeless. Small wonder that faced with such a task, the researcher might throw up his hands. Far better to work with alcoholics alone, measure body fluids, delve into motivations and emotional vicissitudes, probe character style and attempt to modify behavior. But in the end, the simple fact is that alcoholism is a family disorder. The family cannot be ignored simply because it is complex and difficult to study. So let us see what we have learned in the past 35 years.

General Trends in Psychosocial Family Research

Before discussing specific issues related to alcoholism and the family, it might be useful to review some general trends in family research that might help place our assessment in proper perspective. How consonant is the work in our field with family research in general? To what extent have we been able to capitalize on advances in family theory and research methods?

Basically, there have been two major traditions in family research, one associated with family sociology, the second represented by clinician–researchers from the fields of psychiatry, psychology and social work.

The sociological tradition has looked at the family as a group of interacting individuals having as its primary task the socialization of children and the perpetuation of societal values (11). The processes that contribute to these two goals have been studied by survey techniques, placing primary emphasis on such constructs as social class, community relationships, and social roles.

A pivotal event in this line of research was the series of studies carried out by Hill and his colleagues at the University of Minnesota on the family's response to stress, stimulated in large part by an attempt to understand patterns of family coping styles in dealing with the stress of World War II wartime separation and economic deprivation (12). However, although the studies surely had clinical applicability, research techniques remained very much those of traditional sociological study.

A second line of research, the clinical tradition, emerged in the 1950s and 1960s. Not surprisingly, clinical researchers were primarily interested in understanding the role of family factors in pathological processes. Normative processes took a back seat to the attempt to isolate factors in family life that might contribute to the onset and continuance of psychiatric and medical conditions (13, 14). Such conditions as schizophrenia, major affective disorders, psychosomatic illness, and addictive behavior have been intensively investigated from this perspective.

In the past decade the two lines of research have been gradually merging. Family sociologists have shown increasing interest in interactional behavior and in applying developmental concepts to the family life cycle. Clinical researchers have come to realize that their exclusive emphasis on pathological functioning has introduced a bias into their work, especially their tendency to label as abnormal behavior now thought to be part of the family's attempt to cope with stress. Gradually, therefore, each line of research has begun to recognize the relevance of the other's sphere of expertise, incorporating findings and theoretical notions, and has, on rare occasions, even designed cooperative, multidisciplinary studies.

Early Trends in Alcoholism Research

This brief history of the development of family research is very much reflected in some of the dilemmas that plagued the study of alcoholism and the family during the decades of the 1950s and 1960s. During those years the field was dominated by an intensive, and largely unnecessary, debate between sociological and clinical researchers.

The sociological researchers, best represented by the work of Joan Jackson (15, 16), examined family behavior as a response to alcoholism, following the tradition of Hill and his studies of families under stress. Much was made of the fact that the behavior of nonalcoholic spouses should be viewed not as examples of pathology, but rather as adjustment responses to the stress of living with alcoholism. Basically, the debate boiled down to the question of whether nonalcoholic family members were victims of stresses beyond their control, trying to cope in the best way that they could, or perpetrators of alcoholism, victimizing the alcoholic in the family.

The wisdom of hindsight has led us to feel that this debate was largely irrelevant (17). Differences that emerged in studies stemming from the sociological rather than the clinical tradition were largely based on asking different questions of different types of families.

We are now able to find in the literature reports of studies based on carefully thought through theoretical positions and research designs that use state-of-the-art techniques for study of the family.

Recent Research Trends

As work has become more sophisticated, interest in alcoholism and the family has centered around four major questions:

(1) What accounts for the dramatic familial nature of alcoholism (the transmission issue)?

(2) What role does the family play in the maintenance of alcoholism (the chronicity issue)?

(3) What is the impact of alcoholism on the family itself?

(4) Of what use is the family in the treatment of alcoholism?

Familial Nature of Alcoholism. Inheritance is obviously an important issue in family life. In addition to the willful passage through time of worldly goods, family generations pass on characteristics and values of critical importance in determining the behavior of subsequent generations. For the biologist, the process of transmission from one generation to another is a genetic one. For the social scientist, intergenerational transmission is also of interest. But this time what is being transmitted is not only genetic information, but family and cultural values, and patterns of organization of family life.

That alcoholism is transmitted in families has been known for a long time. Yet for many years only novelists and playwrights seemed interested in exploring the issue in any depth. In the last decade, some evidence that the occurrence of alcoholism may be determined in part by the inheritance of a genetic predisposition has appeared. Basically two standard strategies have been employed to examine this question. The first looked at incidence rates in monozygotic vs dizygotic twins; the second studied subjects raised by biological vs nonbiological parents. Both the twin studies and adoption studies have yielded interesting results. A series of studies by Goodwin and his colleagues (18–20) of Scandinavian populations have consistently demonstrated findings supportive of a genetic hypothesis for the familial prevalence of alcoholism. Although Goodwin has cautioned against over-interpretation of the results, pointing out that a heritable component for alcoholism may apply only to a specific subgroup of alcoholics, the findings suggest that a genetic component must be considered as one factor accounting for the familial occurrence of alcoholism.

The transmission issue has also been addressed in an intriguing series of studies that are the psychosocial complements to Goodwin's genetic studies. Wolin and Bennett (21, 22) studied family rituals as a vehicle for understanding why alcoholism is transmitted across generations in certain families but not in others. Using a carefully designed, semi-structured interview schedule and a conjoint interviewing format (interviewing all family members simultaneously), the investigators demonstrated that ritualized behavior in two important areas of life, dinner-time behavior and behavior around holidays, are predictive of whether transmission of alcoholism will occur. The crucial factor appears to be the extent to which alcoholismic behavior forces the family to alter significantly the nature of the organized, symbolically important ritual behaviors connected with dinner time and holidays. In families in which the unique characteristics of rituals are preserved even though alcoholismic behavior might be stressing them severely, children do not appear to develop alcoholism when they reach adult life. On the other hand, when rituals give in to the demands of alcoholism, alcoholism appears to win out in subsequent generations as well.

Family Environment as a Factor Maintaining Alcoholism. Current wisdom suggests that alcoholism is a continuum of disorders, rather than a single disease. To what extent is the family environment a factor determining which position along the continuum individuals will fall? Is the family environment capable of attenuating or exaggerating the consequences of alcoholism? Does the family play a role in subsidizing the condition? Are different family coping styles more or less successful in dealing with alcoholism? Should the family be included in treatment? Several examples of studies in this area will help give a flavor of the range of approaches that have been used to answer these questions.

One approach has been to study the interaction in "alcoholic marriages" or "alcoholic families" in laboratory settings. These studies have relied on systematic measurement of aspects of behavior generated by game-playing or problem-solving situations in a laboratory (that is, controlled) environment in an attempt to identify features unique to alcoholic families.

Every time a study has been designed to contrast alcoholic couples or families with nonalcoholic couples or families, investigators have been able to point to aspects of interactional behavior in the alcoholic group that first, are significantly different from the nonalcoholic group, and second, seem intuitively to bode poorly for the quality of family life. However, the specific behaviors are quite different from one study to another (23, 24) and the differences are not accounted for by differences in measurement technique. They are often frankly contradictory findings.

In reviewing these studies, one conclusion seems clear. The more interesting findings have emerged only when some care has been taken to tease out the independent alcoholism-related variable with the same energy addressed to the construction of the family-interaction dependent variables. When families have been grouped on the basis of the characteristics of alcoholism, a far more interesting picture has emerged than is the case when only a comparison of alcoholic and nonalcoholic families is attempted. This is not to say that these distinctions are sophisticated ones, e.g., it might be the grouping of alcoholic families into those in which the alcoholic is drinking or abstinent (25), or the addition of a third group of families containing members with a chronic illness other than alcoholism (24).

A second strategy has been an attempt to measure family environment using a questionnaire that asks family members about family values, rules, and characteristic styles of dealing with specific situations. One such instrument, the Family Environment Scale (26), is thought to tap into a series of dimensions of the social environment of the family. When applied to alcoholic families, the scale has identified aspects of this family environment presumably shared by most alcoholics, but has also indicated that the nature of the family environment is a prognostic indicator of treatment outcome. It is these links between family-level characteristics and behavior of the individual alcoholic that address most directly our questions about the relative importance of the family in shaping the development and course of alcoholism.

The most extensive series of studies (27–29) addressing the role of family environment in the maintenance of alcoholic behavior has proposed that for many families alcoholism has paradoxically come to assume a stabilizing rather than disruptive force in family life. As is perhaps true in many chronic illnesses, there is a tendency for family life to become organized around the concerns of family members in trying to deal with the consequences of alcoholism. But even more strikingly, we have found that in many families behavior associated with alcoholism proves to have "adaptive consequences" for the family. It is not only that during periods of intoxication a set of interactional behaviors are elicited that seem otherwise unavailable to the family. When family behavior was directly observed during such periods, it was found that many of the behaviors had positive consequences for the family: for example, affective expression was often heightened in families that were otherwise emotionally constricted; or physical and sexual contact increased dramatically; or the family appeared far more assertive in dealing with its external environment. This characteristic—a tendency to cycle between two distinct interactional states—helps account for the remarkable stability in the face of what might otherwise be a potentially destructive condition.

Not all alcoholic families observed responded the same way during periods of intoxication; however, each family seemed to have a characteristic set of interactions that was predictably present when its alcoholic member was drinking. One implication of this finding is that research designs which merely examine the contrast in interactional behavior between alcoholic and nonalcoholic familes would in all likelihood yield inconclusive or potentially contradictory findings.

I have also suggested that a developmental perspective be introduced in studying the alcoholic family. Since it is clear that the course of alcoholism often extends over decades, a "life history" model (30) could be applied as a way of understanding the course that alcoholism might play in the life of the family. The model suggested proposes that the alcoholic family goes through a series of stages tied to the current drinking behavior of its alcoholic member and that each stage in turn is associated with an identifiable pattern of interaction.

We have been most impressed with the power of this model both as a clinical approach and as a conceptual base for interpreting research data. For example, we have been extensively studying a cohort of families using sophisticated measurement techniques of family interaction and have been able to demonstrate that patterns of interactional behavior in three very different settings—the families' own home, the families' behavior in a multiple family discussion group, and family behavior in a laboratory setting—are clearly and statistically sensitive to what we have called the current "family alcohol phase" (31). This developmental perspective was first described by Jackson (15).

The Impact of Alcoholism on the Family. There is now substantive evidence to support the notion that alcoholic families organize significant aspects of their interactional life around the fluctuating vicissitudes of the process of alcoholism itself. Does this mean, however, that the impact or consequence of alcoholism for these families is the same in every case? Obviously not.

A substantive body of data also emerging in the last decade indicates that the differential interactional style of the family around such issues as coping behaviors, family cohesiveness, and family adaptability are predictive of the ultimate impact of alcoholism on the family. The work of Moos et al. (26) examining aspects of family environment as predictors of treatment outcome, the work of Orford and his colleagues (32) on coping styles in wives of alcoholics, and our own work (33) examining levels of symptomatology in alcoholic families are all examples of studies attempting to tease out answers to some of these questions.

Moos and his colleagues have compared data about family characteristics obtained with the Family Environment Scale (26), with treatment outcome for a large sample of alcoholic families. Three measures of family functioning proved to be highly correlated with treatment outcome as measured by alcohol consumption—cohesiveness, conflict, and family organization. No surprises were found in the direction of the correlations; that is, better treatment outcomes were found in families scoring high on cohesiveness, low on conflict, and high on organization. (Low cohesiveness and high conflict scores also correlated with high frequencies of physical symptomatology, and lower levels of social participation by family members.)

The study by Orford et al. (32) examined prospectively the relationship between aspects of marital interaction patterns, demographic factors, personality traits and subsequent treatment outcome. The critical factor proved to be a cohesion dimension which was highly predictive of treatment outcome. When their total sample of 89 couples were divided into "good," "bad," and "equivocal" treatment outcome, the ratio of couples with high cohesion to those with low cohesion in the good treatment outcome group was 22 to 6, while a reverse ratio was found among couples with bad outcome, 9 to 20. Based on the variables that made up this cohesion dimension, Orford et al.'s findings suggest that poor prognosis is associated with a marital pattern which includes affective distance, a relatively negative view voiced by wives about their husbands when husbands were sober, noninvolvement of alcoholic husbands in family tasks (as contrasted with decision-making, which did not appear to be affected), and a generally pessimistic view about the future of the marriage.

Our own work has suggested a clear-cut link between stylistic patterns of interactional behavior of family members in their own homes, and such measures of impact as degree of psychiatric symptomatology and the social and behavioral consequences of alcoholism for the family.

All this work points to the same conclusion: the family plays a critical role in determining whether alcoholism is to be circumscribed or allowed to invade family life in a more malignant fashion. In the first instance, the impact of alcoholism is often delimited to physical consequences for the alcoholic individual. In the second instance, the family's life is insidiously pervaded by the social and behavioral consequences—work and economic disruptions, diminishing social contacts, conflicts with extended family, internal dissension, and finally, incest and domestic violence.

Family Treatment Approaches to Alcoholism. Given the growing evidence that the family both plays a significant role in the maintenance of alcoholism and can influence treatment outcome, it is not surprising that family treatment techniques have been gaining in popularity over the past two decades. In recognition of this trend, the 1974 special report to Congress on alcohol and health called family therapy "the most notable current advance in the area of psychotherapy" (34, *p. 149*).

At the present stage of development, family therapy is a term used to describe a wide variety of therapeutic techniques. Murray Bowen, one of the major figures in this movement, identified at least 12 distinct therapeutic approaches as bonafide forms of family therapy (35); all have been tried at one time or another in the treatment of alcoholism. If one looks only at the published literature reporting case studies or formal outcome data, we find all manner and form of involvement of the family in the treatment process (36). These include concurrent therapy groups for alcoholics and their spouses, conjoint family therapy, multiple family group therapy, and the involvement of family members in self-help groups (Al-Anon, Alateen, etc.). Although this literature probably represents only a small fraction of programs that have systematically incorporated family therapy in treatment, a quick review gives at least a sense of what has been tried and what are some of the problems.

From a historical point of view, a report from Johns Hopkins Hospital (37) is of note because it probably represents one of the first attempts to systematically evaluate a family treatment approach to alcoholism. The approach, concurrent group therapy meetings of men alcoholics and their wives, attempted to adapt group therapy, the most successful psychotherapy of alcoholism at that time, to a family orientation. Although the results were equivocal, it remains instructive because of the range of outcome variables used by the investigators. Instead of using reduction of drinking as the sole criterion of successful or unsuccessful treatment, the research team introduced a range of psychosocial outcome variables, including the psychological status of the nonalcoholic spouse. Even in this early study, therefore, the perspective of the family therapist is clearly seen. The patient is the family, not the identified alcoholic alone, and treatment success must therefore be measured against change manifested in all family members.

A second study (38), also evaluating the use of concurrent group therapy with men alcoholics and their wives, reported two important long-term follow-up findings. First was the finding of a significantly greater persistence in therapy by those alcoholics whose wives were attending a concurrent group psychotherapy session. Second, long-term follow-up indicated significantly improved control of drinking and considerable improvement in marital harmony by men in concurrent group therapy with their wives. Since engagement in therapy has long been known to be a major stumbling block to the successful treatment of alcoholism, the indication that the involvement of family members in the treatment process significantly increased the likelihood of involving the identified alcoholic in treatment represented a finding of major importance.

The use of conjoint family therapy in the treatment of alcoholism has been reported less frequently in the literature, but no less enthusiastically. Meeks and Kelly (39) perhaps provide the clearest description in the literature of how the technique can be integrated into traditonal alcoholism treatment programs. Conjoint family interviews were instituted during the outpatient aftercare phase, and continued for a number of months in conjunction with an intensive 7-week program of individual and group psychotherapy. Therapy focused on interaction, communication, performance, and redefinition of problems in family rather than individual terms. In this sense, the approach parallels traditional conjoint family therapy techniques. Once again treatment evaluation included an assessment of family interaction and family equilibrium, as well as the assessment of drinking by the identified alcoholic.

A third approach, one that has actually received the greatest attention in the published literature, is the use of multiple couples group therapy. One of the few controlled treatment outcome studies of the technique is Cadogan's report (40). Forty couples were recruited and then randomly assigned to one of two groups: an immediate treatment group or a waiting-list group, in which they continued with the traditional treatment program but did not engage in the outpatient, multiple couples group. Six-month follow-up results were striking. Nine alcoholics in the therapy group remained abstinent, 4 were doing some drinking and 7 had relapsed completely. Among the control group, however, only 2 were abstinent, 5 were drinking moderately, and 13 had demonstrated complete relapse.

Finally, we have the use of Al-Anon family groups as adjunct to formal treatment programs. Al-Anon is an indigenous self-help movement which arose spontaneously as a parallel but separate movement to Alcoholics Anonymous in the late 1940s. Well over 5000 Al-Anon family groups now exist, with world-wide distribution. Al-Anon is similar to A.A., a group fellowship of peers sharing a common problem. In the case of Al-Anon the peers are spouses, children, and close relatives of alcoholics who are usually but not necessarily A.A. members. Although infrequently studied by social scientists, those reports that have appeared have included enthusiastic antecdotal reports of favorable clinical outcomes in the families involved.

The literature, although not very satisfying in terms of its sci-

entific quality, clearly reflects the growing enthusiasm among clinicians for family therapy techniques, either as a core feature or an
adjunct to traditional treatment programs. Unfortunately, many of
the clinicians in the field are only marginally trained in family treatment techniques. They come predominantly from alcoholism backgrounds. They have shown the courage to incorporate innovative
techniques in programs, but often do not have the skills to match
this courage. On the other hand, it is also well documented that
traditional family therapists frequently demonstrate an appalling lack
of knowledge about alcoholism and fail to take into account, in their
treatment approaches, the unique consequences and impact of alcoholism on family life.

As a result of these various factors, it is difficult to accurately
assess the efficacy and appropriate place of family treatment techniques in dealing with alcoholism. As we have seen, reports appearing in the literature have been largely anecdotal and, although
enthusiastic, must obviously be taken cautiously. A great deal of
critical work remains to be done before we can arrive at a reasonable
judgment about how to include families in the treatment of alcoholism. Several well-designed treatment evaluation studies examining
specific aspects of family techniques are currently in progress, and
we anticipate being in a far better position to address this issue
within the next several years. There are also a small but growing
number of family therapists who have shown an interest in developing specialized treatment approaches designed specifically for the
treatment of families with alcoholics. The future in this regard therefore looks quite positive.

Comments

Where do we stand regarding our understanding of alcoholism and
the family at this point in time? We have only begun to scratch the
surface. Although one can now point to examples of studies, firmly
grounded in theory, clearly building one on another, and yielding
results that paint an informative picture about aspects of alcoholism
in family life, the examples are few.

Yet of the research questions that seem to capture our interest
as we move into the 1980s, the role of the family comes up over and
over again. The fetal alcohol syndrome, teen-age drinking, domestic
violence, the transmission of alcoholism from generation to generation, the effect of alcoholism on children are all areas of major research
and public policy concern that cannot be divorced from a consideration of family issues.

Although psychosocial research in alcoholism and the family is still at a fledgling stage, our experience in studying the role of the family related both to psychopathology and to the management of chronic illness helps to fill in the picture. By and large, we have come to see the family as one of a series of mediating variables that play a significant role in three major areas: first, whether a clinical pathological condition will actually develop; second, the way the condition will actually express itself (source, course, severity, etc.); and third, the seriousness or effect of the condition once it has emerged. A series of studies have convincingly demonstrated that the question of whether an individual at high risk will in fact become symptomatic can be greatly influenced by the structure and patterns of interaction manifested by the family environment. The same is true in predicting whether the condition has a relatively short life or becomes chronic. And lastly, we have been able to demonstrate, both through direct studies of family process and through family therapy treatment outcome studies, that the family is a powerful factor in determining eventual outcome of the clinical condition.

If family studies in alcoholism continue to keep pace with general trends in family research, we might expect to see emerging in the 1980s a series of studies that emphasize sophisticated assessment procedures for identifying families at risk and examining the longitudinal course of alcoholism. We should also expect to see growing interest in coping styles of the family as well as the individual. And we should be able to answer the question of which types of family therapy are most effective in the treatment of alcoholism.

REFERENCES

1. JELLINEK, E. M. The disease concept of alcoholism. Highland Park, N.J.; Hillhouse Press; 1960.
2. WITTKOWER, E. P. Historical perspective of contemporary psychosomatic medicine. Int. J. psychiat. Med. **5**: 309–319, 1974.
3. ENGEL, G. L. The need for a new medical model; a challenge for biomedicine. Science **196**: 129–135, 1977.
4. CASSEL, J. The contribution of the social environment to host resistance. Amer. J. Epidemiol. **104**: 107–123, 1976.
5. DOHRENWEND, B. S. and DOHRENWEND, B. P., eds. Stressful life events; their nature and effects. New York; Wiley; 1974.
6. PRICE, G. M. Social casework in alcoholism. Quart. J. Stud. Alc. **19**: 155–163, 1958.
7. BAILEY, M. B. Alcoholism in marriage; a review of research and professional literature. Quart. J. Stud. Alc. **22**: 81–97, 1961.

8. CAHALAN, D., CISIN, I. H. and CROSSLEY, H. M. American drinking practices; a national study of drinking behavior and attitudes. (Rutgers Center of Alcohol Studies, Monogr. No. 6.) New Brunswick, N.J.; 1969.

9. U.S. NATIONAL INSTITUTE ON ALCOHOL ABUSE AND ALCOHOLISM. First special report to the U.S. Congress on alcohol and health from the Secretary of Health, Education, and Welfare. (DHEW Publ. No. HSM 72-9099.) Washington, D.C.; U.S. Govt Print. Off.; 1971.

10. WHITEFIELD, C. L. The patient with alcoholism and other drug problems; medical aspects for medical students. Springfield; Southern Illinois University of Medicine; 1975.

11. CHRISTENSEN, H. T. Development of the family field of study. In: CHRISTENSEN, H. T., ed. Handbook of marriage and the family. Chicago; Rand McNally; 1964.

12. HILL, R. Families under stress. New York; Harper; 1949.

13. JACOB, T. Family interaction in disturbed and normal families; a methodological and substantive review. Psychol. Bull. **82**: 33–64, 1975.

14. RISKIN, J. and FAUNCE, E. An evaluative review of family interaction research. Fam. Process, Balt. **11**: 365–456, 1972.

15. JACKSON, J. K. The adjustment of the family to the crisis of alcoholism. Quart. J. Stud. Alc. **15**: 562–586, 1954.

16. JACKSON, J. K. Alcoholism and the family. Pp. 472–492. In: PITTMAN, D. J. and SNYDER, C. R., eds. Society, culture, and drinking patterns. New York; Wiley; 1962.

17. PAOLINO, T. J., Jr. and MCCRADY, B. S. The alcoholic marriage; alternative perspectives. New York; Grune & Stratton; 1977.

18. GOODWIN, D. W., SCHULSINGER, F., HERMANSEN, L., GUZE, S. B. and WINOKUR, G. Alcohol problems in adoptees raised apart from alcoholic biological parents. Arch. gen. Psychiat. **28**: 238–243, 1973.

19. GOODWIN, D. W., SCHULSINGER, F., MØLLER, N., HERMANSEN, L., WINOKUR, G. and GUZE, S. B. Drinking problems in adopted and nonadopted sons of alcoholics. Arch. gen. Psychiat. **31**: 164–169, 1974.

20. GOODWIN, D. W., SCHULSINGER, F., KNOP, J., MEDNICK, S. and GUZE, S. B. Alcoholism and depression in adopted-out daughters of alcoholics. Arch. gen. Psychiat. **34**: 751–755, 1977.

21. WOLIN, S. J., BENNETT, L. A. and NOONAN, D. L. Family rituals and the recurrence of alcoholism over generations. Amer. J. Psychiat. **136**: 589–593, 1979.

22. WOLIN, S. J., BENNETT, L. A., NOONAN, D. L. and TEITLEBAUM, M. A. Disrupted family rituals; a factor in the intergenerational transmission of alcoholism. J. Stud. Alc. **41**: 199–214, 1980.

23. GORAD, S. L. Communication styles and interaction of alcoholics and their wives. Fam. Process, Balt. **10**: 475–489, 1971.

24. KENNEDY, D. L. Behavior of alcoholics and spouses in a simulation game situation. J. nerv. ment. Dis. **162**: 23–34, 1976.

25. STEINGLASS, P. The alcoholic family in the interaction laboratory. J. nerv. ment. Dis. **167**: 428–436, 1979.

26. Moos, R. H., Bromet, E., Tsu, Z. and Moos, B. Family characteristics and the outcome of treatment for alcoholism. J. Stud. Alc. **40:** 78–88, 1979.

27. Steinglass, P., Weiner, S. and Mendelson, J. H. A systems approach to alcoholism; a model and its clinical application. Arch. gen. Psychiat. **24:** 401–408, 1971.

28. Davis, D. I., Berenson, D., Steinglass, P. and Davis, S. The adaptive consequences of drinking. Psychiatry **37:** 209–215, 1974.

29. Steinglass, P., Davis, D. I. and Berenson, D. Observations of conjointly hospitalized "alcoholic couples" during sobriety and intoxication; implications for theory and therapy. Fam. Process., Balt. **16:** 1–16, 1977.

30. Steinglass, P. A life history model of the alcoholic family. Fam. Process, Balt. **19:** 211–226, 1980.

31. Steinglass, P. The alcoholic family at home; patterns of interaction in dry, wet, and transitional stages of alcoholism. Arch. gen. Psychiat. **38:** 578–584, 1981.

32. Orford, J., Guthrie, S., Nicholls, P., Oppenheimer, E., Egert, S. and Hensman, C. Self-reported coping behavior of wives of alcoholics and its associations with drinking outcome. J. Stud. Alc. **39:** 1254–1267, 1975.

33. Steinglass, P. The impact of alcoholism on the family; relationship between degree of alcoholism and psychiatric symptomatology. J. Stud. Alc. **42:** 288–303, 1981.

34. U.S. National Institute on Alcohol Abuse and Alcoholism. Second special report to the U.S. Congress on alcohol and health, June, 1974. Mark Keller, editor. (DHEW Publ. No. ADM 75-212.) Washington, D.C.; U.S. Govt Print. Off.; 1974.

35. Bowen, M. Family therapy after twenty years. In: Ariete, S., ed. American handbook of psychiatry. New York; Basic Books; 1975.

36. Steinglass, P. Experimenting with family treatment approaches to alcoholism, 1950–1975; a review. Fam. Process, Basel **15:** 97–123, 1976.

37. Gliedman, L. H., Rosenthal, D., Frank, J. D. and Nash, H. G. Group therapy of alcoholics with concurrent group meetings with their wives. Quart. J. Stud. Alc. **17:** 655–670, 1956.

38. Ewing, J. A., Long, Z. and Wenzel, G. G. Concurrent group psychotherapy of alcoholic patients and their wives. Int. J. group Psychother. **11:** 329–338, 1961.

39. Meeks, D. E. and Kelly, C. Family therapy with the families of recovering alcoholics. Quart. J. Stud. Alc. **31:** 399–413, 1970.

40. Cadogan, D. A. Marital group therapy in the treatment of alcoholism. Quart. J. Stud. Alc. **34:** 1187–1194, 1973.

19

The Chronic Drunkenness Offender on Skid Row

Earl Rubington

The original *Alcohol, Science and Society* contains a panel discussion entitled "Penal Handling of Inebriates" (1). In that discussion, Maltbie said that the machinery, but not the tools, all ready existed for the reformation of alcoholic offenders, and recommended that the state of Connecticut "establish a single institution to which men of this type can be sent." Banay felt that "we should first attempt to change public opinion" before changing laws dealing with the redemption of the alcoholic offender. And, "in a jail," Bacon said, the alcoholic offender "is going to learn, if that is possible, how to be more unsocialized than he already is."

Almost 35 years have passed since the publication of that discussion. What, if anything, has happened to the chronic drunkenness offender problem? What changes have taken place in the social responses to this pervasive problem? If there have been any changes, what kinds of results have been obtained? Were these results intended or side effects? And what new problems, if any, have become caught up in the traditional chronic drunkenness offender problem?

This chapter seeks to answer these questions. It will attempt to do so by looking at the chronic drunkenness offender on Skid Row at two arbitrary points in time through a social problems framework. It will first look at the situation of the chronic drunkenness offender on Skid Row as it may have been in 1940 and then how it probably is in 1980. In between this overview and comparison, it will note the many efforts of the alcoholism movement to alter the definition of that situation. Then, it will conclude with an assessment of the successes and failures of current policy and suggestions for some variations in that policy.

This research was supported by grant No. AA-02900 from the National Institute on Alcohol Abuse and Alcoholism.

The Problem of Public Drunkenness in 1940

The literature on this subject is vast, complex, and contradictory. The same may be said for any aspect of alcoholism taken seriously. Hence, what I present here is a composite portrait that stresses central tendencies. What it tries to do is to focus on the main outlines of the forest. It may well miss some or even a lot of the trees. But since any forest is more than the sum of its trees, this big picture, with all the advantages and disadvantages of scope, will give us some idea of the terrain below.

Some definitions and qualifications are in order before proceeding. First, a chronic drunkenness offender is any person who has made contact with the criminal justice system (that is, the police-court-jail complex) three or more times in a given year. Second, Skid Row refers to the generally well-established homeless men's quarter in middle-sized and large cities. Third, homelessness and arrests for public drunkenness probably occur more often off rather than on Skid Row. And fourth, although some women live on Skid Row and some women are arrested for public drunkenness regardless of where they live, this chapter deals solely with men.

Skid Row, the natural habitat of many chronic drunkenness offenders, usually exists just off the central business district of the city. In this area are found a variety of institutions which cater to unattached men. These include bars, package stores, inexpensive restaurants, municipal lodging house, cage hotels sometimes called flophouses, second-hand stores, rescue missions, Salvation Army shelters, barber college, tattoo parlor, all-night movie, burlesque theatre, employment agencies, etc. Though the land values are high, the price of goods and services are very cheap, tailored to the pocketbooks of low-income males.

A subculture is the way of life of a group and Skid Row has its own unique and deviant subculture (2). It consists largely of a set of beliefs, values, and norms which have been devised by Skid Row residents for solving the problems of their markedly deviant existence. The two major problems of that existence are how to behave as a social deviant while avoiding its social and psychological consequences. Some elements of this subculture include skills, self-image, norms, and ideology. Skills comprise techniques for "getting by"; self-image consists of the alternate view of the self Skid Row men fashion for themselves as a defense against the hostile rejection of the conventional social world; norms refer to rules on how to act and include a high tolerance of dependency and deviance and a taboo on asking any personal questions; and its ideology justifies independence, nonalignment with the conventional world, and an excess of the appearance, if not the actual possession, of the masculine virtues.

In 1940, there are probably 750,000 unattached men living on an estimated 150 Skid Rows. They either dropped, slid, or drifted into this natural area of the city. Their reasons for coming here are either economic, social, or psychological. The Great Depression brought many men to live cheaply on Skid Row while they looked for work (3). Others came there when they were unable to work because of some kind of disability or handicap. And some came to escape sanctions while pursuing their deviant careers in crime, homosexuality, or alcoholism. Through drinking together, living in shelters or cage hotels, and experiencing social punishments, all came to learn, share, and then transmit the Skid Row subculture.

Heavy drinking is a Skid Row norm, and most residents drink. Alcoholism exists among all Skid Row social types, though not to the degree that the outside world surmises. Situational alcoholism exists largely among the resident workingmen. Reactive alcoholism is more characteristic of the handicapped and the disabled. And essential alcoholism is found most often among those who fled to Skid Row to continue their drinking.

A variety of social control agencies exist including the police, public health institutions, welfare, shelters, and missions. But the main control agency is the police and rates of arrest for public drunkenness are very high on Skid Row. It is well to remember that arrests for public drunkenness occur in other areas of the city as well. In all, in 1940 there probably are 750,000 arrests for public drunkenness both on and off Skid Row. Another important informal social control agency, Alcoholics Anonymous, only has 22 groups (4).

Nels Anderson pointed out in his book *The Hobo* (5) that in 1923 Chicago's Skid Row men all knew the difference between the hobo, the tramp, and the bum. The conventional world thought of Skid Row men as outcasts, derelicts; control agents on Skid Row viewed them as addicted to vice, sinners, or petty criminals.

Maltbie, in the 1945 panel (1), says that the law seeks three objectives; deterrence, segregation, and reformation. And Skid Row police, whether acting on the chief's order, responding to the complaints of citizens, or acting on their own initiative, saw themselves as enforcing and upholding the law. But sometimes, when they were neither deterring, segregating, or reforming, they were simply acting out a policy of retribution. At other times, they used their powers of arrest to solve the immediate presenting problem as it appeared to them (6). But very often they arrested Skid Row men as well as chronic drunkenness offenders off Skid Row and charged them with being drunk when in fact they were sober. In effect, they often made public drunkenness a crime of status. But, in arresting men because of their condition or status rather than overt behavior, they negated the official policies of deterrence and reformation.

Official social control policies had results markedly different from those intended. Enforcing the law that enjoined drunkenness in public either stabilized or increased the rates of public drunkenness rather than reducing those rates. In effect, then, social control perpetuated the very problem it had sought to reduce if not eliminate (7). On discharge from correctional institutions, men felt that they owed themselves a drink whether they could or could not control themselves in the act of drinking (8). Men returned to Skid Row, got drunk, and got arrested again, sometimes in the space of only a few hours. This cycle of drunkenness-arrest-jail-release-drunkenness-arrest later came to be known as the "revolving door" (9). The process labeled some men as intemperate drinkers and marked them for special attention by the police; others became dependent on the process to "dry out," thereby becoming institutionalized in the bargain. Thus, arresting and jailing chronic drunkenness offenders was only a short-term solution that compounded itself into a long-term and massive social problem.

The Problem of Public Drunkenness and the Alcoholism Movement

It would seem relatively easy to define social problems. For example, it seems obvious that alcoholism, bastardy, crime, delinquency, drug addiction, homosexuality, mental illness, prostitution, and witchcraft are all social problems. But such is not the case at all. Sometimes they are and sometimes they are not. Unless certain social requirements are met, no problem exists. To qualify as a social problem, four requirements must be met: a condition, real or imagined, must exist; it exists when a group defines it as a social problem and calls for action. When groups continually say, "We've got to do something about it," the problem has begun to exist (10).

Many of the conditions that give rise to social problems as defined above involve people who make trouble for themselves as well as for others. Groups make such people a social problem when they call for organization and action on the case which disturbs them. Oftentimes, they have a completely new definition of the situation with a set of actions for changing it. They redefine the entire situation when they give a new *definition of clients* (the problem people) by suggesting that old as well as new *control agencies* should follow new or different *policies* to achieve the *results* they intend.

By 1940 a coalition of groups had already formed or was beginning to form and had begun to call for action on alcoholism. This loose coalition of groups comprises the alcoholism movement. Alcoholism became and still is a social problem to participants in the movement because they defined alcoholism as an illness and the client as an alcoholic or sick person, they specified that public health control agencies deal with these clients, they argued for a policy of treating illness, and they intended that the results of their proposed policies would either be the reduction if not the complete elimination of the problems of alcoholism.

A brief history of the American alcoholism movement will show how its work came to have real bearing on subsequent attempts to solve the problem of public drunkenness. In the beginning, the Yale Center of Alcohol Studies, the National Council on Alcoholism, and Alcoholics Anonymous all joined hands to redefine the alcoholism situation for the American people. These three groups both talked and acted on the four points Dwight Anderson had set down in an article (11) in 1942 when he said: (1) That the problem drinker is a sick man, exceptionally reactive to alcohol; (2) That he can be helped; (3) That he is worth helping; and (4) That the problem is therefore a responsibility of the healing professions, as well as of the established health authorities.

Maltbie, Banay, and Bacon were only the first in a long line of people both in and out of the alcoholism movement who began to raise the question about the appropriateness of jailing alcoholics. Health and correctional officials began expressing similar doubts in print. And soon afterwards studies on the jailing policy and its consequences began appearing. Pittman and Gordon's *The Revolving Door* (9) was perhaps the best known of these studies, all of which pointed to the obvious conclusion that imprisonment did not reduce public drunkenness.

In the early 1960s, action appeared on the legal front. Appellate courts ruled in both the Driver and the Easter cases that alcoholism was a defense against the crime of public intoxication (12). And even though the Supreme Court refused a similar finding for Powell, the tide of public opinion was clearly turning in favor of values the alcoholism movement had been espousing for years, namely, that "jail is no place for the alcoholic." A year after the Driver and Easter cases, the first civil detoxification facility appeared in St. Louis to be followed rather quickly by one in Washington, D. C., and another in New York (13). The following year the Cooperative Commission on Alcoholism recommended in its report (14) that alcoholics no longer be jailed for public drunkenness. And later that same year, President Johnson's Task Force on Drunkenness recommended that public drunkenness be abolished as a crime, that persons found drunk in public view be brought to civil detoxification facilities rather than jails, and that the police be freed of the onerous burden of arresting drunks so that they could concentrate on crime (15).

Soon organizations emerged for taking action on these definitions of the alcoholism movement. First, a small office called the National Center for the Prevention of Alcoholism opened within the National Institute of Mental Health. Within three years the Alcoholism Treatment and Rehabilitation Act was passed establishing the National Institute on Alcohol Abuse and Alcoholism. NIAAA had been granted equal status with the National Institute of Mental Health and given a mandate at the federal level for developing and carrying out a national policy on alcoholism in four areas: research, treatment, education and prevention.

Within two years of that move, a group met to develop a model Alcoholism and Intoxication Treatment Act that could become the basis for similar policies at the state-wide level (16). This model legislation defined the alcoholic as a sick person, not a criminal, forbade penal handling of inebriates, required treatment to be voluntary rather than mandatory, established in law the principle of continuity of care, required that all alcoholics entering the system of comprehensive alcoholism services be equipped with a treatment plan that would follow them as they moved through the system, and, finally, required the development of comprehensive alcoholism services that would include an integrated network of detoxification centers, intermediate treatment centers, residential care, and outpatient clinics. Twenty-five of 50 states as of this writing have adopted such legislation.

The Problem of Public Drunkenness in 1980

In 1940, the population of the U.S. was 125 million; in 1980, it rose to 230 million. In 1940 Haggard and Jellinek estimated that there were 2½ million intemperate drinkers in the United States, 750,000 of whom were alcoholics; in 1980, NIAAA estimated that there were 11 million problem drinkers in the country, 6 million of them being alcoholics (17, 18). Thus, while the total population had almost doubled in 40 years, the number of problem drinkers had quadrupled, while there were 13 times more alcoholics in 1980 than there were in 1940. And, finally, the ratio of male to female alcoholics in 1940 was said to be 6:1; there were even those in 1980 who were claiming that it was 4:3.

Many significant changes in economy, culture and social structure took place in the 40-year period under review, and several of them affected the chronic drunkenness offender–Skid Row situation. The mechanization of agriculture, changes in the age and ethnic composition of the population, the rise of a service economy, the ending of the depression, the development of the welfare state, and the rise of the alcoholism movement all had their impact on the Skid Row scene and the public drunkenness problem.

Skid Row, the natural habitat of chronic drunkenness offenders among others, underwent some drastic changes in between 1940 and 1980. Urban renewal shrunk the number of well-established Skid Rows from 150 to 35, while the total population of all Skid Rows dropped from 750,000 to 135,000 (19, 20). Meanwhile, the composition of the population changed markedly. Formerly an almost all-White deviant enclave, Skid Row in 1980 had become bimodal, a place where older Whites sought to survive alongside of younger Blacks. And, on several large Skid Rows, as for example New York's Bowery, Blacks actually came to outnumber Whites. In 1980, the average age of Skid Row Whites had risen to 55.

Grafted onto the older set of beliefs, values, and norms that made up Skid Row subculture was a newer set of skills, self-image, norms, and ideology. Now more skills were aimed at capitalizing on dependency and making it a virtue rather than a vice; more and more there was an acceptance of the self-image of a drug-dependent person entitled to help rather than disrespect; the band of norms broadened to include rightfully demanding what was due one by reason of dependency or infirmity; and finally the ideology expanded to a justification of dependent status.

Deinstitutionalization of mental hospitals increased the numbers of mentally ill persons living in or near the Skid Row sector (21). The number of drug addicts as well as multiple drug users similarly increased. And the proportions of alcoholics in the total Skid Row population actually rose while the distribution of types of alcoholics shifted dramatically. If one third of the Skid Row population were alcoholics in 1940, by 1980 two thirds of that population were alcoholics. In 1940, the distribution of types of alcoholics were rank-ordered as situational, reactive, and essential. In 1980, this order was now reversed; essential alcoholics outnumber the reactive alcoholics and the reactive alcoholics outnumber the situational alcoholics (22).

In 1940, the main route of entry into Skid Row was by way of extreme economic impoverishment. Physical handicap and social disability was next to be followed by psychological disturbance. In 1980, these routes reversed themselves. The undersocialized came to outnumber the desocialized. And before where the main agencies of Skid Row socialization had been drinking establishments, shelters, and jails, in 1980, the situation of treatment of alcoholism and drug dependency came to be the major stimulus of socialization.

Throughout the country, police, missions, and shelters continued as social control agencies, but now public health and welfare had become more important as agencies of formal social control. In the country at large, there had been a great expansion of general hospitals with specialized alcoholism treatment wards at both federal, state, and municipal levels. Similarly, a network of specialized alcoholism treatment centers, as envisioned by the planners who drew up the model Alcoholism and Intoxication Treatment Act, had come into being. For instance, by 1980 there were probably 1000 detoxification centers, 250 residential intermediate care programs, 500 halfway houses, and 375 outpatient clinics; most, if not all of these alcoholism treatment centers, were parts of state alcoholism programs financed partly or wholly by NIAAA. A sizeable minority of chronic drunkenness offenders throughout the country were becoming clients of this network of alcoholism treatment agencies.

Decriminalization of public drunkenness existed in some states either through police underenforcement or the passage of the Model Act (23). Other states had not changed the law enjoining public drunkenness. As a result, in 1980 there were 1½ million arrests for public drunkenness, whereas in 1940 there were probably only 750,000 such arrests. At the same time, the actual number of chronic drunkenness offenders rose; in 1940 there were an estimated 125,000 drunkenness offenders, whereas in 1980 the absolute number of such offenders had grown to 375,000. The increase in offenders in 1980 was probably due to the growth and expansion of chronic drunkenness offenders living off rather than on Skid Row.

Another sign of the times, of course, was the phenomenal growth of A.A. As noted earlier, there were only 22 groups in the entire United States in 1940. In 1970, the number of A.A. groups had grown to 9541 (4). And by 1980 that number had reached an estimated 12,500. The chances of chronic drunkenness offenders coming into contact with A.A. influences in 1940 were quite remote; in 1980, the chances of escaping those influences were remote.

One consequence of the turbulent 1960s probably was an

increased tolerance for deviance in the society at large. Perhaps more importantly for chronic drunkenness offenders was the lessening of the social stigma attached to alcoholism and wider public acceptance of the illness concept (24). Thus, the public opinion change that Banay had called for before changes in the law may well have come about largely through the combined efforts of the three wings of the alcoholism movement. It is rather likely that the most significant of the three probably was the word-of-mouth information on the success of A.A. In any event, the definition of the chronic drunkenness offender in 1980 was no longer that of a sinner or criminal; he had now become accepted as a sick person. This acceptance, of course, was not without its overtones of moral ambivalence (25).

The policies of the dominant social control agencies had been spelled out in the Model Treatment Act and included the switch from crime to illness, the shift from mandatory to voluntary treatment, the emphasis on integration in a comprehensive alcoholism program in which clients would participate according to an individualized treatment plan. This principle was called continuity of care. Through it all the policy was based on locating the client in a given treatment niche depending on where he was in his treatment career. A model of stages of recovery was implicit in the general policy and that in some way there would be steps in treatment that paralleled Jellinek's famous phases in alcoholism (26) which the alcoholism movement had popularized so widely and so successfully. The model implied that if there was a progression into alcoholism, so similarly was there a reverse progression out of alcoholism and that integration into the body of alcoholism treatment agencies was perhaps the best way to speed alcoholics along that rehabilitation route.

The new definition of the clients, according to the alcoholism movement, had to coexist alongside of the earlier more traditional conception of the chronic drunkenness offender as an unreclaimable outcast. "The alcoholic is a sick person" existed alongside the earlier accepted statement that "once a drunk, always a drunk." Nonetheless, some of the results of the new social control policies were intended, although, of course, there were a set of side or unintended effects. For example, evidence from a number of public opinion studies over the past 10 years or so suggests that more members of the general public accept the dictum of the alcoholism movement that the alcoholic is a sick person (24, 25). Less empirically based though strongly believed, particularly by personnel working in the new system of alcoholism treatment agencies, is the notion that a small, but significant percentage of offender-clients of the new treatment system have actually recovered from alcoholism. On the other hand, treatment evaluations of the new detoxification–halfway house circuit in cities like Boston, Minneapolis, New York, Seattle, St. Louis, and Washington, D. C., suggest that the revolving door has moved out of the criminal justice system and into the new circuit of alcoholism treatment agencies (27–30).[1] And by virtue of the voluntary character of the new treatment system, all of these doors were now revolving even faster, so much so that one commentator has referred to it as a "spinning door" (31).

Ironically, a comment on the "revolving door" included in President Johnson's Commission on Law Enforcement and Administration of Justice might stand as a summary of the accomplishments of the new detoxification–halfway house system if the words "criminal justice system" are changed to "system of comprehensive alcoholism services" (15): "The criminal justice system appears ineffective to deter drunkenness or to meet the problems of the chronic alcoholic offender. What the system usually does accomplish is to remove the drunk from public view, detoxify him, and provide him with food, shelter, emergency medical service, and a brief period of forced sobriety. As presently constituted, the system is not in a position to meet his underlying medical and social problems."

Unlike its predecessor, the alcoholism treatment system is voluntary, not mandatory. Many detoxification centers, for example, aspire to hold their clients for a 5-day period; on average, their clients stay for 2.5 days (30).[1] Under the old system, some offenders compiled 50 or more arrests as a lifetime record in one jurisdiction, now some clients compile a record of 50 or more detoxifications a year. It is difficult to regard such a high frequency of relapse as a sign of social rehabilitation. Perhaps one word that best summarizes the results of the new system is "containment" (32). This pattern of many short stays, frequent relapse and return does shorten the length of each drinking bout while containing the client in one system of detoxification. This may well be a significant change over the earlier system of social control and may well have some hidden advantages as well as all of the obvious disadvantages.

Successes, Failures, and Some New Directions

On balance, the new social control policy of 1980 with chronic drunkenness offenders can claim some modest success. First of all, so far as definitions are concerned, it seems that not only has there

---

[1]Also, RUBINGTON, E. The social organization of relapse. Presented at the annual meeting of the Society for the Study of Social Problems, New York, 26 August 1980.

been a wider appreciation of alcoholism as an illness among the general public but that among the clients of the treatment system itself, namely, chronic drunkenness offenders, there is an increased willingness to regard alcoholism as a form of illness rather than a sign of weak willpower, moral failing, or bad habit (33).

The switch from a police to a public health response has resulted in much better medical care for clients of the system. And unquestionably the treatment which clients experience in the detoxification–halfway house system is considerably more humane than what they would have experienced had they been sentenced to serve time for public drunkenness. There is also some evidence suggesting that in some instances lives have been saved while in others they have actually been lengthened.

The police have been relieved of a great burden. In those jurisdictions where detoxification centers exist, police contacts with drunkenness offenders have diminished considerably (28). A network of alcoholism treatment agencies has been established. Probably a few remarkable recoveries from alcoholism have been successfully accomplished. Meanwhile, the further development and integration of these agencies continues.

It is much easier to point to the failures of the new system of social control of public drunkenness. To begin with, innovations always induce unrealistic aspirations and the decriminalization–detoxification package is no exception. As one highly-placed Boston police official put it, the police expected the new system "to wipe out alcoholism" (33). And many people who first went to work in detoxification centers knowing nothing about alcoholism felt that they personally were going to sober up many alcoholics. Both police and new staff rapidly became disaffected when there was both a shortage of beds to go around and when most of these beds began to fill up rapidly and repeatedly with clients who had just been in for detoxification the week before, in some instances the day before.

The extremely high rates of recidivism disillusioned many workers rather early and staff "burn-out" (low morale, high apathy, rapid turnover) became the staff reciprocal of the "revolving door" (34). A great deal of money had been invested in the system and many people began to wonder if the meager benefits justified the enormous expense. And, in addition, critics began to point out that the system was institutionalizing as many if not more alcoholics than the previous criminal justice system. In some cases, they claimed that clients had now become as dependent on the detoxification as they were on alcohol. In still other cases, people argued that a triple dependency had come into being. Now there were clients who in addition to being dependent on detoxification and on alcohol had now become through frequent detoxifications addicted to chlordiazepoxide (Librium) or diazepam (Valium).

The most systematic treatment evaluation of the detoxification–halfway house circuit has been done in Canada (27). The results are not encouraging. To begin with, detoxification had been viewed by system planners as the agency that would detoxify, then refer clients to the next agency in the system, intermediate care, halfway houses, or outpatient clinics. In Canada, however, researchers found that clients did not stay as long as necessary, made frequent returns, and accepted few referrals to halfway houses. Of those who did accept referrals, most stayed in the houses only a very short time, got drunk and ended up in detoxification centers again (35).

There are not too many suggestions one can make. To begin with, the failures of the new system are as nothing when compared with its predecessor. It is not the case that there will be a return to the criminal justice response to public drunkenness but rather that both systems will continue to exist side by side. Part of the Canadian problem is that public drunkenness continues to be a crime alongside of a quasi-voluntary detoxification–halfway house system. But that is also true still in many parts of the U.S., particularly in places where public drunkenness has been decriminalized but where there is no center for citizens or police to take public inebriates.

Costs and drug dependency can both be reduced by establishing more social setting model detoxification centers (36). Most of the centers in the U.S. are based on the medical model (doctors are responsible while registered nurses supervise). This means that staff manages withdrawal from alcohol by dosing clients with chlordiazepoxide or diazepam usually on a three-day reduction cure. The regimen is costly in two ways. First, medical personnel have high salaries. And, second, the procedure increases the risk of dependency on tranquilizing drugs. Social setting detoxification, wherever it exists, uses fewer staff who simply talk clients down from any anxieties they experience during withdrawal from alcohol. Clients need not stay as long, but as long as they do stay, they are not caught up in the "give-me-a-pill-for-pain" syndrome. Such clients will come to learn that alcoholism does not go away by the efforts of medical people.

Summary and Conclusions

In 1940, penal handling was the principal social control response to chronic drunkenness offenders on or off Skid Row. The main effect of this kind of handling was that offenders came to view themselves as degraded social objects. People who view themselves in these terms have little reason to stop drinking, every reason to continue. As a result, the jailing policy perpetuated rather than eliminated or reduced the problem of public drunkenness. Then the rising alcoholism movement began calling attention to the futility and the inhumanity of penal handling and suggested the public health response in its place. The public health response, in the form of the civil detoxification facility, has only recently made its appearance nation-wide. Although considerably more humane, because of its medical trappings, clients come to view themselves as social objects who can control most situations by means of chemicals. People who look at themselves in these terms have little reason to change, particularly as these are the same terms in which agents of social control who are trying to help them also see them. Perhaps the civil detoxification facility, without the medical trappings, may go the farthest in helping its clients to slowly rid themselves of their alcohol dependency. For the social setting model encourages clients to view themselves as social objects who can control themselves and the situations in which they find themselves. People who come to view themselves as self-controlling social objects may have a better chance in their struggle against alcoholism. This would seem to be the case because this kind of agency of social control does not insist that its clients become dependent on it, its philosophy, or any of its agents, chemical or social.

REFERENCES

1. MALTBIE, W. M., BANAY, R. S., and BACON, S. D. Penal handling of inebriates; a panel discussion. Pp. 373–385. In: Alcohol, science and society; twenty-nine lectures with discussions as given at the Yale Summer School of Alcohol Studies. New Haven; Quarterly Journal of Studies on Alcohol; 1945.
2. WALLACE, S. E. Skid Row as a way of life. Totowa, N.J.; Bedminster Press; 1965.
3. SUTHERLAND, E. H. and LOCKE, H. J. Twenty thousand homeless men. Philadelphia; Lippincott; 1936.
4. LEACH, B. and NORRIS, J. L. Factors in the development of Alcoholics Anonymous (A.A.). Pp. 441–543. In: KISSIN, B. and BEGLEITER, H., eds. The biology of alcoholism. Vol. 5. Treatment and rehabilitation of the chronic alcoholic. New York; Plenum; 1977.
5. ANDERSON, N. The hobo. Chicago; University of Chicago Press; 1923.
6. BITTNER, E. The police on Skid Row; a study of peace-keeping. Amer. sociol. Rev. 32: 699–715, 1967.

7. PITTMAN, D. J. and GILLESPIE, D. G. Social policy as deviancy reinforcement; the case of the public intoxication offender. Pp. 106–124. In: PITTMAN, D. J., comp. Alcoholism. New York; Harper & Row; 1967.

8. SPRADLEY, J. P. You owe yourself a drunk; an ethnography of urban nomads. Boston; Little, Brown; 1970.

9. PITTMAN, D. J. and GORDON, C. W. The revolving door; a study of the chronic police case inebriate. (Rutgers Center of Alcohol Studies, Monogr. No. 2.) New Brunswick, N.J.; 1958.

10. SPECTOR, M. and KITSUSE, J. I. Constructing social problems. Menlo Park, Calif.; Benjamin–Cummings; 1977.

11. ANDERSON, D. Alcohol and public opinion. Quart. J. Stud. Alc. 3: 376–392, 1942.

12. HUTT, P. B. The legal control of alcoholism; towards a public health concept. Pp. 124–128. In: PITTMAN, D. J., comp. Alcoholism. New York; Harper & Row; 1967.

13. NIMMER, R. T. Two million unnecessary arrests. Chicago; American Bar Foundation; 1971.

14. COOPERATIVE COMMISSION ON THE STUDY OF ALCOHOLISM. Alcohol problems; a report to the nation. Prepared by T. F. A. Plaut. New York; Oxford University Press; 1967.

15. U.S. PRESIDENT'S COMMISSION ON LAW ENFORCEMENT AND ADMINISTRATION OF JUSTICE. TASK FORCE ON DRUNKENNESS. Task Force report; drunkenness. Washington, D.C.; U.S. Govt Print. Off.; 1967.

16. U.S. NATIONAL INSTITUTE ON ALCOHOL ABUSE AND ALCOHOLISM. First special report to the U.S. Congress on alcohol and health from the Secretary of Health, Education and Welfare. (DHEW Publ. No. HSM 72-9099.) Washington, D.C.; U.S. Govt Print. Off.; 1971.

17. HAGGARD, H. W. and JELLINEK, E. M. Alcohol explored. New York; Doubleday; 1942.

18. U.S. NATIONAL INSTITUTE ON ALCOHOL ABUSE AND ALCOHOLISM. Alcohol and health; third special report to the U.S. Congress from the Secretary of Health, Education, and Welfare, June 1978; technical support document. Ernest P. Noble, editor. (DHEW Publ. No. ADM 79-832.) Washington, D.C.; U.S. Govt Print. Off.; 1979.

19. BAHR, H. M. The gradual disappearance of Skid Row. Social Probl. 15: 41–45, 1967.

20. BOGUE, D. T. Skid Row in American cities. Chicago; University of Chicago Family Study Center; 1963.

21. ROONEY, J. F. Societal forces and the unattached male; an historical review. Pp. 13–38. In: BAHR, H. M., ed. Disaffiliated man; essays and bibliography on Skid Row, vagrancy, and outsiders. Toronto; University of Toronto Press; 1970.

22. RUBINGTON, E. The changing Skid Row scene. Quart. J. Stud. Alc. 32: 123–135, 1971.

23. GIFFEN, P. J. and LAMBERT, S. Decriminalization of public drunkenness. Pp. 395–440. In: ISRAEL, Y., GLASER, F. B., KALANT, H.,

POPHAM, R. E., SCHMIDT, W. and SMART, R. G., eds. Research advances in alcohol and drug problems. Vol. 4. New York; Plenum; 1978.

24. HABERMAN, P. W. and SHEINBERG, J. Public attitudes toward alcoholism as an illness. Amer. J. publ. Hlth **59:** 1209–1216, 1969.

25. RIES, J. K. Public acceptance of the disease concept of alcoholism. J. Hlth social Behav. **18:** 338–344, 1977.

26. JELLINEK, E. M. Phases of alcohol addiction. Quart. J. Stud. Alc. **13:** 673–684, 1952.

27. ANNIS, H. M. The detoxication alternative to the handling of public inebriates; the Ontario experience. J. Stud. Alc. **40:** 196–210, 1979.

28. FAGAN, R. W., JR. and MAUSS, A. L. Padding the revolving door; an initial assessment of the Uniform Alcoholism and Intoxication Treatment Act in practice. Social Probl. **26:** 232–247, 1978.

29. WESTERMEYER, J. and LANG, G. Ethnic differences in use of alcoholism facilities. Int. J. Addict. **10:** 513–520, 1975.

30. REGIER, M. C. Social policy in action; perspectives on the implementation of alcoholism reforms. Lexington; Lexington Books; 1979.

31. ROOM, R. Comment on "The Uniform Alcoholism and Intoxication Treatment Act." J. Stud. Alc. **37:** 113–144, 1976.

32. WISEMAN, J. P. Stations of the lost; the treatment of Skid Row alcoholics. Englewood Cliffs, N.J.; Prentice-Hall; 1970.

33. RUBINGTON, E. Top and bottom; how police administrators and public inebriates view decriminalization. J. drug Issues **5:** 412–425, 1975.

34. MASLACH, C. The client role in staff burn-out. J. social Issues **34:** 111–124, 1978.

35. ANNIS, H. M. and LIBAN, C. B. A follow-up study of male halfway-house residents and matched nonresident controls. J. Stud. Alc. **40:** 63–69, 1979.

36. WEIL, M. Social-setting withdrawal from alcohol. M.P.H. thesis, Yale University, 1975.

20

Special Populations

Edith Lisansky Gomberg

The Third Special Report to Congress from the National Institute on Alcohol Abuse and Alcoholism in 1978 (1) contained a chapter, "Special Population Groups." These groups were defined in terms of demographic variables: age, gender and race. Thus the chapter includes a discussion of the drinking problems of young people, women, the elderly, American Indians, Spanish-speaking and Black Americans. As the term is currently used, special populations do not include White ethnic groups, e.g., Irish-Americans or Polish-Americans, although the drinking patterns and problems of the groups are discussed in national surveys and, occasionally, in special studies. This categorization does not include alcoholics classified by *socio-economic status*, thus the drunkenness offender on Skid Row is excluded. And, although there is a good deal of space devoted to industrial and governmental employee assistance programs, the classification of alcoholic persons as homeless, poor, blue collar, middle class, etc., does not appear. Finally the term, special populations, as currently used, does not divide alcoholics into subgroups of *other diagnostic features*.

The 1981 report to Congress on alcohol added Native Alaskans, Asian Americans, and the Gay Community to the age, gender and racial groups which appeared in the earlier report.

These special populations, so designated in 1978 and 1981, express the concerns of the last two decades. The subgroupings which appear in the 1945 *Alcohol, Science and Society* are quite different. It was a different time with a different view of the world, one focused on the White male alcoholic. As one reads the lectures given 35-odd years ago, two major groupings of alcoholics emerge. First, alcoholics are seen to be different in terms of other diagnostic features, i.e., alcoholics were viewed psychologically or psychiatrically in terms of clinical symptoms like psychopathy (or sociopathy), neurosis, psychosis or impaired characterological development. Second, alcoholism and drinking behavior itself are seen in terms of social class differences.

The Other Diagnostic Features of Alcoholism

Classification of the other diagnostic features of alcoholism did not originate in *Alcohol, Science and Society*. In 1942, Bowman and Jellinek (3) had summarized in tabular form the different suggested classifications of "abnormal drinkers" which reached back historically to the turn of the century. These included intemperate drinkers, addicts, cyclothymic and schizoid types, intermittent, regular and paroxysmal dipsomania, and many others. The table included classifications of alcoholics in terms of related diagnostic features and in terms of patterns of drinking.

In *Alcohol, Science and Society*, Landis, who was lecturing on "the alcoholic personality" (4) chose the classification by Cimbal (5) as most useful. Cimbal had grouped alcoholics as decadent drinkers, impassioned drinkers, stupid drinkers, and self-aggrandizing drinkers. Landis suggested that this classification was both theoretically and clinically useful.

Several of the classifications listed in Bowman and Jellinek's table have survived and still appear to have clinical and research viability. R. P. Knight (6) and W. C. Menninger (7) had classified patients as essential alcoholics, reactive alcoholics, neurotic characters, psychotic personalities, and symptomatic inebriates. These classifications are rooted in psychoanalytical theory and linked to psychiatric nosology, and the essential vs reactive differentiation has a counterpart in the classification of the schizophrenias. In a current attempt to classify women alcoholics in terms of other psychiatric clinical features, the differentiation of primary and secondary alcoholism is made in terms of *antecedent* psychiatric symptomatology (8).

The attempts to classify alcoholics continued; Jellinek's *Disease Concept of Alcoholism* (9) described the alcoholisms in terms of drinking patterns, characterizing each with a Greek symbol, thus alpha alcoholism seems to be what we might call a problem drinker whereas gamma alcoholism is closer to that observed in mental hospital patients where chronicity, withdrawal symptoms, and loss of control may indeed be present. It is not clear just how useful such a classification is. More recent attempts to describe "personality types" have used sophisticated statistical manipulation, e.g., factor analysis. In one study, Partington and Johnson (10) argue that "pure types" of alcoholics do not exist but that alcoholics represent mixtures of five types characterized by rebelliousness, thought disturbance, neuroticism, defensiveness, and a combination of these behaviors.

Social Class Differences

In Bacon's lecture in *Alcohol, Science and Society* (11), he notes that a complex, highly specialized modern industrial society produces a social stratification system, and in a lecture applying the system of social stratification to drinking behavior, Dollard (12) predicted the drinking behavior of the different social classes. First he pointed to the constraints placed on drinking, constraints derived from the self and from society. Then, following the work of Warner and Lunt (13) and other social anthropologists who were exploring the American class system, Dollard described the drinking and attitudes toward drinking of upper, middle and lower classes. The data of epidemiological research have, by and large, supported the descriptions given by Dollard, with one notable exception: the description of the debauchery of the lower income groups has not stood up as well with time.

The issue of social-class status and alcohol problems is also linked in Reverend Murphy's discussion of "alcohol and pauperism" (14) and in the terms used by Jellinek, e.g., "alcoholic derelicts." The lecturers in *Alcohol, Science and Society* were quite aware of the chronic drunkenness offender and Skid Row; their task was to convince their colleagues of the existence of respectable alcoholics.

One attempt to classify alcoholics both by social-class status and drinking patterns was made by Straus and McCarthy (15). Contrasting the alcoholic seen by clinicians and therapists, "the addictive drinker," with homeless men living on Skid Row, they termed the latter "nonaddictive pathological drinkers." While fundamentally a differentiation based on drinking patterns, Straus and McCarthy were distinguishing between different societal groups and emphasizing the heterogeneity of pathological drinking.

The Special Populations

Current usage has defined the special populations of alcoholic persons as the young, the elderly, women, and the minority groups. What follows is a brief summative description of the state of knowledge about the drinking patterns and problems of each of these groups.

Age

Adolescents. The National Institute on Alcohol Abuse and Alcoholism describes "youth" as teen-agers. We are not speaking here of young adults. As early as 1945, those persons who treated alcoholics knew that young adults did indeed appear at various treatment facilities (it is likely that the military and the veterans' facilities were quite familiar with young adult men alcoholics). In Fleming's lecture on treatment in *Alcohol, Science and Society* (16), he commented: ". . . the young drinker is much more difficult to treat than the older one" *(p. 398).*

But the 1945 book did not mention drinking patterns and problems of *adolescents.* The concern with drinking by junior and senior high-school students, with minimum age of legal purchase, with youth and traffic accidents came after World War II. Analysis of the 120 surveys of teen-age drinking practices from 1941 to 1975 shows that the proportion of teen-age drinkers rose steadily from the earlier surveys until about 1965 and it has leveled off. The proportion of teen-agers who reported experiencing intoxication continued to rise: 19% reported such experience in 1966 and 45% in 1975. The discrepancy between boys' and girls' drinking has lessened although it is still true that more boys drink and consume greater quantities than girls. The major factors which influence teen-age drinking are peers, family and parents, sociocultural background, contexts of drinking, and the personality of the drinker. Heavier drinking among adolescents is apparently related to drinking outside the home, with age peers, and—more rarely—drinking alone (1, 2, 17).

Although the increase in drinking and intoxication among adolescents is viewed with moralistic alarm, a large real social problem is the relationship between alcohol consumption and automobile accidents. One estimate of the economic costs of alcohol-related traffic accidents runs over $5 billion, and a disproportionate number of these accidents occur among young men. The number of adolescents seen in alcoholism treatment facilities remains small but it may very well be that youngsters who have problems associated with their drinking, e.g., objections from significant others, are more likely than others to develop alcoholism in their twenties, thirties and forties.

Adolescent problem drinking is associated with child–parent relationships characterized by rebelliousness, less parental disapproval of drinking and more parental heavy drinking, and with lack of interest and affection from parents. Adolescent problem drinkers are more responsive to peers than to parents and frequently have friends who drink heavily and show other problem behaviors. Such friends are supportive of heavy drinking. The psychological characteristics reported in adolescent problem drinkers include unhappiness, boredom, aggressiveness, frustration and dissatisfaction. Since so many of these characteristics are shared with unhappy adolescents in general, early intervention programs should look to (*1*) the drinking behavior itself; and (*2*) associated problem behaviors, particularly antisocial behaviors. Heavy drinking among adolescents has been linked to precocious sexual behavior, poor school performance, acting out behavior in the classroom, family problems, truancy, and high drop-out rates from school (1, 2, 18).

Although this is obviously an age group for whom prevention may have great economic and social benefits, most preventive programs have been limited to educational materials about alcohol. The military services have had an "employee assistance" treatment program for many years but have only recently become motivated to expand into prevention programs which deal with older adolescents separated from home, and their concern is with the efficient performance of recruits. It would seem that educators, military personnel and church groups might be effective in banding together and exchanging ideas and methods.

The Elderly. The adolescent years, we know, are a time of decision about drinking—whether to drink, where, how much. And, interestingly enough, it appears that the later years are also a time of decision or perhaps redecision about drinking. The number and proportion of abstainers increases, the proportion of heavy drinkers decreases among both sexes at about age 50 and there is a second decrease among men around age 65. At the same point in the life cycle, there is evidence that some people who have not necessarily had drinking problems before, do drink problematically, we think, for the first time. A distinction has been made between older problem drinkers with early onset and a long history of problems who survive into their sixties ("survivors") and older problem drinkers for whom the onset of alcohol problems is relatively recent ("reactives"). And there may be still others with a history of occasional episodes of problem drinking who become alcoholics in their older years (19, 20).

During the period in which *Alcohol, Science and Society* appeared, and for a long time afterwards, developmental study meant study of the infant, child and adolescent. Only in the last 10 to 20 years has interest expanded to the whole life cycle and there is now debate about which cognitive functions survive and which are impaired with aging and about the question of personality change throughout the lifetime. Mid-life crises, the "empty nest," and gerontology are now familiar terms. But the early theories about alcoholism began and ended with events in early life and the theories reflected the lack of interest in the events and stresses of adult life which was characteristic of psychological and psychiatric theorizing of the time.

With an increasing proportion of the population (at least in Western societies) living to age 65 and over, interest in gerontological issues has grown. We do not have 120 surveys as we do of adolescent drinking but there are hints and trends in the published literature. Moderate social drinking among older persons is associated with good health and with social networks which encourage drinking (as is true among younger persons). Heavy, problematic drinking, when of recent origin, is likely to be associated with depression, loss and isolation. One survey reported that older men, recently retired and recently widowed, are at high risk for alcohol problems (21), and for suicide as well. One unexplored problem is the interaction between medication and alcohol in the elderly; although they represent approximately 11% of the U.S. population, older people receive about a quarter of all prescriptions written.

When older men manifest alcohol problems, they are less likely to be visible than younger men simply because fewer are in the workplace. Older people who drink excessively at home tend to be overlooked, particularly if they have withdrawn from family and friends. There is less belligerence among older alcoholics than among younger ones but, interestingly enough, older men alcoholics do tend to be picked up by the police, but primarily for public intoxication rather than for being drunk and disorderly. Older men alcoholics are most visible in hospital wards and emergency rooms, or as public intoxication offenders. Elderly women alcoholics appear to be invisible—except for family, neighbors, or visiting welfare workers who may spot them.

Generally, older alcoholics are probably most visible to spouse, family, friends and neighbors, and those networks appear to be a primary source of casefinding. Health problems associated with excessive drinking loom large so that physicians' offices, emergency rooms, hospital wards and services may also be significant casefinding sites.

Many agencies and therapists are not particularly responsive to the needs of older alcoholics although there is evidence that older alcoholics are very responsive to help, particularly to the social therapies. It is believed that shame and denial play a significant role in keeping older problem drinkers away from sources of help; this is consistent with their underrepresentation at community mental health centers.

The drinking behaviors of older persons is a new research frontier. Explanations of the increase in abstinence include economics, the physiological effects of alcohol, generational attitudes. What can be predicted about the drinking behaviors of the older populations of the future: Will the increase in adolescent drinking be carried through into later life? Gerontological alcoholism is a new research area: What are the diagnostic and behavioral cues we should know? How do we differentiate alcoholism from depression or dementia? What are the best techniques of treatment and of prevention? All these questions wait upon the interest and activity of alcohol researchers.

Sex Differences

The woman alcoholic appears once in *Alcohol, Science and Society*. In Baker's discussion of social case work with alcoholics, one of the six cases presented was of a woman alcoholic (22). In 1945, there were two papers on female alcoholism, one from Bellevue Hospital and one from Bloomingdale Hospital, both in New York. When my first study of alcoholic women appeared in 1957 (23), it was possible to summate all that had been published in seven hypotheses:

1. Women alcoholics are "much more abnormal" or "worse" than male alcoholics.
2. Women alcoholics are more variable, i.e., "the problem is more highly individual."
3. Female alcoholism is more associated with a precipitating, stressful antecedent event than is true of male alcoholism.
4. Female alcoholism and sexual promiscuity are somehow related.
5. The onset of alcoholism is somewhat later in women alcoholics but the shift from controlled to uncontrolled drinking is more rapid.
6. There is a relationship between female alcoholism and hormone function.
7. Medical complications of alcoholism are more frequent among women.

The 1957 report presented a comparison of the clinical records of men and women alcoholics attending an outpatient clinic, and a comparison of outpatient women and women in a state prison attending an alcoholism treatment unit. The men and women alcoholics of the outpatient clinic turned out to be quite similar in socioeconomic status and family history except for one item: the frequency with which alcohol problems had appeared in early family history was much higher among the women who, in general, reported more disruptive experience in early life. The most significant differences between men and women alcoholics appeared to lie in their drinking histories: onset, duration, drinking contexts and consequences. On the other hand, socioeconomic status differences were striking in comparing the outpatient women and the prison women. The former were "more respectable" and still maintained a good degree of social integration, and the latter, whose drinking becomes a police matter because it is more publicly visible, were more often multiproblem women whose drinking compounded their problems, had far fewer personal resources, fewer interested people, and fewer strengths. It was suggested that the outpatient women resembled Straus and McCarthy's "addictive drinkers" while the prison women resembled the "nonaddictive pathological drinkers."

Very little interest was shown in the issue of female alcoholism for about 15 years after this report, but interest exploded in the 1970s, a by-product of feminism and a response to the greater numbers of women who began appearing at treatment facilities. The rate of female alcoholism may or may not have increased (the accuracy of the previously accepted male : female ratio of 6:1 was questioned) but there was no question but that women were joining Alcoholics Anonymous and coming into treatment more than they had before. The burst of interest in female alcoholism has given rise to much research activity, most of it summarized in literature reviews (e.g., 24).

A recent summary of the state of the art (25) includes the epidemiological information that rates of social drinking rose among women for several decades after World War II but leveled off in the 1970s; the most striking differences are age differences—among younger women, the proportion who drink is very high but it drops rapidly among women over age 50. The issues of intoxication and the double standard of response to drunken comportment are still important and there is no indication that the stigma associated with female intoxication has lessened.

Most studies of women alcoholics compare them with men alcoholics, usually at the same treatment facility. Some studies compare women alcoholics with nonalcoholics with or without psychiatric diagnoses. More recent data indicate the following about the points raised in the 1957 report:

1. There is no consistency in reports of treatment outcome and the question of differential prognosis is not resolved. There is some limited evidence that women alcoholics are more deviant from female norms than men alcoholics are from male norms.

2. Women alcoholics are a variable group, differing in terms of social class, ethnicity, occupational category, etc. Men alcoholics are also a variable group.

3. Women cite stressful events as precipitating their alcoholism more frequently than do men, which may be due to defensiveness, sex differences in reporting, or greater vulnerability to environmental stress and to loss.

4. Little evidence has been found for the association of female alcoholism and sexual promiscuity. Women who drink in public places may be more vulnerable to sexual assault. The relationship between alcohol and enhancement of sexual experience has not been well studied among women alcoholics or any other group of women.

5. Women begin drinking later, experience their first intoxication later, develop alcoholism later, and come to treatment with briefer histories than do men. There are some recent data which suggest that these sex differences are minimal when younger alcoholics are compared.[1]

6. The relationship between alcoholism and hormonal status is unclear. There is some research which suggests that blood alcohol concentration varies in different stages of the menstrual cycle and there is a good deal of published material about "the fetal alcohol syndrome."

7. Medical complications of alcoholism are a significant issue and it would appear that women alcoholics are more vulnerable to several medical complications (26). Health concerns are important motivations for women to seek treatment.

In addition, research since 1957 would add the following:

8. Comparison of men and women alcoholics shows that an alcoholic spouse or lover occurs much more frequently in female alcoholism. Women alcoholics are often married to alcoholics but the histories suggest that transmission is most usually male to female.

9. Women drink alone much more than do men, partly because of social role and guilt, partly because of stigmatization. When women alcoholics do not drink alone, they are most likely to be drinking with a significant other. Far fewer women alcoholics drink in public places than do men alcoholics.

10. Women in general are more often prescribed drugs than men

[1]GOMBERG, E. S. L. and HNAT, S. Age comparisons of men and women alcoholics. [In preparation.]

and more often take psychoactive drugs, and significantly more women alcoholics appear to be polydrug misusers. Drug histories may be more similar among younger men and women alcoholics than among middle-aged alcoholics.

11. Recent studies suggest risk indicators for women may include the following: recent marital disintegration experienced by women under age 35; "dual role stress" for women of all age groups who deal with double responsibilities, working outside and working at home; non-White young women who are the head of the family and raising young children with little financial and social support; and the recently widowed.

It should be noted that a gloomy view of prognosis prevailed but that interest and attention to issues about female alcoholism, to a large extent, have countered that view. The effectiveness of treatment is often a self-fulfilling prophecy and the recent interest in female alcoholism has had a positive effect on treatment outcomes.

Minorities

American Indians and Native Alaskans. The drinking and alcohol problems of American Indians is discussed in this volume by Lemert. I will therefore touch on only a few points.

Alcohol has been involved in the problems of American Indians and Native Alaskans since earliest contact with Europeans. Alcohol-related problems are still a major source of difficulty and debates as to whether Native Americans are addicted, alcoholismic or problematic in their use of alcohol seem academic. Life expectancy is shorter than for other groups and the death rate from alcohol-related causes is high. Alcohol-related suicide rates among American Indians are considerably higher than they are for the general U.S. population. Alcohol figures prominently in arrest records: Los Angeles Police Department data from a 15-year period show that of all adult Indian arrests, 90% were for intoxication (2). Native people in Alaska, including Eskimaux, Indians and Aleuts, comprise 17% of the population of Alaska but account for 60% of the deaths due to alcoholism (2).

The image of "the drunken Indian" is a widely held stereotype, and American Indians themselves see excessive drinking as one of their major social problems. Recent research at a reservation school indicated that 58% of the boys and 30% of the girls reported "regular drinking" before age 12 (27). The author noted that although there was the possibility of getting into trouble, the 7th and 8th graders he studied expressed positive attitudes toward drinking and that although the boys reported the beginning of drinking at an earlier age than girls, both boys and girls reported regular drinking by the age of 13.

The study of American Indian drinking and alcohol problems is complex. American Indians include many different peoples with unique languages and cultures and with different histories of contact with alcohol. Research among American Indians must also distinguish between populations on the reservation and those which have moved off the reservation, most frequently to large urban centers. A recent discussion (28) of alcohol use by Indians as "a community-based phenomenon" argues that alcohol problems are not psychological but based on the impact of urban, industrial living on Indians who lack "the necessary sacred social controls." Alcohol problems among Native Alaskans, too, seem rooted in the breakdown of traditional group customs. It is perhaps in response to these analyses that recent trends in alcoholism therapy with native peoples have moved toward emphasis on native culture and customs.

Black Americans. Surveys in the 1960s and more recent surveys have shown little difference in the drinking patterns of Black and White men: 38% of Blacks and 31% of Whites reported abstinence, heavy drinking by 19% and 22% (1). Whether Black heavy drinkers and alcoholics are less visible is a question we have not yet answered. Another question on which we have little information is the Black community's perception of and response to heavy drinking and alcohol problems.

Among Black women, 51% were abstainers, compared with 39% of the White women, but the proportion of heavy drinkers is in the opposite direction: 11% of the Black women and 4% of the White women reported heavy drinking (1). Interestingly, though, heavy drinking Black women reported a greater likelihood of a heavy drinker at home and they also reported drinking with a spouse or family member, patterns also reported by heavy drinking White women. Compared with other Black women, heavy drinking Black women are less oriented toward respectability and less apt to be regular churchgoers, and they are more permissive about drinking by men, more apt to drink in public places, and drink more often for reasons of escape (29).

Summarizing the characteristics of Black drinking drawn from the data of national surveys, Harper (30) suggests the following:

1. Blacks tend to be group rather than solitary drinkers.
2. Blacks tend to drink more frequently and heavily on weekends.

3. Blacks often use alcohol for social celebrations and for sexual activity.

4. Black and White adolescents in the South have similar drinking rates.

5. Many Blacks drink in public or in outdoor places, e.g., street corners, automobiles, in front of stores or residences.

6. Black college men often drink heavily on weekends in group settings.

7. The young Black urban male who drinks heavily often shows medical consequences and an arrest record. Urban Blacks tend to drink more than rural Blacks, and in the 15 to 30 age range, the Black male is most likely to be a victim of alcohol-related homicide. Violent consequences of alcohol use are much more serious for Blacks, particularly Black men.

8. Black alcoholics tend to be younger than White alcoholics.

9. Black alcoholics utilize treatment facilities less than do White alcoholics although it is a question whether that would be true for facilities designed for outreach in the Black community.

10. There is a strong association of problem drinking with health and social problems, particularly in crowded Black ghettos. There is a clear relationship between alcohol and homicide among Black men and women homicide offenders.

11. The life styles of Black and White Skid Row alcoholics differs: Blacks are younger, less likely to be homeless, less likely to get jobs or utilize welfare resources, and they tend to identify with the Black community.

12. Compared with Whites, Black alcoholics admitted to hospitals and clinics show stronger motivation for treatment, fewer complaints and more cooperation during the treatment process.

There is a high rate of liver disorders among Black women alcoholics, and recent research indicates serious biomedical consequences of excessive drinking among Blacks of both sexes. The generally found high rates of hypertension among Blacks compound the medical problems associated with heavy drinking. More information is needed about the relationship of drinking patterns, alcohol problems, health consequences, and the lower life expectancy of Blacks.

The data we have is pretty much limited to epidemiological and medical studies. Ethnographic studies of Black communities, such as *Tally's Corner* (31) or *Soulside* (32), could explore the role of alcohol in Black social life and community structure. What Black communities perceive to be alcohol problems needs to be understood if there is to be effective intervention and prevention. Black communities need to be viewed as complex, socially stratified structures: studies of drinking patterns of the Black poor, middle class, and well to do, are needed.

Hispanic Americans. Generalizations about Hispanic-American drinking patterns must be qualified: the Hispanics include sizable populations drawn from different countries and cultures. The most numerous are those who come from Mexican rural backgrounds and are most represented in farm labor in the U.S. The next larger subgroup, the Puerto Ricans, also come largely from agricultural and rural backgrounds but are concentrated more in service occupations in the northeast. The heaviest concentration of Mexican Americans is on the west coast and in the southwest. Both groups tend to be younger than the general population, and are characterized by lower educational achievement and income than the general population. Cubans and persons from South America were drawn largely from middle class backgrounds although the most recent influx of Cubans has changed that middle class character.

To complicate matters even further, there are variations in the number of years in the U.S. and in the proportion of second generation American born. Mexican Americans have one of the highest birthrates in the U.S. The first wave of Cuban refugees after the Castro revolution, on the other hand, contained a greater proportion of elderly people than the general U.S. population. Most Hispanics live in metropolitan areas and Mexican Americans are the largest group living outside metropolitan areas.

There appear to be high rates of alcohol problems among Mexican Americans and among Puerto Ricans living on the mainland, consistent with reports of alcohol problems at the points of origin. The rate of alcoholism in Mexico is reported by the World Health Organization to be among the highest of the countries on which it has information, and the rates of alcohol problems on the island of Puerto Rico are considered to be quite high (1).

The high rates of alcohol problems among these groups in the U.S. are indicated by statistics of arrests for driving while intoxicated, public drunkenness and alcohol-related offenses of all kinds. The statistics may be qualified by the differential likelihood of arrest as lower status people are more likely to be arrested by the police and both Mexican Americans and Puerto Ricans are overrepresented in the low income brackets. However, the greater incidence of alcohol problems among these Hispanic groups is supported by mortality statistics. Alcocer (33) reported, in his review, higher rates of death from cirrhosis among Hispanics than in the general U.S. population. Alcohol-related automobile accidents are also higher among Hispanics. Other indicators of high rates of drinking problems are the above-average concentrations of alcohol outlets in Spanish-speaking communities, high ratings of alcoholism as a community problem by the Spanish-speaking communities, and some evidence that acculturation stress and higher rates of alcoholism are linked (1). Discrimination, prejudice and low social status have been linked to alcohol problems and a high degree of stress is theorized among those who show a pattern of abandoning their root culture with little opportunity to achieve status and identify with the dominant culture.

There are many questions which need research. One finding reports a lower proportion of drinkers among young Hispanics than young Anglos and the question is, What produces the shift from relative moderation among the young to higher rates of alcohol use and problems among Hispanic adults? Another question relates to reports which consistently indicate a high proportion of abstainers among Hispanic women. Few Hispanic women apparently develop alcohol problems but there is some suggestion that increasing acculturation is associated with the breakdown of traditional sanctions against women drinking.

Still another question may be raised about cultural differences in the perception of alcohol problems and cultural differences in willingness to utilize treatment facilities. A few surveys suggest that Hispanics are more likely to view alcoholism as moral weakness, not as a condition requiring treatment. The underrepresentation of Hispanics in treatment facilities is consistent with their underutilization of mental health facilities in general and probably indicates an unwillingness to use the institutionalized, psychological approach of most treatment programs. The few outcome studies available suggest that Spanish-speaking patients do at least as well and generally better than Anglo patients. An unresolved question is whether outreach and therapy with indigenous personnel are more effective; it would certainly make sense to have at least part of the staff from similar cultural background in an agency designed to reach a particular population. Finally, there are family, religious and Spanish-speaking community support systems which must be utilized for effective intervention.

Asian Americans. Asian Americans, like Hispanic Americans, are comprised of a number of different groups: recent immigrants and first, second and third generation peoples from China, Japan, Korea, the Philippines, the Pacific islands, and southeast Asia. Generally, Asian Americans are concentrated in Hawaii and the west coast. Since they are relatively small in number and maintain "a low profile," what we know of Asian American use of alcohol tends to fit stereotypes: e.g., they have cultural taboos against heavy drinking and low rates of alcohol problems and they may have an allergic sensitivity to alcohol, the so-called cutaneous "flushing response" (34).

Research indicates that, on the whole, Asians drink less than Europeans and that more Asian Americans, both adolescents and adults, are abstainers than other Americans. There is a relatively low level of alcohol consumption in general but, among adolescents of Oriental background, it appears that those who drink often drink heavily (17); these may be acculturated, multiproblemed young people.

With increasing assimilation, there are shifts. For example, Issei (first generation immigrant Japanese) are considerably less permissive about drinking than Nisei (second generation) or Sansei (third generation). Both Chinese Americans and Japanese Americans have low arrest rates in general, but if a member of these groups is arrested, it is very likely to be an alcohol-related offense. It is possible that alcohol misuse occurs along with other deviant behaviors.

The idea of counseling services is alien to a traditional Oriental emphasis on the role of the family in dealing with its members' problems, and there is also shame attached to a family member's deviant behavior. Asian Americans underutilize community mental health services in general and Kitano (34) has raised the question of special treatment facilities and a needed emphasis on prevention. Urbanization, changes in family structure, economic mobility, cultural conflict, and acculturation are likely to push drinking and problem rates upward. Prevention programs for a group which has offered relatively few alcohol problems up to now seems a most rational course.

Discussion

The definition of "special populations" represents some real progress in recognizing the complexities of alcohol problems. Although the differentiations made among alcoholics in *Alcohol, Science and Society* are valid and important, the recognition of unique issues relating to alcoholics in different age and racial groups and of different gender is useful and good. The term has come to mean those groups who have special treatment needs and who have been underserved. If we are to talk of treatment needs, however, it would be well to consider the 1945 emphasis on classification by different types of drinking behaviors and different diagnostic features.

If we are ever to be effective in prevention, however, we must know a great deal more than we do now about the drinking customs and social uses of alcohol in the special populations. As a matter of fact, there is no reason why our study of social drinking patterns should be confined to the officially designated special populations. The study of drinking mores among some White ethnic groups (Italians, Irish, Jews, Armenians, etc.) contributes much to our understanding of the role played by alcohol among these groups and in the total society. The information we have about normative drinking in the various social classes, ethnic groups, and special populations, needs expansion, if for no other reason than to serve the ends of prevention.

REFERENCES

1. U.S. NATIONAL INSTITUTE ON ALCOHOL ABUSE AND ALCOHOLISM. Alcohol and health; third special report to the U.S. Congress from the Secretary of Health, Education and Welfare, June 1978; technical support document. Ernest P. Noble, editor. (DHEW Publ. No. ADM 79-832.) Washington, D.C.; U.S. Govt Print. Off.; 1979.

2. U.S. NATIONAL INSTITUTE ON ALCOHOL ABUSE AND ALCOHOLISM. Fourth special report to the U.S. Congress on alcohol and health from the Secretary of Health and Human Services, January 1981. John R. DeLuca, editor. (DHHS Publ. No. 81-1080.) Washington, D.C.; U.S. Govt Print. Off.; 1981.

3. BOWMAN, K. M. and JELLINEK, E. M. Alcohol addiction and its treatment. Pp. 1–80. In: JELLINEK, E. M., ed. Effects of alcohol on the individual; a critical exposition of present knowledge. Vol. 1. Alcohol addiction and chronic alcoholism. New Haven; Yale University Press; 1942.

4. LANDIS, C. Theories of the alcoholic personality. Pp. 129–142. In: Alcohol, science and society; twenty-nine lectures with discussions as given at the Yale Summer School of Alcohol Studies. New Haven; Quarterly Journal of Studies on Alcohol; 1945.

5. CIMBAL, W. Trinkerfürsorge als Teil der Verwahrolostenfürsorge. Allg. Z. Psychiat. 84: 52–86, 1926.

6. KNIGHT, R. P. The psychoanalytic treatment in a sanitarium of chronic addiction to alcohol. J. Amer. med. Ass. 111: 1443–1448, 1938.

7. MENNINGER, W. C. The treatment of chronic alcohol addiction. Bull. Menninger Clin. 2: 101–112, 1938.

8. SCHUCKIT, M. A., PITTS, R. N., JR., REICH, T., KING, L. J. and WINOKUR, G. Alcoholism. 1. Two types of alcoholism in women. Arch. gen. Psychiat. 20: 301–306, 1969.

9. JELLINEK, E. M. The disease concept of alcoholism. Highland Park, N.J.; Hillhouse Press; 1960.

10. PARTINGTON, J. T. and JOHNSON, F. G. Personality types among alcoholics. Quart. J. Stud. Alc. 30: 21–34, 1969.

11. BACON, S. D. Alcohol and complex society. Pp. 179–200. In: Alcohol, science and society; twenty-nine lectures with discussions as given at Yale Summer School of Alcohol Studies. New Haven; Quarterly Journal of Studies on Alcohol; 1945.

12. DOLLARD, J. Drinking mores of the social classes. Pp. 95–104. In:

Alcohol, science and society; twenty-nine lectures with discussions as given at the Yale Summer School of Alcohol Studies. New Haven; Quarterly Journal of Studies on Alcohol; 1945.

13. WARNER, W. L. and LUNT, P. S. The social life of a modern community. New Haven; Yale University Press; 1941.

14. MURPHY, A. J. Alcohol and pauperism. Pp. 239–249. In: Alcohol, science and society; twenty-nine lectures with discussions as given at the Yale Summer School of Alcohol Studies. New Haven; Quarterly Journal of Studies on Alcohol; 1945.

15. STRAUS, R. and McCARTHY, R. G. Nonaddictive pathological drinking patterns of homeless men. Quart. J. Stud. Alc. **12:** 601–611, 1951.

16. FLEMING, R. Medical treatment of the inebriate. Pp. 387–401. In: Alcohol, science and society; twenty-nine lectures with discussions as given at the Yale Summer School of Alcohol Studies. New Haven: Quarterly Journal of Studies on Alcohol; 1945.

17. RACHAL, J. V., MAISTO, S. A., GUESS, L. L. and HUBBARD, R. L. Alcohol use among adolescents. In: U.S. NATIONAL INSTITUTE ON ALCOHOL ABUSE AND ALCOHOLISM. Alcohol consumption and related problems. (Alcohol and Health Monogr. No. 1.) Rockville, Md. [In press.]

18. BRAUCHT, N. G. Problem drinking among adolescents; a review and analysis of psychosocial research. In: U.S. NATIONAL INSTITUTE ON ALCOHOL ABUSE AND ALCOHOLISM. Special population issues. (Alcohol and Health Monogr. No. 4.) Rockville, Md. [In press.]

19. GOMBERG, E. S. L. Drinking and problem drinking among the elderly. Ann Arbor; University of Michigan Institute of Gerontology; 1980.

20. GOMBERG, E. S. L. Patterns of alcohol use and abuse among the elderly. In: U.S. NATIONAL INSTITUTE ON ALCOHOL ABUSE AND ALCOHOLISM. Special population issues. (Alcohol and Health Monogr. No. 4.) Rockville, Md. [In press.]

21. BAILEY, M. B., HABERMAN, P. W. and ALKSNE, H. The epidemiology of alcoholism in an urban residential area. Quart. J. Stud. Alc. **26:** 19–40, 1965.

22. BAKER, S. M. Social case work with inebriates. Pp. 419–435. In: Alcohol, science and society; twenty-nine lectures with discussions as given at the Yale Summer School of Alcohol Studies. New Haven; Quarterly Journal of Studies on Alcohol; 1945.

23. LISANSKY, E. S. Alcoholism in women; social and psychological concomitants. I. Social history data. Quart. J. Stud. Alc. **18:** 588–623, 1957.

24. GOMBERG, E.S.L. Problems with alcohol and other drugs. Pp. 204–240. In: GOMBERG, E. S. L. and FRANKS, V. eds. Gender and disordered behavior; sex differences in psychopathology. New York; Brunner/Mazel; 1979.

25. GOMBERG, E. S. L. Women, sex roles, and alcohol problems. Professional Psychol. **12:** 146–155, 1981.

26. HILL, S. Biological consequences of alcoholism and alcohol-related problems among women. In: U.S. NATIONAL INSTITUTE ON ALCOHOL

ABUSE AND ALCOHOLISM. Special population issues. (Alcohol and Health Monogr. No. 4.) Rockville, Md. [In press.]

27. COCKERHAM, W. C. Drinking attitudes and practices among Wind River Reservation Indian Youth. J. Stud. Alc. **36:** 321–326, 1975.

28. THOMAS, R. K. The history of North American Indian alcohol use as a community-based phenomenon. J. Stud. Alc., Suppl. No. 9, pp. 29–39, 1981.

29. STERNE, M. W. and PITTMAN, D. J. Drinking patterns in the ghetto. 2 vol. St. Louis; Washington University Social Science Institute; 1972.

30. HARPER, F., ed. Alcohol abuse and Black America. Alexandria, Va.; Douglass; 1976.

31. LIEBOW, E. Tally's corner; a study of negro streetcorner men. Boston; Little, Brown; 1967.

32. HANNERZ, U. Soulside; inquiries into ghetto culture and community. New York; Columbia University Press; 1969.

33. ALCOCER, A. Alcohol use and abuse among the Hispanic American Population. In: U.S. NATIONAL INSTITUTE ON ALCOHOL ABUSE AND ALCOHOLISM. Special population issues. (Alcohol and Health Monogr. No. 4.) Rockville, Md. [In press.]

34. KITANO, H. Alcohol drinking patterns: the Asian Americans. In: U.S. NATIONAL INSTITUTE ON ALCOHOL ABUSE AND ALCOHOLISM. Special population issues. (Alcohol and Health Monogr. No. 4.) Rockville, Md. [In press.]

21

Alcohol and the Church

Rev. David C. Hancock

The source and background of the religious community's teachings and preachings about the substance alcohol make up the basic material covered by Dr. Roland H. Bainton in *Alcohol, Science and Society*'s chapter "The Churches and Alcohol" (1). Starting with the Old and New Testaments, he traced the church's concern with the rightness or wrongness of drinking down through Roman Catholic and Protestant church history, Prohibition and Repeal.

But what has happened since then? What has the church world been saying and doing about alcohol and alcohol problems during these last 35 years?

It is important to note at the outset that there are very few clergy active in the churches today who remember much about Prohibition or who were involved in ministry when the U.S.A. was a "dry" nation. Practically all present-day pastors and rabbis have served only congregations whose adult members could drink legally if they chose to.

Also, very few present-day clergy know much of the history and facts of the controversy which divided the churches and the nation between 1840 and 1933, but which has now pretty much subsided; namely, is drinking alcohol a sin?

Today 68 to 71% of the population (depending on whose statistics you use) have occasion to use beverage alcohol. This means that in most of the major religious denominations a large number of the members do drink. And the proportion of clergy who drink follows only a little behind that of the laity.

A survey of the United Presbyterian Church in the U.S.A. in June, 1979, found that 83% of its members drink, and 76% of its pastors (2). Among the ministers who were not abstainers, 43% said that they drank at least once a month; 25% said that they drank from one to three times a week; 8% four to six times a week; and 6% drank everyday; 15% of the pastors said that they drank "rarely."

Whatever a similar survey of other denominations might show, I think we can safely assume that the use of alcohol by both church members and clergy has increased steadily over the years since Prohibition, just as it has in the general population.

Along with the waning of the wet–dry controversy, and surely influenced by the increased use of alcohol in the United States, has come a lessening of concern about most societal problems related to the misuse of alcohol. Save for a few skirmishes here and there across the nation—local option elections, some law enforcement issue, or legislating the age of purchase of alcoholic beverages—the churches today are just not fighting much about alcohol-related issues.

Nevertheless, there still remains a great deal of ambivalence about alcohol. George Gallup reported in 1979 that nearly half of our population "basically disapprove of drinking even though some of them, of course, are drinking. And one in five would like to see a return to prohibition."[1] As someone put it a few years ago, we as a nation have not really decided whether drinking is good, bad, or indifferent.

It is not surprising, then, to find that the American religious community is still largely found among the ranks of what Selden Bacon has called "the avoiders," those who want nothing to do with alcohol problems.

But this lack of interest and action is not limited to the churches. It is reflected in many other disciplines as well: law, medicine, social work, education. Dr. Joseph Pursch has said that most physicians suffer from the "four-two-one syndrome: four years of medical school with only two hours of lectures on how to deal with America's number one health problem, alcoholism." By the same token, most clergy suffer from the "three-zero-one syndrome."

There is another current attitude which has affected the churches vis-à-vis alcohol problems. It is rooted in the influence which Alcoholics Anonymous has had in our society. A.A. and its effectiveness have become so well known that the churches, along with the general public, have bought the idea: "If you've got a drinking problem, go to A.A." Period! So this becomes for many if not most clergy the extent of their responsibility. Consequently, they show little concern about other societal problems related to the misuse of alcohol: drunken driving, lack of adequate treatment facilities, lack of effective alcohol education in schools and churches, and careless, unthinking public attitudes which more and more enlarge and

[1]GALLUP, G., Jr. An address before the Alcohol Information and Action Network of the United Presbyterian Church in the U.S.A., at its General Assembly, Kansas City, Missouri, 26 May 1979.

multiply the uses of alcohol as the sine qua non of grace and culture, and de rigueur for any kind of social entertaining.

But there are also some good signs. The church is beginning to see that there is much more to society's problems with alcohol than getting the problem drinker to A.A. There are some events and actions which show promise that the church is moving away from its old battle about drinking and not drinking to a much broader concern which sees that the chief issue is not just about a chemical, but a whole range of problems and attitudes, philosophies and values which affect and are affected by our use of alcohol. We shall try to trace and interpret some of this movement.

Just as much of the advancement in our nation's dealings with alcohol problems can be traced back to the Yale (now Rutgers) Center of Alcohol Studies and its Summer Schools, so, much of what has happened in the religious community goes back to the same beginnings.

Prodded by some graduates of the Yale Summer School of Alcohol Studies, the Presbyterian Church in 1946 issued a paper officially recognizing the alcoholic as an ill person and calling for a special ministry to him or her.

Informed by the same Yale experience, Roman Catholic clergy in 1948 organized the National Catholic Clergy Association as a fellowship to help with problems of alcoholism within its priesthood; and it is still providing valuable leadership and service to the church.

In 1951, the Reverend David A. Works, an Episcopal parish priest in the little village of North Conway, New Hampshire, attended the Yale Summer School of Alcohol Studies. Later that year he, together with other graduates of the Summer School, some members of the Yale Center of Alcohol Studies staff, representatives of the New Hampshire Council of Churches, and of the Episcopal Diocese of New Hampshire, formed the North Conway Institute (NCI).

Now based in Boston, NCI, more than any other institution in the religious world, has been the leader and catalyst in getting the American religious community aroused and moving to do something about alcohol problems. Since its 1951 founding this ecumenical, interfaith, interdisciplinary organization has been working diligently and effectively toward reducing and preventing problem drinking. Indeed, the story of "Alcohol and the Church" over the past three decades cannot be written apart from NCI. Most of the significant ecumenical actions and many of the denominational policies and actions on alcohol problems have grown directly out of or have been influenced by NCI's spirit and work.

Beginning in 1955, NCI has each year convened a summer "assembly" of religious leaders from a wide variety of denominations and faiths, both laity and clergy, together with representatives of many disciplines and professions—from the most dedicated temperance organizations to representatives of the alcoholic beverage industries. And through the years NCI has demonstrated that persons of radically different feelings about the substance alcohol can work together toward solving the problems in society which are related to its misuse.

NCI's pioneering efforts are demonstrated in its first four themes: "The Churches and Alcoholism," "What Shall We Say to Young People?" "The Churches and Social Drinking," and "The Church's Concern for the Alcohol Education of its People." Other early topics were: "Pastoral Care of Alcoholics and Their Families," "Inebriety, Alcoholism and the Church," and "The Church's Role in the Prevention of Alcoholism."

NCI had a large part in helping the National Council of Churches prepare its landmark 1958 statement, "The Churches and Alcohol," which called on its member denominations to minister to alcoholics and their families, to provide education about alcohol for their members and for the general public, and to work for more effective social and legal controls to help reduce the damage caused by problem drinking.

The year 1958 produced another landmark statement on alcohol. It was the brave publication of the Protestant Episcopal Church, *Alcohol, Alcoholism, and Social Drinking* (3). Here was the first time that a major denominational publication dealt honestly with the issue of social drinking. This pamphlet sought to interpret for its members the theological basis for the church's official position on alcohol, and it offered guidance to members for making their personal decisions about drinking or abstaining. It is one of the earliest of contemporary references in religious literature to "responsible use of alcoholic beverages" along with specific suggestions on the responsibility of a host or hostess when offering alcoholic beverages.

Another noteworthy advance in the history of the church's dealing with alcohol problems (also influenced by NCI) was the work in 1965 by an interfaith group in the Boston area called The Ecumenical Council on Alcohol Problems. These church leaders, after citing the serious societal problems related to the misuse of alcohol, declared (4):

"These problems are not new but they are acute, and are made more so by an attitude of complacency and irresponsibility on the part of the general public in whose hands the final determination of alcohol policy lies. It is urgent that churchmen and others concerned with human needs and the moral foundations of our society endeavor to create a more responsible public attitude toward drinking.

"We believe that we all may unite on the ground of the virtue of sobriety. This can be practiced in two ways. One is by total abstinence from beverage alcohol for religious motives. The other is by true moderation in the use of alcohol, also for religious motives. On this common ground the virtue of sobriety may be practiced both by abstainers as well as by those who drink moderately.

"Although differences of conscientious conviction in relation to certain current drinking customs exist among us, the area of our agreement with regard to drunkenness and alcoholism is sufficiently large and significant as to enable us to unite our best efforts for the alleviation and ultimate solution of these alcohol-related problems."

The Ecumenical Council also called on all religious bodies to help their members examine carefully "the personal and social issues involved in drinking," and to "understand *the role of alcohol in society* [italics mine] and the gravity of the problems connected with its use; to help persons understand their own motivations for drinking or abstaining so that an individual choice may be made free from the necessity to conform" (4).

The American Lutheran Church in 1968, in a pamphlet written by three Yale Summer School of Alcohol Studies graduates, said, "It is time for the churches to shift the discussion of alcohol problems out of the realm of dogmatic rigidities into the openness of Christian freedom and responsibility" (5); i.e., stop trying to legislate the rightness or wrongness of drinking. It further called on Lutherans to help Americans face and come to grips with the whole range of alcohol problems.

In the United Methodist Church, from the mid-1950s onward, staff members from its Board of Temperance (later renamed The Department of Alcohol and Drug Concerns) were strong supporters of and participants in NCI while the Methodists were re-examining their church policy which required Methodists and their clergy to abstain from beverage alcohol. Sparked by their church's study of the report of the Cooperative Commission on the Study of Alcoholism titled, *Alcohol Problems; A Report to the Nation* (6), the church spent nearly a year trying to resolve the drinking issue.

Finally, the Methodist General Conference in April, 1968, voted to eliminate all "legal" requirements for abstinence in both members and clergy. The church recommended abstinence, but no longer required it, thus making the choice of whether to drink a matter of individual conscience.

Two years later the Presbyterian Church in the United States (sometimes referred to as the "Southern Presbyterian") took similar action citing the same principle of freedom of conscience before God.

Another milestone in the church's journey toward a broader, more realistic approach to alcohol problems is found in its response to the publication of *Alcohol Problems; A Report to the Nation* in October, 1967. In December that year the National Council of Churches appointed a Task Force on Alcohol Problems to study the report, and to recommend to the churches approaches and actions for dealing with the whole range of alcohol problems which the report had identified.

This Task Force, made up of 34 church and alcohol authorities representing various denominations, disciplines, and points of view about alcohol, worked for more than a year. In January, 1969, it presented its report to the General Board of the National Council at its meeting in Memphis. The Task Force's report, titled simply *Problem Drinking* (7) could be studied with great profit by every church member or clergyperson concerned about the subject.

Problem Drinking's chief significance is that it lifts the issue of alcohol problems above narrow parochial interests and places them before the entire religious community as a concern which all American society should face and deal with. Further, it spells out some specific interpretations and recommended actions for the churches to take in the six areas of problems which were identified by *Alcohol Problems; A Report to the Nation*.

Incidentally, the Task Force was the only group in the U.S.A. which made any kind of formal response to the Cooperative Commission report.

In *Problem Drinking* the Task Force calls the religious community "to open up for discussion and responsible action" the entire range of problems related to the use of alcohol in American society. Not only the illness or disorder of alcoholism, but also the problems of chronic drunkenness offenders, drinking and driving, legal controls around alcohol, youth education about alcohol, and public attitudes and customs which determine how alcohol is used. These issues are addressed from the perspective of the Judeo-Christian heritage. By deliberately avoiding the pitfalls of the old wet–dry controversy, *Problem Drinking* removes most of the thorns from what has been a painful area too long avoided by most of the religious community.

In 1972, under a contract with the National Highway Traffic Safety Administration of the U.S. Department of Transportation, the

North Conway Institute carefully studied five Alcohol Safety Action Programs around the nation, then prepared a guide book for the religious community outlining specific, practical steps which churches and synagogues could take to help reduce the fatalities caused on our nation's highways by drunken drivers. Unfortunately, however, the churches still remain largely silent on the issue. For example, have you ever heard a sermon on drunken driving? Yet it kills more Americans in two or three years than died in Vietnam in ten years!

In the late 1960s and continuing on to the present the NCI summer assemblies began to address the role and responsibility of the churches in the primary prevention of alcohol problems. At its 1973 assembly (funded by a special grant from the National Institute of Mental Health) it identified many of the basic issues which are involved in the prevention as well as the treatment of alcohol (and other drug) problems. Most of those identified issues are easily recognized by the churches as lying within their area of special expertise; e.g., personal goals and values, the need for meaningful human relationships and healthy family life, the role of the peer group, developing skills and resources for coping with life's problems, the search for self-understanding, for God, for reality.

Continuing its emphasis on prevention, the NCI assembly in 1977 carefully reviewed the findings and recommendations of the *Final Report* of the Education Commission of the States' Task Force on Responsible Decisions about Alcohol (8). The Task Force recognized that the church plays an important role in the formation of character and the capacity for decision-making.

In the published papers of that 1977 assembly (9) NCI reported to the churches that one of the most important issues confronting the church and the nation is to find a way "to get the prevention of alcohol-related harm on the agenda of the country." The churches, said the assembly, have "a crucial leadership role" to play in helping devise and implement "a national policy on alcohol" because they are concerned "about the spiritual values and behaviors inherent in the use and nonuse of alcohol beverages."

One of the conclusions which can be drawn about the work of the churches in relation to alcohol problems over the last 35 years is that more and more of them are moving from concern primarily about the chemical, alcohol, to a comprehensive, holistic approach to the whole range of societal problems associated with its misuse; and then on beyond that to help persons live their lives in such ways as to avoid trouble with alcohol in the first place.

One of the great truths demonstrated by A.A. is that recovery from alcoholism requires more than just "putting the cork in the bottle." Recovery requires a reliance on a "Higher Power" and a re-ordering of one's life. So also the religious community is beginning to see that alcohol and drug problems involve not just chemicals but the total person and his or her relationships to "a living God at work in the world in which we live, work, play and worship."

The church is beginning to see that its mission is not to be a pressure group, a social agency or a political party. Its role becomes, rather, that of a catalyst, a motivator to stimulate and to challenge the whole of society to examine the moral and spiritual aspects of critical issues, and whatever in society denies to God's people the abundant life.

As Dr. Francis Sayre, Dean of the Washington Cathedral, put it a few years ago: "Perhaps the church's peculiar 'thing' just now may be the humble job of just being what it is—a symbol and a sign of the Lord, a reminder to men that there is another dimension of life beyond our own making. Perhaps this is a time in the church's life when we are called simply to be a landmark in a lost land."

The Yale (now Rutgers) Summer School of Alcohol Studies (SSAS), recognizing the significant role of the churches in dealing with alcohol problems, has from its earliest days offered one or two courses designed especially for clergy and others interested in the pastoral counseling of alcoholics and their families, and in helping the churches become more effective in dealing with alcohol prob-lems at the parish and community level. Hundreds of these SSAS alumni can be found across the nation working in other schools of alcohol studies, in colleges, theological seminaries, in alcoholism treatment centers, alcohol councils, training institutes, etc., helping provide opportunities for church leaders to increase their under-standing and improve their skills in dealing with alcohol problems.

There is encouraging evidence that the church is beginning to awaken, both to the seriousness of these problems, and to the importance of its own unique role and opportunity.

But, alas, there is also discouraging evidence of a considerable gap between the seriousness of the problems and the church's efforts to deal with them. This is pointed up in the findings of the 1979 survey on the church and alcohol problems by the United Presby-terian Church (2).

Based on an 80% return of a four-page questionnaire mailed to nearly 4000 Presbyterians, both laity and clergy, we found that 44% of the laity and 72% of the clergy said that "the consumption of beverage alcohol was a problem" in their community.

To the question, "Do you feel that the abuse of alcohol is a problem area in which the church should be involved," 46% of the members and 83% of the clergy said yes. But 81% of the lay members and 76% of the clergy said their churches were *not* currently involved in programs which address the issue of alcohol misuse.

Of those who said that their church was not involved in doing something about alcohol misuse, we asked, Why? Of several suggested possible reasons which might account for their church's lack of activity, 75% checked "other congregational programs have higher priorities," and 65%, "alcohol problems have not surfaced as an issue in the congregation." And about two-thirds said that doing something about alcohol problems "should be left to other community agencies, or to A.A. and Al-Anon."

We also asked whether fear of causing controversy within their congregations, or fear of alienating church members who drink had anything to do with the lack of involvement with alcohol problems. But only a fourth to a third of the clergy marked these fears as reasons for their avoidance of alcohol issues.

Because the Gallup Poll in 1978[2] had found that 24% of the U.S. population reported that liquor had "caused trouble" in their families, we wanted to know whether these families were seeking help from their clergy. Gallup had asked: "If you or someone else in your family had a drinking problem, where or to whom would you turn for help?" "To Alcoholics Anonymous," was the response of 42% of the general public; 9% listed God, the Lord, prayer, or the Bible; 8% said they would go to a priest, minister, or rabbi; and 8% said they would seek help from the church and its trained personnel. Combining these last three categories of help results in a total of 25% of Americans who would turn to some kind of religious resource for help with their own or a loved one's drinking problem.

Borrowing a question which Harold A. Mulford used in his 1963 survey of Iowa clergy, we asked the Presbyterian ministers, "During the past twelve months how many persons have sought help from you specifically about their own drinking problems or those of a family member?"

One-fourth of the Presbyterian ministers said that no one had sought help about any kind of a drinking problem; 44% said that from one to three persons had come about a drinking problem; and 23% had seen from four to ten persons with such problems in a year.

So, if people with alcohol problems are not exactly beating down the door of the pastor's study, can or should the pastor do anything about the nearly one-fourth of the nation's families where

[2]AMERICAN INSTITUTE OF PUBLIC OPINION. Gallup Poll news releases, Princeton, N.J., 2, 3 July 1978.

"liquor is causing trouble"? Remember that 83% of the Presbyterian clergy respondents had said that the church should be doing something about alcohol problems.

As one of the final questions, then, we asked the clergy, "During the past twelve months have you preached about any subject related to the use of alcohol?" and "Have you spoken in public about any subject related to the use of alcoholic beverages?" We wondered what kind of initiative pastors were taking to deal with this set of social and personal problems, even if only a few persons were seeking their direct help.

Sixty-three per cent of Presbyterian ministers had not discussed from the pulpit any subject related to the use or misuse of alcohol. And 71% had not spoken about the subject in any kind of public setting. Yet 83% had said that the church should be doing something.

This is how it was with the two-and-a-half-million member United Presbyterian Church in the U.S.A. and the problems of alcohol in the summer of 1979. How is it with other denominations? It would be helpful to know. But my guess is that the picture would be largely similar among most of them.

But even though the churches are not as highly involved in these issues as some of us would wish, it is encouraging to note the large numbers of members and clergy who feel strongly that the church *must* get to work on these problems. And increasing numbers are finding that they can deal constructively with alcohol and drug problems regardless of their personal views about the use or nonuse of any particular chemical.

The religious leaders, both lay and clergy, who had assumed their responsibility in the alcohol–drug problems area learned from experience that the place where we all have to begin is with the victims of problem drinking, both the afflicted and the affected, the alcoholics and their families, ministering to them with love and compassion and understanding.

Pastors and church members have a responsibility to learn about alcoholism—its dynamics, how it affects people, how to recognize it, and how to intervene in a constructive way. And how to help the alcoholic's family even though the drinker may not yet be ready to seek help. A friend of mine who directs a refuge center for families in crisis said to me not long ago, "Why do pastors so often miss this illness among their members?"

One reason, at least, may be our wrong attitudes. In one of my early parishes I watched a young woman my age die of alcoholism without realizing what was happening to her. For whenever I made pastoral calls in the home, she had a drink in her hand. Because I considered her drinking a sin it was impossible for me to relate to her in a loving or pastoral manner—plus the fact that I knew absolutely nothing about alcoholism.

For church leaders to deal effectively with alcoholism does not mean that they must "cure" all the alcoholics in the congregation. But it does mean that they will make problem drinking a concern of high priority and begin to learn how to deal with it. And one of the very first things to do is to purchase some A.A. and Al-Anon literature for the church's literature rack, so that those who need it can at least get some help through the printed word.

A next step is to find a few persons in the congregation who are interested or can be interested in learning more about alcohol problems. There *are* such persons in every congregation if you'll take the trouble to find them: educators, physicians, nurses, social workers, counselors, personnel workers, parents, young people, persons working in law enforcement, and persons who have close personal experience with their own or someone else's alcoholism. Such a group can become the church's formal or informal alcohol–drug concern and action committee. They can begin to educate themselves, then discover and get acquainted with the other community agencies which are at work on these problems: Al-Anon, A.A., treatment centers, outpatient clinics, mental health centers, crisis phone lines, etc. In this way the church can become a member of the community "team" and pick up on its unique opportunity and responsibility for helping to interpret the inescapable spiritual and moral dimensions of alcohol problems.

A.A., Al-Anon, and Alateen, which can be found in most communities of any size, have "open" meetings from time to time, meetings at which anyone is welcome, whether there is a drinking problem or not. Attend these open meetings. You'll be cheered by the warm reception and amazed at the understanding you'll gain; and you'll find their excellent literature a great help with your education about alcoholism and about how to help alcoholics and their families.

I feel that the church has a responsibility to call all persons, young and old, to examine the role and meaning of alcohol in our society. And if we drink, we need to examine most carefully the place and purpose of drinking in our own lives. Whether we like it or not, use it or not, alcohol is a part of our culture, probably here to stay. But how well do we understand it: what it is, what it does to persons, and what drinking really means?

From a good many years' experience working in this field I have seen the following hypothesis proven true in countless persons, and I'm about ready to call it Hancock's Law; namely, The more important your drug of choice is to you, the more it means to you, the more you feel that it does for you, the more trouble you are going to have with it.

Every person who drinks should take a very careful look at his or her drinking. The little grandmother who has a glass of wine on Christmas Eve, and that's it for the year, is not going to have much trouble with alcohol. But the man who feels he must drink to be a man among men, to be accepted, or to project the kind of image he wants to have; or who drinks to deal with boredom, frustration, loneliness, anxiety; who drinks to cope or to solve his problems or to change the way he feels; or who cannot have a good time or be comfortable without alcohol—that man is headed for trouble sooner or later.

Without condemning church members who drink, nor canonizing those who abstain, and without trying to lay down rules, the church in its role as prophet can rightly call us all to look most carefully at our use or nonuse of alcohol. The National Council of Churches's Task Force in 1969, called for "widespread public discussion which would help people clarify their own attitudes, and enable them to see how the feelings and practices of a society are significant factors in the development of alcoholism and other alcohol problems" (7). They went on to say, "The religious community is uniquely equipped and strategically placed to develop new attitudes within society which will enable society to throw off its indifferences and engage in those preventive and therapeutic measures which aim at enabling persons to make responsible personal and social choices about the shape their lives will take."

In truth, the church is the only group in this country which can call us to assess honestly our own use or nonuse of all potentially harmful substances, and to evaluate the real place, role, meaning, and significance of alcohol and other drugs in one's life and in society at large.

The church, in grappling with alcohol problems, must of necessity begin with ministry to persons who are hurting because of a drinking problem. But we cannot stop there. We must go beyond the illness of alcoholism and try to head off problem drinking before it starts harming the drinker or loved ones.

Therefore, we must get into the area of public attitudes, and look at drinking customs and the practices in which alcohol misuse occurs. It is here that we can begin to do something toward the prevention of alcohol problems. As Howard Clinebell has put it, the church "can no longer ignore the socially destructive effects of alcohol," even in situations where addiction is not involved, nor can it ignore "the dangerous patterns of so-called 'social drinking' which provide the soil in which the seeds of addiction easily take root."[3]

The church should be able to help people to look honestly at both the positive and the negative aspects of alcohol use—the possible benefits and the possible dangers—and ask themselves some serious questions: Why do I drink? Why do I abstain? What does drinking really mean to me? What risks do I want to take with it? What am I doing, what am I saying with my drinking? How does my drinking affect the quality of my relationship with my family? What kind of pattern do I set for my children or my neighbor's children? And what effect does my drinking have on my spiritual life, my relationship with God?

And the church should not hesitate to ask some questions about the appropriate role of alcohol in our society. What is its true place and meaning? Does the drug alcohol have to be such an important substance? Do we have to put it on a pedestal and make the ability to drink the end-all and be-all of sociability, or the proof of sophistication and adulthood? While recognizing that there are some positive values to the safe and proper use of alcohol, the church still has the right to challenge some of our common attitudes which have cast alcohol in the role of panacea, problem-solver, prime social lubricant, and hallmark of hospitality.

Nor can the church forget its responsibility to the nearly one-third of our nation's citizens who choose not to drink. We need to insist broadly and loudly that their abstaining does not make them strange, peculiar, anti-social, or maladjusted, any more than it makes them angels. It is time for all of society, not just the church and A.A., to champion the idea that it's okay not to drink, and that abstinence is a perfectly respectable, socially gracious, and in some cases life-saving option which should be accepted as being equally as adult, warm, sophisticated, and responsible as drinking.

The church can also call to society's attention a very, very important truth which is not generally recognized; namely, that one does not have to be "an alcoholic" to have a problem with alcohol. All too often in our culture the risk potential or the growth-inhibiting capacities of the drug alcohol are ignored or denied. Ernest Shepherd pointed out years ago that many people who are far from being alcoholics still use alcohol inappropriately, many as a crutch, and, consequently, do not develop their fullest capabilities.

[3]CLINEBELL, H. J. A multi-dimensional model of prevention in American churches. Presented at the International Congress on Alcohol and Alcoholism, Washington, D.C., 19 September 1968.

There seems to be a myth abroad in our land that if you're not a falling-down drunk, then you don't ever need to look at your drinking. Just drink up! "Grab the gusto!" Enjoy! In 1978 Gallup found[2] that 47% of the nation's drinkers had no specific rules or guidelines for their own use of alcohol. And 51% had no guidelines for their children's use of it.

The church, in calling society to evaluate the place and meaning of alcohol in our culture, will do well to remind us of two cold, hard facts of which we must forever be aware. First, alcohol is a drug. It is a mind-altering, mood-changing drug, a central nervous system depressant similar to ether and chloroform. It is an intoxicant. It can produce not only welcome effects, relieve stress, reduce anxiety, create feelings of euphoria; but it can also, even in relatively small amounts, produce harmful effects such as impairment of conscience, judgment, reason, insight, memory, speech, hearing, vision, and muscular coordination. And taking alcohol in combination with other drugs, even legal prescription or over-the-counter drugs, can be dangerous, sometimes fatal.

The second hard fact about alcohol is that it is potentially addicting. You can become dependent on it, "hooked" on it. As you come to use it more and more regularly you can gradually and subtly reach the point where you are uncomfortable without it, where you rely on it, where you need it to cope, to function. And ultimately you have to have it to live. Or so you think. And that is alcoholism. There is an old saying in A.A. that the chains of addiction are seldom felt until they are too strong to break. Yes, alcohol can produce addiction. Everyone who drinks should be constantly aware of that fact, and should regularly examine and reassess his involvement with alcohol.

In light of these facts about the nature of alcohol the church can now suggest to those who would use it safely and wisely that they should never, never use it to deal with their problems of living, never use it as a medicine, or for coping. To put it a different way, *never* take a drink when you "need" it! Using a drug like alcohol to manage your troubles can actually interfere with your own native ability to conquer them. The temporary euphoria or escape it gives can sabotage your efforts to deal with troubles more realistically. There are no adequate chemical cures for "humanitis"—the ordinary, everyday whips and scorns of daily living.

What happens to my strength as a person when I depend on a mood-altering chemical to help me face a crisis or manage my life? Can alcohol really deliver all the rewards we have been led to expect of it?

There is, finally, one other way in which the church can help its members avoid problems with alcohol. That, as we pointed out earlier, is in helping them to understand some of the ethical, moral, and spiritual dimensions of alcohol use.

For example, what is the social impact of my drinking? How does my use of alcohol affect others? Even smoking can have serious social impact; for as G. B. Shaw observed, "It is impossible for the smoker and the nonsmoker to be equally free in the same railway car." How does my drinking affect my spouse, my children, my parents, my neighbor, or other drivers on the highway? Drinking is one of the rights which adults have in our society. But it is a right which carries the responsibility to drink in such a way that it does not harm me or anyone else. If a man's only problem with drink is his wife's objecting to it, then he must decide whether drinking is more important than his marriage; and if it is, then he most certainly has a drinking problem.

And then there are the spiritual dimensions. How does my drinking affect my relationship with God? What does it do to my conscience, my spiritual perception and sensitivity; how does it square with my value or belief system? It is with good reason that the Bible again and again condemns drunkenness, even though it does not forbid the drinking of wine. We are called to love God with all our heart, soul, mind and strength. But the misuse of God's good gifts can impair our ability to do so. Susanna Wesley, John Wesley's mother, had this definition of sin: "Whatever weakens your reason, impairs the tenderness of your conscience, obscures your sense of God, or takes off the relish of spiritual things—that is sin to you."

It is important to realize that alcoholism involves philosophical and theological issues, not just chemicals or physical symptoms. As someone has said, "You don't help the alcoholic just by getting his enlarged liver back inside his pants. You have to help him change his way of living!" And this is what the church is all about. This is its reason for being.

Indeed, the religious community is the only resource, the only institution or fellowship in our society whose leaders are especially trained in helping people to find meaning in life, to understand its spiritual dimensions, its non-physical, non-chemical aspects.

And the church is to be found everywhere. Its multitudes of educated clergy are experienced in working with children, youth, adults, and families; and they are welcomed into their members' homes without invitation or warrant. The church's opportunity to educate its people and to shape public opinion is matched only by schools and the media. Amazingly, it operates very successfully without any tax dollars; and there are no fees for its services. It has access to persons and families who are hurting, who are in trouble, to offer them God's love and compassion, acceptance and healing.

Obviously the religious community does not have all the answers to the prevention and treatment of problem drinking, but it does have a significant contribution to make far greater than it has yet made. It must now begin to take its rightful place as a member of the community team.

REFERENCES

1. BAINTON, R. H. The churches and alcohol. Pp. 287–298. In: Alcohol, science and society; twenty-nine lectures with discussions as given at the Yale Summer School of Alcohol Studies. New Haven; Quarterly Journal of Studies on Alcohol; 1945.

2. UNITED PRESBYTERIAN CHURCH IN THE U.S.A. RESEARCH DIVISION. Attitudes toward alcohol use and abuse. New York; 1979.

3. PROTESTANT EPISCOPAL CHURCH IN THE U.S.A. JOINT COMMISSION ON ALCOHOLISM. Alcohol, alcoholism, and social drinking. Greenwich, Conn.; Seabury; 1958. (Reprinted by North Conway Institue, 1979.)

4. NORTH CONWAY INSTITUTE. Alcohol and the American churches. Boston; 1967.

5. AMERICAN LUTHERAN CHURCH. COMMISSION ON RESEARCH AND SOCIAL ACTION. Alcohol problems—in bits or as one? Minneapolis; 1968.

6. COOPERATIVE COMMISSION ON THE STUDY OF ALCOHOLISM. Alcohol problems; a report to the nation. Prepared by T. F. A. Plaut. New York; Oxford University Press; 1967.

7. NATIONAL COUNCIL OF CHURCHES OF CHRIST IN THE U.S.A. Problem drinking; report of the task force on alcohol problems. New York; 1969.

8. EDUCATION COMMISSION OF THE STATES. TASK FORCE ON RESPONSIBLE DECISIONS ABOUT ALCOHOL. Final report. Denver; 1977.

9. NORTH CONWAY INSTITUTE. Responsible decisions about alcohol; a report to the churches. Boston; 1977.

22

Alcohol, Science and Social Control

Robin Room

Alcohol, Science and Society was both a beginning and an ending. In its pages we can glimpse the beginning of the publicly-oriented alcoholism movement, a movement which eventually brought into being alcoholism agencies in every state, the National Institute on Alcohol Abuse and Alcoholism, and the publicly-funded treatment systems for alcoholics which now treat perhaps 1 million cases a year in the U.S. I term it the "publicly-oriented alcoholism movement" to distinguish it from another intertwined social movement, Alcoholics Anonymous. The existence of A.A. was important in shaping the publicly-oriented alcoholism movement, and many people were active in both movements, but the aims have been very different. A.A. directs its attention inward, at self-help and mutual support. While outwardly-directed activity ("twelfth-step work") is encouraged, it is seen as a personal matter by individuals rather than as a collective activity.

The public health alcoholism movement's attention, on the other hand, is directed outwards, towards persuading the larger society not only to take a different attitude toward alcoholics but also to expend public resources on treatment. Dwight Anderson propounded the position in the pages of the *Quarterly Journal of Studies on Alcohol* in 1942 (1), and repeated it at the first Yale Summer School in 1943. In *Alcohol, Science and Society*, the argument appears again, tacked on to Anderson's talk on a quite different topic: "Here are four postulates: . . . first, that the problem drinker is a sick man, exceptionally reactive to alcohol; second, that he can be helped; third, that he is worth helping; fourth, that the problem is therefore the responsibility of the healing profession, as well as the established health authorities, and the public generally" (2, *p. 367*).

Beyond an argument, however, in 1944 the nascent movement was in process of developing an organization that eventually became

Preparation of this chapter was supported by alcohol research center grant No. AA-03524, from the U.S. National Institute on Alcohol Abuse and Alcoholism.

the National Council on Alcoholism; the Yale (now Rutgers) Center of Alcohol Studies was in the process of becoming the mecca of the movement; and the Summer School, of which *Alcohol, Science and Society* is a record, was already becoming an occasion of pilgrimage which validated one's authority in the movement. Within the next few months, the Center of Alcohol Studies would start the Yale Plan Clinics and become involved with the precedent-setting Connecticut State Commission on Alcoholism. In a portent of this future, Anderson announced: "There is an organization being formed . . . for the purpose of education on the subject of alcohol*ism*. It is to be called, The National Committee for Education on Alcoholism. . . . Mrs. Marty Mann, who is a student here with you, is to be Executive Director" (2, *p. 368*).

The story of the brilliant success of the subsequent campaign, by a combination of talented recovered alcoholics with public-relations backgrounds and versatile scholar-entrepreneurs from Yale, joined eventually by many others, has not yet been fully told, although it has its early chronicles (3, 4) and its more recent exegeses (5–8). It would have taken an unusually prescient prophet to have predicted in 1945 where the movement has brought us today.

In *Alcohol, Science and Society*, then, we can already discern around the margins of the main action of the Summer School, a very significant beginning. But the book is also a kind of ending. From some perspectives, it is the last significant glimpse of a world before it was occluded. The world of research and thought about alcohol which succeeded it is enormously larger in its population of scientists and more developed in much of its knowledge. As Anderson had argued to the research constituency (1), one effect of the success of the alcoholism movement was enormously increased support for alcohol research. But in some ways the world of alcohol studies in the era following *Alcohol, Science and Society* was foreshortened and constrained compared to the world it replaced.

Alcohol studies as mirrored in *Alcohol, Science and Society* has a sense of history. Participants knew about the turn-of-the-century research of the Committee of Fifty to Investigate the Liquor Problem (9, *pp. 240–243*), the origin of the drunkenness offense in ecclesiastical law (10, *pp. 301–305*), Father Mathew's temperance crusade among the Irish (11, *p. 295*), and Yandell Henderson's plea for dilution at the time of Repeal (12, *p. 41*). In the succeeding era, this sense of history was lost; indeed, I believe that many people in the field today might place *Alcohol, Science and Society* itself in an era of prehistory.

In *Alcohol, Science and Society*, alcohol problems are seen in the perspective of structural and institutional factors as well as in terms of the individual problem drinker. Consideration is given to the interaction of the individual and the social agency, to the role of alcohol in complex society, to inebriety's social aspects, and to the laws controlling the availability of alcoholic beverages. The extent to which structural and interactional sense of the definition of alcohol problems has faded is apparent in a comparison of the table of contents of *Alcohol, Science and Society* with that of the present volume.

Perhaps most crucially, the field of discourse in *Alcohol, Science and Society* is defined as "alcohol studies," and not "alcoholism research." The sparseness of references to "alcoholism" partly reflects the book's appearance at a time of terminological transition. Indeed, when "alcoholism" is used, it is sometimes in the older sense of the health consequences of long-term drinking. The term often used in the book instead is the then already obsolescent "inebriety." But inebriety was meant in a broad sense of problems of drunkenness, rather than in the restricted sense of gamma alcoholism. For most of the lecturers in *Alcohol, Science and Society*, there was more to alcohol studies than research on alcoholism.

The breadth of vision of *Alcohol, Science and Society* reflects the breadth of interests and expertise of the research staff of the early Yale Center of Alcohol Studies, and particularly of the Director of its Summer School, E. M. Jellinek. This same breadth of perspective on alcohol studies can be seen in the early volumes of the *Quarterly Journal of Studies on Alcohol*. The historical depth of *Alcohol, Science and Society*'s view of alcohol studies also reflects the presence among the staff and students at the Summer School of moderate temperance forces, many of whom brought substantial knowledge to bear in the dialogue with the Yale researchers, and gently reminded the audience of the historical antecedents for a disease conceptualization (e.g., 13, *p. 417*).

After the appearance of *Alcohol, Science and Society*, the focus of research and attention narrowed to center on "clinical alcoholism." Despite his own wide erudition, Jellinek himself had presaged this narrowing in his influential "Outline of Basic Policies for a Research Program on Problems of Alcohol" (14), resisted at the time by Bacon (15). Looking back in 1960, Jellinek acknowledged ruefully that "it must be admitted that in America, the scientific literature and the public and private agencies concerned with 'alcoholism' have concentrated to such a degree on the true alcohol addict and the problem drinker that other important problems arising from the use of alcoholic beverages have been neglected" (16, *p. 174*).

The shift in focus in the pages of the *Quarterly Journal* after 1945 reflected a change in the direction of the Yale Center of Alcohol Studies. With the foundation of the Yale Plan Clinics and of what became the National Council on Alcoholism, what had been a research organization also became the nucleus of a rehabilitation and social movement. Yale Center staff embarked on a heady round of nationwide speaking tours, radio network broadcasts, and promotional projects (5). The shift to an activist stance coincided with and helped precipitate an estrangement from even the "moderates" in the temperance movement (17). The broader agendas of *Alcohol, Science and Society* were submerged in the flood of alcoholism movement concerns.

As early as 1960, social scientists were calling for a redistancing of research from alcoholism movement agendas (e.g., 18). Twenty years after this new beginning, vigorous traditions of historical research from primary sources are flourishing.[1] Economic and political economic research has broadened in several directions (e.g., 21).[2] The anthropological, sociological and psychosocial literatures have burgeoned beyond the possibility of coherent summarization (15). Although the over-all weight of the literature is still heavily oriented toward the individual (22), sociological, and more generally, social science research in alcohol studies are no longer constrained within the perspectives of the alcoholism movement.

One aspect of the newer lines of research has been a renewed interest in social controls and drinking behavior. In the titles of the lectures in *Alcohol, Science and Society*, "control" makes only one appearance, in Baird's magisterial lecture (10) on legal controls of drinking. But the broader sociological sense of "social control" is mentioned in Bacon's lecture (23) on "Alcohol and Complex Society," and this broader sense is implicit in the contributions of others.

As I have noted elsewhere (24), the concept of "control" has a number of meanings in the alcohol literature. In the framework alcoholism movement, "control" has referred primarily to the "pathognomic symptom" of the classic disease concept of alcoholism, loss of control over drinking, and thus over one's life (25). Levine (26)

[1]See, for example, Blocker's bibliographies (19, 20).

[2]Also, BUNCE, R. The political economy of California's wine industry. University of California Social Research Group working paper F88, revised January 1980 [Contemporary Drug Problems, forthcoming]; and COOK, P. The effect of liquor taxes on drinking, cirrhosis and auto accidents. Prepared for the National Academy of Sciences' panel on alternative policies affecting the prevention of alcohol abuse and alcoholism, May 1980 [In: MOORE, H. H. and GERSTEIN, D. R., eds. Alcohol and public policy; beyond the shadow of prohibition. Washington, D.C.; National Academy Press; 1981].

suggests that this concern over personal control over drinking is an expression of more generalized concerns over the self-control of behavior, first fully expressed in the early 19th century as middle-class interests came to the fore, a suggestion in line with Bacon's discussion of self-control in complex society (23). Today, another individual-level meaning of "control" derives from recent work in behavioral psychology on "controlled drinking" (e.g., 27).

On a societal level, "control" has also had a special meaning. This meaning derives from an elite-based and very successful small social movement which preceded the alcoholism movement but has been largely forgotten. "Alcohol control" as a philosophy of state control of alcoholic beverage production, distribution and sales emerged early in the 20th century in the U.S., drawing on British and Scandinavian antecedents, as a technocratic alternative to prohibition. Instead of prohibiting alcohol, the argument went, the state could manage the alcohol trade so as to minimize alcohol problems. In the depths of the Depression, with Prohibition losing popular support and taxes on alcohol seeming an attractive source of revenue, the well-connected Association against the Prohibition Amendment and the Rockefellers financed a number of studies of "alcohol control" which determined the shape of the regulations adopted at Repeal (28).[3] Although the systems paid lip-service to "promotion of temperance," a major aim was to keep order and peace between segments of the relegalized industries. Recently, interest has been renewed in studying the inception, operation and potential of these systems, which still exist in broad outline (e.g., 29).[4]

With a fresh infusion of Scandinavian and Canadian influence, in the form of such locutions as "alcohol control policies," the limited concerns of earlier U.S. discussions of ABC systems have been broadened to encompass a "public health model," in which regulatory controls are seen, along with other public health measures, as a means of minimizing alcohol problems (30, 31). This broader sense of "control" is very close to the sense used by Baird, in *Alcohol, Science and Society*, to cover all legal controls of drinking behavior, including public drunkenness and minimum age of purchase of alcoholic beverages (10).

Even this broader usage, however, is more narrowly defined

[3]Also, LEVINE, H. and SMITH, D. A selected bibliography on alcohol control, particularly before and at Repeal. University of California Social Research Group working paper F71, December 1977; and LEVINE, H. The Committee of 50 and the origins of alcohol control. University of California Social Research Group working paper F129, August 1980 [Journal of Drug Issues, forthcoming].

[4]Also, MORGAN, P. The evolution of California alcohol policy. University of California Social Research Group working paper F92, April 1979.

than the classic sociological conception of "social control." In the remainder of this chapter I shall examine alcohol and social control in this broad sociological sense, picking up some loose threads in existing literatures and suggesting some directions for their interweaving in the future.

As Pitts notes (32) in his discussion of the concept in the sociological literature, "social control" is "essentially an American term," first popularized in 1901 by Edward Ross. Ross defined social control "as concerned with that domination which is intended and which fulfills a function in the life of society," acting through a variety of material and ideological social institutions. From the first, then, the concept was concerned with both intentions and functions in the existence and exercise of power in society. However, in many sociological discussions the emphasis has been on what Lemert terms "passive" social control, using a "Sumnerian assumption of automaticity in the control process" operating through "reified . . . mores, social norms and social laws" (33, pp. 53, 95). In the hands of such theorists as Talcott Parsons, the concept also became associated with an assumption that deviance derived solely from individual impulses (32, p. 394), so that social control was seen only in terms of society's constraints on individual deviant behavior.

Lemert (33) places social control in the context of a more dynamic, purposive and complex view of societal functioning. Social control is not only a matter of "conformity to traditional norms," but also "a continuous process by which values are consciously examined, decisions made as to those values which should be dominant, and collective action taken to that end" (pp. 53–54). It is not an automatic manifestation of a *conscience collective*, but represents the efforts of "human beings [to] define, regulate and control behavior of other human beings" (p. 29). It is not simply a matter between the individual and the society, but involves also, as both objects and agents of social control, a wide variety of intervening social and cultural collectivities. Whereas such writers as Parsons and Pitts recognize the existence of subcultures, using such terms as "secondary institutions" and "fringe organizations" (32, pp. 384, 393–394), their organismic vision of society (34, p. 235) tends to consider only the functional significance of subcultures for the society at large. Lemert's conception recognizes the pervasiveness of social conflict and change, whether "functional" or not, and the often considerable autonomy of subcultures. Thus subcultures and social organizations can, irrespective of individual impulse, exercise a control on the individual which may put him or her at odds with the larger society. Thus, too, social control, rather than operating always on individuals, is often a matter between collectivities, as in legal controls on corporations or in imperial domination of colonies. That individuals are often used as hostages in such power relations or conflicts does not negate their nature as efforts at social control between collectivities.

Besides his more general sociological discussions of social control, Lemert has made significant contributions on the topic specifically of alcohol and social control, in a widely-disseminated essay (35) and in a 1958 conference presentation recently reprinted (36). The discussions reflect Lemert's flexible general perspective on social control: it is recognized that there are positive functions of drunkenness in some collectivities (35, *p. 554;* 36, *p. 46*), and there is explicit attention to the relation of institutionalized drinking groups to social control (36, *pp. 47–48*). But the main emphasis is on social control in the narrower, public health sense of societal "alcohol control policies"; in fact, issues in the social control of the "addictive drinker" are excluded from the discussion (35, *p. 569*). In my view, Lemert's general contributions on social control as well as his specific discussions of alcohol and social control need to be taken into account in future work.

So far, however, there has been little further literature explicitly on alcohol and social control, although Bruun (37) commented on and offered an alternative to Lemert's typology of alcohol control policies, and Sargent's recent analysis (38) of drinking in terms of "power relations" draws on Lemert's general discussions of social control. Instead, a recent increase in interest in social control and alcohol has been expressed in the form of several so far separate literatures. The best-developed of these is the previously-mentioned line of general discussions of "alcohol control policies," which tend to take social control issues rather for granted in discussions of alternative policy options. This literature seems to be fast attaining serious consideration in the alcohol field (e.g., 31; 39; 40, *pp. 313–325*).[5] Another area of concern relates alcohol to the general "interests of the state" (41, 42)[6]—fiscal, commercial, social policy and public order interests—in different conditions of societal organization. A third area is a growing "social constructionist" literature, using such terms

[5]Also, the National Academy of Sciences' forthcoming report of the panel on alternative policies affecting the prevention of alcohol abuse and alcoholism. [MOORE, M. H. and GERSTEIN, D. R., eds. Alcohol and public policy; beyond the shadow of prohibition. Washington, D.C.; National Academy Press; 1981.]

[6]Also, MORGAN, P. Examining United States alcohol policy; alcohol policy and the interest of the state. Presented at the meeting of the Ninth World Congress of Sociology, Uppsala, Sweden, 13–20 August 1978; and MÄKELA, K., ROOM, R., SINGLE, E., SULKUNEN, P., WALSH, B. with 13 others. Alcohol, society and the state; I. A comparative study of alcohol control. Toronto; Addiction Research Foundation. [Forthcoming.].

as "ideological hegemony" and "governing image" to emphasize the importance of social definitions in the creation and maintenance of "public domination" (6, 43–46).

So far the literatures have tended to concentrate on specific and often separate sectors of the terrain of alcohol and social control. The control-policy literature is specifically concerned with the control of alcohol-related problems, while the state-interests literature considers alcohol issues in the context of more general issues of social control—production, reproduction, security, etc. In both literatures, the emphasis is on the state's interests in and powers of control. The social-constructionist literature is oriented rather to the definition and handling of deviance by a variety of social institutions, with a particular emphasis on alternative social rubrics of alcohol-related deviance, but with consideration also of a broader framework in which the alcohol attribution is itself an aspect of the social control choices. Only Sargent (38) has attempted to combine most of these areas of interest in one study.

While it seems appropriate to call for more cross-fertilization between these literatures, it may be premature to attempt any full synthesis. For the moment, it may suffice to draw some distinctions and point to some directions for development. "Alcohol and social control" can clearly have at least two meanings: the social control of alcohol-related problems, or the role of alcohol in social control in general. While much material relevant to the latter topic is available in the anthropological literature, there are few general discussions of the processes involved.

Alcohol, like many other drugs, is potentially an instrument of social control for both the dominant and the subordinate side of power relations, and in relations both at system levels and in face-to-face interactions. At system levels, while denial of alcohol to subordinate statuses is a potential instrument of domination (47),[7] alcohol has at least as often been useful to rulers as an opiate for the masses (36, *pp. 46–47*; 45). It is a reflection of a formerly widely-used means of social control (e.g., 48, *p. 140*) that the International Labor Convention Protocol No. 95 on the Protection of Wages contains provisions prohibiting the payment of wages in the form of strong alcoholic beverages (49, *p. 214*). On the other hand, drinking is intimately associated with various revolutionary or countercultural movements (50). Taverns were major centers for seditious activities

---

[7]Also, ROOM, R. Alcohol as an instrument of intimate domination. Presented at the meeting of the Drinking and Drugs Division of the Society for the Study of Social Problems, New York, 26 August 1980.

by the colonials in the events preceding the American Revolutionary War (51). Lurie (52) has argued that American Indian drinking patterns can be interpreted as a permanent "protest movement" against the dominant Anglo culture.

Many of these system-level relations are impersonal in nature; alcohol is not necessarily involved directly in the relations of the dominant and the subordinate. But alcohol is also potentially an instrument of social control in face-to-face and intimate relations. The greater acceptance of drinking by men and other dominant statuses, and the cultural belief in alcohol's power to disinhibit, make alcohol a useful instrument of domination in intimate relations.[7] On the other hand, alcohol is also a potential tool for the subordinate. Trice and Belasco (53) have related how immediate subordinates often benefit by keeping the company president well plied with liquor, and Lurie (52) and Stivers (48, *pp. 155 ff*) portray drunkenness as a major element in roles oppressed ethnicities played out to some extent to their own advantage.

There are two directions in which studies of the social control of alcohol problems might be particularly fruitful. The first is attention to the interaction of social control and the "social worlds" of heavy drinking (54). On the one hand, there has been little systematic discussion of the processes of social control in drinking groups. Bruun's pathbreaking study (55) of drinking behavior in small groups of workmates, with its evidence of repeated pressures to drink more but of none to drink less, has not yet been followed up. Burns' (56) account of a night of "getting rowdy with the boys" in working-class Boston is evocative in its description not only of the processes of hell-raising but also of the pressures on the narrator to join and stay with the occasion. Mass-observation's account (57) of the exact matching with which the draining of glasses proceeds, sip by sip, in groups in British pubs suggests how fine-grained the social controls of drinking can be. The most systematic accounts available of controls within drinking groups are of Skid Row "bottle gangs" (58–60).

On the other hand, there is also little but anecdotal evidence of the processes by which the larger society exercises control over drinking groups and the social worlds of heavy drinking, except, again, for Skid Row (60–62). Few discussions of teen-age drinking follow Sower's lead (63) in viewing it as group behavior and studying the dynamic processes of adult attempts at control and the teen-age world's patterned evasions. In this general area of social control, again Lemert offers some starting-points for analysis (36, *pp. 47–48*).

The second general direction of study is the social handling of those identified as having alcohol problems—already a major concern of the social-constructionist literature, as mentioned above. In fact, the strength of this literature, in line with general sociological trends, has been its insistence in viewing therapy, supports, and rehabilitation efforts as serving—among other functions—as instruments of social control. But so far, the discussion tends to have been focused around one thematic dimension, the "medicalization of deviance": the general shift in recent decades from legal and penal to medical and therapeutic models and institutions to deal with alcohol and a variety of other problems. Although there has been some recognition of the potential to shift in the opposite direction, as in the case of drugs during the 1920s (64), or towards redefinition away from being a "public problem" (65), as happened during the 1970s to homosexuality (46), there has been in my view too single-minded an emphasis on the medicalization dimension, which provides only a part of the recent history of social control and handling of those with alcohol problems. For example, already in California in 1950, before the publicly-oriented alcoholism movement had made its influence felt, two of the three major social institutions handling those with alcohol problems—mental hospitals, county general hospitals, jails—operated under a medical rubric. An interpretation of the subsequent history requires attention not only to medicalization and the efforts of the movement but also to such other general dimensions of change in the exercise of social control as: "deinstitutionalization" in the handling of deviant populations; transfer of financing of social handling to state and federal levels in connection with recurrent fiscal crises in local government; the individualization of the definition of problems, particularly in the 1950s and 1970s; differentiation of social services in the postwar era; and increases in insurance coverage and scope.[8]

The years since *Alcohol, Science and Society* have seen a great deal of mapping of the social terrain of alcohol use and problems, using surveys, observations, indirect measures, social statistics, etc. But much of the mapping has treated the terrain as a flat surface, with no vertical axis. As Bruun (66) noted in a review of Jessor et al.'s analysis of drinking in a tri-ethnic community, there has been a miss-

[8]BUNCE, R., CAMERON, T., MORGAN, P., MOSHER, J. and ROOM, R. California's alcohol control experience, 1950–1980. University of California Social Research Group working paper F108, revised November 1980 [To be published as: California's alcohol experience; stable patterns and shifting responses. In: SINGLE, E., MORGAN, P. and DE LINT, J., eds. Alcohol, society and the state; II. The social history of alcohol control experiences in seven countries. Toronto; Addiction Research Foundation; forthcoming]; and ROIZEN, R. and WEISNER, C. Fragmentation in alcoholism treatment services; an exploratory analysis. University of California Social Research Group Rep. C24, October 1979.

ing dimension, that of power: "one could ask if it really is fruitful to define deviance in purely statistical terms, without consideration of content and without thought of who are the agents of social control, who define deviance, and how do they operate."

Maps with no topological indication are not much use in hilly terrain. We need to add the contour-lines of the social control dimension to our understanding of drinking practices and problems.

REFERENCES

1. ANDERSON, D. Alcohol and public opinion. Quart. J. Stud. Alc. 3: 376–392, 1942.
2. ANDERSON, D. Analysis of wet and dry propaganda. Pp. 355–372. In: Alcohol, science and society; twenty-nine lectures with discussions as given at the Yale Summer School of Alcohol Studies. New Haven; Quarterly Journal of Studies on Alcohol; 1945.
3. HIRSH, J. The problem drinker. New York; Duell, Sloan & Pearce; 1949.
4. ANDERSON, D. with COOPER, P. The other side of the bottle. New York; Wyn; 1950.
5. JOHNSON, B. H. The alcoholism movement in America; a study in cultural innovation. Ph.D. dissertation, University of Illinois, Urbana–Champaign; 1973.
6. ROOM, R. G. W. Governing images of alcohol and drug problems; the structure, sources and sequels of conceptualizations of intractable problems. Ph.D. dissertation, University of California, Berkeley; 1978.
7. BEAUCHAMP, D. E. Beyond alcoholism; alcohol and public health policy. Philadelphia; Temple University Press; 1980.
8. WIENER, C. The politics of alcoholism; building an arena around a social problem. Edison, N.J.; Transaction Books; 1980.
9. MURPHY, A. J. Alcohol and pauperism. Pp. 239–249. In: Alcohol, science and society; twenty-nine lectures with discussions as given at the Yale Summer School of Alcohol Studies. New Haven; Quarterly Journal of Studies on Alcohol; 1945.
10. BAIRD, E. G. Controlled consumption of alcoholic beverages. Pp. 299–319. In: Alcohol, science and society; twenty-nine lectures with discussions as given at the Yale Summer School of Alcohol Studies. New Haven; Quarterly Journal of Studies on Alcohol; 1945.
11. BAINTON, R. H. The churches and alcohol. Pp. 287–298. In: Alcohol, science and society; twenty-nine lectures with discussions as given at the Yale Summer School of Alcohol Studies. New Haven; Quarterly Journal of Studies on Alcohol; 1945.
12. HAGGARD, H. W. Metabolism of alcohol. Pp. 31–44. In: Alcohol, science and society; twenty-nine lectures with discussions as given at the Yale Summer School of Alcohol Studies. New Haven; Quarterly Journal of Studies on Alcohol; 1945.

13. McPEEK, F. W. The role of religious bodies in the treatment of ine-
    briety in the United States. Pp. 403–418. In: Alcohol, science and
    society; twenty-nine lectures with discussions as given at the Yale
    Summer School of Alcohol Studies. New Haven; Quarterly Journal
    of Studies on Alcohol; 1945.
14. JELLINEK, E. M. An outline of basic policies for a research program
    on problems of alcohol. Quart. J. Stud. Alc. 3: 103–124, 1942.
15. ROOM, R. Priorities in social science research in alcohol. J. Stud. Alc.,
    Suppl. No. 8, pp. 248–268, 1979.
16. JELLINEK, E. M. The disease concept of alcoholism. Highland
    Park, N.J.; Hillhouse Press; 1960.
17. RUBIN, J. L. Shifting perspectives on the alcoholism treatment move-
    ment 1940–1955. J. Stud. Alc. 40: 376–386, 1979.
18. STRAUS, R. Research in the problems of alcohol; a twenty-year per-
    spective. Pp. 28–31. In: CALIFORNIA. ALCOHOLIC REHABILITATION
    DIVISION. Multidisciplinary programming in alcoholism investiga-
    tion; proceedings of a conference. Berkeley; 1960.
19. BLOCKER, J. S., JR., ed. Alcohol, reform and society; the liquor issue
    in social context. (Contributions in American history, No. 83.) West-
    port, Conn.; Greenwood Press; 1979.
20. BLOCKER, J. S., JR. [Bibliography update.] Alc. & temperance Hist.
    Grp Newslett. 2 (Fall): 5–8, 1980.
21. MÄKELÄ, K. and ÖSTERBERG, E. Notes on analyzing economic costs
    of alcohol use. Drinking & drug Pract. Surveyor, Berkeley, No. 15,
    pp. 7–10, 1979.
22. GREGG, G., PRESTON, T., GEIST, A. and CAPLAN, N. The caravan rolls
    on; forty years of social problem research. Knowledge, Beverly Hills
    1: 31–61, 1979.
23. BACON, S. D. Alcohol and complex society. Pp. 179–200. In: Alcohol,
    science and society; twenty-nine lectures with discussions as given
    at the Yale Summer School of Alcohol Studies. New Haven; Quarterly
    Journal of Studies on Alcohol; 1945.
24. ROOM, R. Introduction: drug and alcohol problems; social control and
    normative patterns. J. drug Issues 10: 1–5, 1980.
25. JELLINEK, E. M. Phases of alcohol addiction. Quart. J. Stud. Alc. 13:
    673–684, 1952.
26. LEVINE, H. G. The discovery of addiction; changing conceptions of
    habitual drunkenness in American history. J. Stud. Alc. 39: 143–174,
    1978.
27. SOBELL, M. B. and SOBELL, L. C. Behavioral treatment of alcohol
    problems; individualized therapy and controlled drinking. New York;
    Plenum; 1978.
28. KYVIG, D. E. Repealing national Prohibition. Chicago; University of
    Chicago Press; 1979.
29. MEDICINE IN THE PUBLIC INTEREST. The effects of alcoholic beverage
    control laws. Washington, D.C.; 1979.
30. BEAUCHAMP, D. E. Public health; alien ethic in a strange land? Amer.
    J. publ. Hlth 65: 1338–1339, 1975.

31. BRUUN, K., EDWARDS, G., LUMIO, M., MÄKELÄ, K., PAN, L., POPHAM, R. E., ROOM, R., SCHMIDT, W., SKOG, O.-J., SULKUNEN, P., and ÖSTERBERG, E. Alcohol control policies in public health perspective. (Finnish Foundation for Alcohol Studies, Vol. 25.) Helsinki; 1975.

32. PITTS, J. R. Social control; the concept. Pp. 281–396. In: SILLS, D. L., ed. International encyclopedia of the social sciences. Vol. 16. New York; MacMillan; 1968.

33. LEMERT, E. M., ed. Human deviance, social problems, and social control. 2d ed. Englewood Cliffs, N.J.; Prentice-Hall; 1972.

34. BOCK, K. E. Evolution, function and change. Amer. sociol. Rev. **28:** 229–237, 1963.

35. LEMERT, E. M. Alcohol, values and social control. Pp. 553–571. In: PITTMAN, D. J. and SNYDER, C. R., eds. Society, culture, and drinking patterns. New York; Wiley; 1962.

36. LEMERT, E. M. An interpretation of society's efforts to control the use of alcohol. Drinking & drug Pract. Surveyor, Berkeley, No. 15, pp. 19–22, 45–50, 1979.

37. BRUUN, K. Implications of legislation relating to alcoholism and drug dependence; government policies. Pp. 173–181. In: KILOH, L. G. and BELL, D. S., eds. Proceedings of the 29th International Congress on Alcoholism and Drug Dependence. Melbourne; Butterworths; 1971.

38. SARGENT, M. Drinking and alcoholism in Australia; a power relations theory. Melbourne; Longman Cheshire; 1979.

39. WORLD HEALTH ORGANIZATION. Problems related to alcohol consumption; report of a WHO expert committee. (WHO Technical Report Ser., No. 650.) Geneva; 1980.

40. U.S. NATIONAL INSTITUTE ON ALCOHOL ABUSE AND ALCOHOLISM. Alcohol and health; third special report to the U.S. Congress from the Secretary of Health, Education and Welfare, June 1978; technical support document. Ernest P. Noble, editor. (DHEW Publ. No. ADM 79-832.) Washington, D.C.; U.S. Govt Print. Off.; 1979.

41. MÄKELÄ, K. and VIIKARI, M. Notes on alcohol and the state. (Social Research Institute of Alcohol Studies, No. 103.) Helsinki; 1976.

42. PARKER, D. A. Alcohol control policy and the fiscal crisis of the state. Drinking & drug Pract. Surveyor, Berkeley, No. 13, pp. 3–6, 1977.

43. GUSFIELD, J. R. Moral passage; the symbolic process in public designations of deviance. Social Probl. **15:** 175–188, 1967.

44. BRUUN, K. Finland; the non-medical approach. Pp. 545–559. In: KILOH, L. G. and BELL, D. S., eds. Proceedings of the 29th International Congress on Alcoholism and Drug Dependence. Melbourne; Butterworths; 1971.

45. HIMMELSTEIN, J. The social labeling of psychoactive drugs; an interim report. Drinking & drug Pract. Surveyor, Berkeley, No. 11, pp. 36–42, 54, 1976.

46. CONRAD, P. and SCHNEIDER, J. W. Deviance and medicalization; from badness to sickness. St. Louis; Mosby; 1980.

47. GUSFIELD, J. R. Symbolic crusade; status politics and the American temperance movement. Urbana; University of Illinois Press; 1963.

48. STIVERS, R. A hair of the dog; Irish drinking and American stereotype. University Park; Pennsylvania State University Press; 1976.

49. MOSER, J., comp. Prevention of alcohol-related problems; an international review of preventive measures, policies, and programs. (Published on behalf of the World Health Organization by the Alcoholism and Drug Addiction Research Foundation.) Toronto; 1980.

50. STAUFFER, R. B. The role of drugs in political change. New York; General Learning Press; 1971.

51. EARLE, A. M. Stage-coach and tavern days. New York; Dover; 1969. [Orig., 1900.]

52. LURIE, N. O. The world's oldest on-going protest demonstration: North American Indian drinking patterns. Pp. 127–145. In: MARSHALL, M., ed. Beliefs, behaviors and alcoholic beverages. Ann Arbor; University of Michigan Press; 1979.

53. TRICE, H. M. and BELASCO, J. A. The aging collegian; drinking pathologies among executive and professional alumni. Pp. 218–233. In: MADDOX, G. L., ed. The domesticated drug; drinking among collegians. New Haven; College & University Press; 1970.

54. ROOM, R. Normative perspectives on alcohol use and problems. J. drug Issues 5: 358–368, 1975.

55. BRUUN, K. Drinking behavior in small groups; an experimental study. (Finnish Foundation for Alcohol Studies, Vol. 9.) Helsinki; 1959.

56. BURNS, T. F. Getting rowdy with the boys. J. drug Issues 10: 273–286, 1980.

57. MASS-OBSERVATION. The pub and the people; a Worktown study. Welwyn Garden City: Seven Dials Press; 1970. [Orig., 1943.]

58. ROONEY, J. F. Group processes among Skid Row winos; a reevaluation of the undersocialization hypothesis. Quart. J. Stud. Alc. 22: 444–460, 1961.

59. RUBINGTON, E. The bottle gang. Quart. J. Stud. Alc. 29: 943–955, 1968.

60. ARCHARD, P. Vagrancy, alcoholism and social control. New York; MacMillan; 1979.

61. SPRADLEY, J. P. You owe yourself a drunk; an ethnography of urban nomads. Boston; Little, Brown; 1970.

62. WISEMAN, J. P. Stations of the lost; the treatment of Skid Row alcoholics. Englewood Cliffs, N.J.; Prentice-Hall; 1970.

63. SOWER, C. Teen-age drinking as group behavior; implications for research. Quart. J. Stud. Alc. 20: 654–668, 1959.

64. MUSTO, D. F. The American disease; origins of narcotic control. New Haven; Yale University Press; 1973.

65. GUSFIELD, J. R. The culture of public problems; drinking–driving and the symbolic order. Chicago; University of Chicago Press; 1981.

66. BRUUN, K. [Review of Jessor, R., Graves, T. D., Hanson, R. C. and Jessor, S. L. Society, personality, and deviant behavior; a study of a tri-ethnic community.] Quart. J. Stud. Alc. 29: 1074–1076, 1968.

23

The Federal Role in Alcoholism Research, Treatment and Prevention

Jay Lewis

The federal alcoholism effort was conceived in the twilight years of the Great Society and, after a gestation period of several years, was actually born following the political demise of its natural parents, the Kennedy–Johnson Administrations.[1]

[1]There was no organized, coherent federal role prior to the 1960s. Activity was confined to some research awards in the alcoholism area by the National Institute of Mental Health. One of the most significant of these was a grant in the late 1950s to establish treatment relations with alcoholics to Massachusetts General Hospital's Alcohol Clinic, headed by Dr. Morris Chafetz, who later become first director of the National Institute on Alcohol Abuse and Alcoholism. The foundation of the federal alcoholism effort was laid during the late 1960s with a series of studies, court decisions, legislation and bureaucratic moves, all aided and abetted by constituency groups. In 1967, the Cooperative Commission on the Study of Alcoholism issued its *Report to the Nation* (1), declaring that more than 20% of the men admitted to state mental hospitals and men discharged from general hospitals were diagnosed alcoholics, and otherwise highlighting the problems of alcoholism. Also in 1967, the Task Force on Drunkenness of the President's Commission on Law Enforcement and Administration of Justice recommended that public intoxication be eliminated as a criminal offense, and that public detoxification centers be established (2). Landmark court decisions in *Easter v District of Columbia* (3), *Driver v Hinnant* (4) and *Powell v Texas* (5) affirmed the status of alcoholism as a disease. In Congress, leading up to the Comprehensive Alcohol Abuse and Alcoholism Prevention, Treatment, and Rehabilitation Act of 1970 (the "Hughes Act") (6) were a number of prior bills, including the Highway Safety Act of 1966 (7), which required a report on alcohol and traffic safety; the Economic Opportunity Amendments of 1967 (8) and 1969 (9), which created the first federally-funded alcoholism treatment programs, under the Office of Economic Opportunity; and the Alcoholic Rehabilitation Act of 1968 (10), which declared that alcoholism was a "major health and social problem" and led to the Community Mental Health Centers Amendments of 1970 (11), which authorized direct grants for special alcoholism treatment projects. Within the federal bureaucracy in 1965 there was but one identifiable alcoholism specialist at the National Institute of Mental Health (NIMH). A small National Center for the Prevention and Control of Alcoholism was created within NIMH in 1967, but funding was never more than $3 million. The National Council on Alcoholism and the North American Association of Alcoholism Programs (later the Alcohol and Drug Problems Association of North America) played catalytic roles in all these precursor developments.

With President Richard M. Nixon as its reluctant midwife, the federal effort came into statutory existence with a stroke of the Nixon ballpoint on New Years Eve, 31 December, 1970. The Hughes Act was one of the last of a progeny of federal categorical programs spawned in an era of boundless social and economic optimism, when it was believed that all of the nation's ills and shortcomings, from poverty to the heartbreak of psoriasis, could be vanquished by bureaucratic phalanxes of social and health programs backed by billions in appropriations.

Coming as it did two years after the Executive Branch was occupied by avowed opponents of such governmental activism, the federal alcoholism program was saddled with stepchild status from its beginning, and its history during the next decade is largely one of surviving against continuing attempts to eliminate it through a variety of bureaucratic, legislative or appropriations devices in the name of the New Federalism.

But survive it did—and even thrive, to some extent—when other Great Society programs were falling victim to either the New Federalism early in the decade, or later to Proposition 13 fever. In fact, the National Institute on Alcohol Abuse and Alcoholism (NIAAA), the federal alcoholism agency created by the Hughes Act, benefited from one of the early casualties of the Nixon Administration's war against the categorical programs of the Great Society era, the antipoverty programs of the Office of Economic Opportunity. In 1972, NIAAA absorbed nearly 200 alcoholism treatment programs as the latter agency was being dismantled (12).

NIAAA's stepchild status was underscored in its first bureaucratic placement, a result of a Congressional compromise to attain passage of its enabling legislation late in 1970. The Hughes Act created NIAAA as an institute within an institute—the National Institute of Mental Health (NIMH)—and, as such, it was somewhat of an organizational anomaly.

Sen. Harold Hughes (D-Iowa), the recovered alcoholic and chief legislative architect of the federal alcoholism effort, and the alcoholism constituency had pressed for independent institute status within the Public Health Service (PHS), as provided in the Senate version of the original bill. The historical antipathy of the alcoholism field to mental health interests was one reason. The other arose from a desire to accord NIAAA the highest possible visibility in the bureaucratic scheme of things (13). (At the time, drug problems enjoyed the ideal location in the White House as the Special Action Office on Drug Abuse Prevention.)

NIAAA's ability to more than hold its own was amply demonstrated two years after its inception, when the institute came out from under the wings of NIMH and emerged as a co-equal institute in the structure of the new Alcohol, Drug Abuse and Mental Health Administration (ADAMHA), successor to the Health Services and Mental Health Administration. NIAAA's programmatic and policy-making autonomy was spelled out by statute in the 1974 reauthorizing legislation that created ADAMHA and, incidentally, gave both NIMH and the National Institute on Drug Abuse (NIDA) statutory existence for the first time (14, 15).

It might be pointed out that the degree to which NIAAA has been able to achieve strength and policy-making autonomy in the face of attempts to hamstring it or merge it with mental health, drug problems or other interests has a direct relationship not only to the nature of the federal role in alcoholism research, treatment and prevention, but indeed, to the very existence of such a role.

Since its placement in ADAMHA, NIAAA's struggle within the parent Health, Education and Welfare (HEW) bureaucracy has been to resist attempts by bureaucrats at the ADAMHA level and above to erode its status as an independent policy-making entity. Under the law, the programs and policies of the federal alcoholism effort are to be carried out by the Secretary of HEW, acting through the institute. The administrator of ADAMHA is given a coordinating and administrative role over the three institutes.

ADAMHA was envisioned by its legislative architects—Rep. Paul Rogers (D-Fla.), Chairman of the House Health Subcommittee, and Senator Hughes, of the Senate Alcoholism and Narcotics Subcommittee—as constituting no more than a thin bureaucratic layer over the three institutes, and the administrator, in the dictionary sense of the word, more involved in administering than setting policy. Rogers said, before final passage of the legislation, that he saw the administrator as being a person with bureaucratic experience, rather than someone drawn from the disciplines (16). James D. Isbister, who was ADAMHA administrator for two years until January, 1977, was a nonphysician with academic and professional credentials in the area of public administration—precisely the kind of official Rogers had in mind.

The most dramatic chapter in NIAAA's fight to maintain its autonomy opened early in 1978 when Isbister's successor, Gerald L. Klerman, a research psychiatrist from Harvard Medical School, advanced a proposal to centralize initial review of grants throughout the three institutes within the office of the ADAMHA administrator.

Under the peer review system at the time, the initial review groups were controlled by the individual institutes and their operating divisions. Genesis of Klerman's proposal was a recommendation by the President's Commission on Mental Health to separate the initial review system from program management as a way of eliminating possible bias.

Klerman's proposal, once revealed, generated a maelstrom of opposition from the alcoholism field and other constituencies embraced by ADAMHA. NIAAA director Ernest P. Noble did not hide his opposition and more than tacitly encouraged the strident attacks of constituency groups on the plan, which was seen as stripping the institute of control over the nature of projects it could fund—one of the major mechanisms in setting policy directions. Seldom have the diverse parts of the alcoholism constituency been so aroused about, and so united on, an issue (17).

Whether it was the strength of the opposition from the alcoholism field or the clout of the opposition from one lawmaker, Sen. Harrison A. Williams (D-N.J.), Chairman of the Senate Labor and Human Resources Committee, which forced Klerman to back down on the initial review proposal, may never be known. At a Capitol meeting in late April, 1978, between then HEW secretary Joseph A. Califano, Jr., Klerman, and Williams, it was agreed that, rather than centralizing initial review at the ADAMHA level, it should be retained within the offices of the three institute directors. Noble could not be saved, however, and he was fired by Califano and Klerman a day after the decision on the grants review proposal, serving as an object lesson that agency directors should not personally man the barricades in constituency rebellions (18).

Early on, NIAAA earned a reputation for being a maverick agency. Its first director, Morris Chafetz, from the Massachusetts General Hospital–Harvard Medical School complex, was not a typical public health service officer, schooled in the virtues of protecting flanks, keeping a low profile and following the Civil Service manual. The feistiness and unorthodoxy of NIAAA and its first director, a kind of mirror of the constituency they embraced, were qualities which both helped the institute survive and, at the same time, made it the special target of attacks by bureaucrats from above.

With the Hughes Act, the alcoholism cause was accorded a small place in the Washington scheme of things, one of some 300 categorical programs within just one department. With the creation of ADAMHA, it was a component of one PHS agency which, itself, was just one component of HEW.

In a very important sense, the difficulties NIAAA has faced in its struggle for survival reflect those in the alcoholism field at large to achieve awareness and recognition of a long-neglected public health problem. The institute has been confronted with the same dynamics of denial that have characterized society's attitude toward alcoholism. Its placement in PHS, achieved only after considerable effort, has not resulted in genuine acceptance of alcoholism as a major public health problem, on a par with cancer, diabetes or even mental illness within PHS itself, much less other realms of HEW.

Benighted attitudes toward alcoholism have been as much of a problem within HEW as in society in general. It took a directive from the secretary of HEW, Joseph A. Califano, Jr., the first in his position to evince personal interest and commitment in alcoholism, to move the department toward an effective program to assist its own employees in trouble with alcohol. A fiat from the secretary was also required to prod the Health Care Financing Administration into looking for ways to provide more appropriate coverage for alcoholism treatment under Medicare and Medicaid (19).

As a result of its problems, NIAAA was often forced to conduct its business in ways representing a marked departure from the accepted practices of other PHS agencies, resulting in increased disapproval from the ADAMHA level to the Office of Management and Budget (OMB) in the Executive Office of the President. The classic example of NIAAA unorthodoxy was a series of grants made to national organizations in 1974–1975 under the directorship of Chafetz. Some $23 million in awards to organizations, including both alcoholism field groups and associations representing broader constituencies, were dispensed in a fashion which was later criticized as highly irregular and for purposes seen as constituency-building in nature. The 16 grants were awarded to 14 organizations for a wide range of projects. Recipients included major alcoholism organizations, such as the National Council on Alcoholism (which had three large grants), the Association of Labor–Management Administrators and Consultants on Alcoholism, Council of State and Territorial Alcoholism Authorities, and the Association of Halfway House Alcoholism Programs.

Other national associations included the National Congress of Parents and Teachers, United States Jaycees, League of Cities–Conference of Mayors, Boys' Clubs of America, Legis 50 (formerly the Citizens Conference for State Legislatures), Education Commission of the States, National Association of Counties, Airline Pilots Association, Group Health Association of America and the National Council on Aging.

At the time, the awards were seen as an astute maneuver, at once strengthening the alcoholism constituency and broadening its base by financing the involvement of large national organizations (with considerable political clout) in alcoholism activities. Since many of the grants were dedicated to prevention activities, they were seen as providing an effective mechanism for raising awareness of alcohol problems by using existing organizational structures with access to large general constituencies throughout the nation.

The fact that the lure of federal dollars was needed to spur interest in alcohol programs within the national associations is a telling commentary, of course, on where alcohol problems stood, not only in their program priorities but in society in general. (Many of these grant recipients later dropped their alcohol programs when the awards were terminated.)

At the time of the grants, NIAAA was confronted with a sudden embarassment of riches in the form of some $80 million in funds previously impounded by the Nixon Administration and freed by court order as a result of a class-action suit filed by various alcoholism and mental health interests (20).

NIAAA had a staff of less than 100 because of persistent denials of additional personnel slots by HEW and OMB. Having a choice between awarding several hundred grants to community projects, with inadequate staff to administer and monitor the grantees, and making a handful of large grants to national organizations with proven track records and established accountability, the institute saw the latter option as the prudent administrative decision. Moreover, OMB had also refused to charter additional initial review groups for NIAAA (the committees of outside experts convened to screen grant applications), limiting the institute's ability to process awards in an orderly fashion. At the time, NIAAA had only two formal initial review groups: one for training and one for research grants. Grants in other areas were reviewed by ad hoc groups.

On the whole, the national organization grants appeared to be accomplishing their objectives during their initial periods. For the first time, an impressive array of prestigious national associations testified alongside the traditional alcoholism field organizations at Senate hearings on the institute's reauthorization early in 1976 (21). The same organizations, for the first time, were also passing resolutions at their annual meetings underscoring the gravity of the nation's problems with alcohol and pledging new efforts in the area.

Then, in May, 1976, the House Commerce Committee, in its report on the NIAAA renewal bill, said it was "distressed by the nature" of some of the large awards and asked that the institute review them (22). A task force was convened by NIAAA that summer and it completed its review toward the end of 1976. In its final report to Congress, NIAAA conceded that its handling of the grants was "highly irregular" (23, *p. 1*) in many cases and pledged to tighten procedures in the future. At the same time, the institute recommended against further funding of most of the grants, for which grantees were expecting renewals. The report was critical of the review procedures, terming them "inconsistent in relation to the formal peer review process utilized for other institute programs of a more traditional and well-defined nature in the training and research areas." There was evidence, the report said, that the office of the NIAAA director, where many of the grants originated, "sometimes ignored or circumvented accepted policies and procedures." Chafetz, director at the time of the awards, had resigned in September, 1975.

Only a handful of the national grants escaped the onus of criticism and received further funding, notably those to the Airline Pilots Association and Group Health Association of America. The National Council on Alcoholism managed to hold onto only one of its three large grants, the Labor–Management Task Force project. The remainder were either phased out or terminated at the end of their first project period.

Many questions remain unanswered in this unusual chapter in NIAAA's history. The review and recommendations can be regarded as amounting to overkill. House staffers who had a hand in the original directive to NIAAA have said privately they had no such far-reaching outcome in mind when the language was inserted in the May, 1976, report. (Indeed, the House report's language was rather ambiguous.)

At any rate, the actions taken in early 1977 against the various national grantees had a convulsive effect on the affected alcoholism constituency groups and served to alienate many of the broader national organizations. Poisonous fallout from the national grants episode continued to affect the institute's political and programmatic being at the end of the decade, although many of the players, at NIAAA and on the constituency scene, may not be fully aware of it.

The national grants debacle is worthy of exploring in such depth because it represented so many of the problems NIAAA faced, as well as the means it was often forced to use to cope with them. The institute's own report to Congress on the grants took note of some of these problems when it related that NIAAA's early years were marked by "rapid growth in budget resources unaccompanied by a commensurate growth in staff and complicated by vetoes, impoundments and continuing resolutions."

Moreover, the report stressed the significance of the historical context and circumstances at the time the awards were made, stating: "It (NIAAA) entered the public health arena committed to the rapid amelioration, if not the resolution of a long-neglected, serious, and poorly understood problem whose roots were deeply embedded in cultural habits. The Institute was encouraged in this role by an articulate constituency who projected an enthusiasm brought about by recognition that their cause was now given national visibility and credibility through federal law and commitment" (23, *pp 1–2*).

These special problems compelled NIAAA throughout its history to improvise and adopt procedures at variance with those of the more established agencies. The practices accentuated the institute's maverick status in the eyes of those higher in HEW and the Administration, who tend to watch closely for signs of straying from accepted bureaucratic behavior. All in all, it has amounted to a "Catch 22" situation in its classic dimensions: A stepchild to begin with, the institute had to rely on unorthodox means to survive, and in so doing earned a reputation for being unorthodox, which in turn made it more difficult to survive.

All this is not to say that NIAAA's energies were entirely devoted to the tough bureaucratic in-fighting necessary for survival during its first decade. The federal alcoholism effort, as represented by the institute's expenditures alone, resulted in $1,328,500,000 in grants and contracts for research, training, treatment, prevention, and formula grants to the states in the fiscal years 1971 through 1980 (24). In addition, NIAAA spent about $88.5 million for administrative overhead. (Alcoholism activities by other federal department and agencies, including the Departments of Defense and Transportation and the Veterans Administration, were funded roughly at the same level, making a total over-all federal outlay of more than $2 billion in the first 9 years of the effort.[2])

Considering the magnitude of the dollars involved, and the paucity of staff to administer them, the fact that no public scandal involving gross waste of federal funds ever touched the institute, as in the case of NIDA, is remarkable, and a credit to NIAAA. The worst brush with publicity occurred in 1974–1975 when columnist Jack Anderson wrote a series of articles dealing with some questionable travel vouchers and other activities of Chafetz and some top aides (25). The flap over the national grants received little or no general media coverage.

In its first 10 years, NIAAA obligated $136.1 million for research activities; $69.7 million for training; $654.4 million for project grants and contracts, including both prevention and treatment programs;

[2]MACRO SYSTEMS, INC. Report on FY 1978 federal activities. Unpublished report to NIAAA. The document projects a total of $351 million spent by all federal agencies, of which $168.7 million was by NIAAA.

and $468.3 million in grants to states, which in turn spent most of these formula grant funds for treatment services.

Thus the major thrust of the federal effort—in dollar terms more than 1 billion, counting the formula grant funds—has been the funding of treatment programs with research and training taking far back seats. Prevention activities, lumped budgetarily with treatment, have been funded at lower levels than even research and training, prompting Congress in 1979 (26) to direct the institute to earmark at least 8% of its project grant and contract appropriations to prevention programs in 1980, and 10% in 1981.

The rationale for spending the major portion of the federal monies for treatment was not seriously questioned until the latter part of the decade when constituencies in the research and prevention areas began to pressure for larger shares of the largesse. The provision of treatment services for the millions of alcoholics throughout the nation where none existed was seen as the most crucial need in 1971, and the policy began to shift only at the end of the decade.

One reason for the shift in policy away from the emphasis on direct funding of treatment projects was the success of federal programs in generating funds from other sources for treatment. In early 1980, the institute reported data on 4219 treatment units nationwide, estimated to be about 90% of the total number. Of these, only 586 (less than 14%) reported receiving assistance from NIAAA in the form of project grants. Annual expenditures of all the programs was on the order of $795 million, of which NIAAA project grant funds represented between 6 and 7%.

In a January, 1980, report to the NIAAA Advisory Council, James Lawrence, Associate Director for Program Implementation, said that the data did not suggest treatment resources were yet equal to the need and the institute therefore retreat from providing further support to treatment. But the data did show that direct assistance was playing a relatively small part in the total service system, and that state and local governments and third party insurance carriers had become major sources of funds. As a result, NIAAA at the end of the decade was in the process of shaping a new role, emphasizing support for demonstration projects and innovative mechanisms for collaboration with states and large units of local government.[3]

The focus of the federal effort on the urgent needs in the treatment arena furnishes one important explanation of why, after

[3]Prepared remarks by James D. Lawrence, NIAAA Associate Director for Program Implementation, to the National Advisory Council on Alcohol Abuse and Alcoholism, 28 January 1980, and summarized in the 8 February 1980 issue of *The Alcoholism Report* (27).

nearly a decade of intense activity and expenditure of well over a billion dollars by NIAAA alone, no over-all strategy or consensus regarding problems with alcohol has emerged. Indeed, even in the treatment sector, it was not until early 1980 that the Institute began shaping standards for the programs it funds and articulating policy regarding the projects it should be supporting. This is not to say that a coherent national policy could or should have been framed earlier. It has taken 10 years to achieve a begrudging and qualified acceptance of alcoholism as a public health problem (and the treatment network that came into being as a result of the federal effort) and to win a limited and probationary membership in the general health care delivery system.

The historic message in the alcoholism field which preceded and furnished rationale of the federal effort was that alcoholism is a health problem of such dimensions that it should be addressed by public agencies with public funds. Moreover, it is a health problem that is treatable. The first 10 years of the Hughes Act were probably necessary to demonstrate these propositions.

The reasons for a lack of a national consensus on alcohol problems (of which alcoholism is just one) are manifold. First, the constituency that helped create the federal effort has been largely treatment oriented, and this constituency has, if anything, been strengthened by the federal emphasis on treatment programs. The major constituency organization is aptly named the National Council on Alcoholism (not alcohol problems), and in January, 1978, its board approved a position that more research was needed to determine whether such issues as alcoholic beverage control laws, health warning labels, dedicated taxation and other control measures are relevant to the disease of alcoholism. The position was reaffirmed in October, 1979, when the organization's board declined to take a stance on the warning label issue.

Second, the federal effort has been subject to policy breaks and fluctuations (some of which arose from NIAAA's problems as a fledgling agency as recounted earlier). Looking over the first 10 years, a certain evolutionary development in NIAAA's leadership can be discerned. The first five years, under the highly articulate Morris Chafetz, gave NIAAA and the cause it represented much needed public visibility and was, generally speaking, a public relations success. With his successor, Ernest P. Noble, a bench research scientist, the institute had a much lower public profile and the focus was more parochial, the accent on problems with the bureaucracy and the constituency. Noble's tenure was marked by a new emphasis on research, resulting in the establishment of the national alcohol research center program, as well as a fundamental change in the institute's prevention policies. It was, in a general sense, a period of consolidation and regrouping after the Chafetz reign. John R. DeLuca, who was named in the spring of 1979 following a year's interregnum during which Loran Archer, the former California alcoholism chief, was acting director, brought to the post a background in public administration, relative youth (age 35), a sharp intelligence, and a formidable ability to perform on the bureaucratic ropes—qualities that should prove of good stead to both the institute and the field in preparing for the problems of the 1980s.

Third, the leadership changes at NIAAA, as well as those at HEW (there have been five Secretaries since the institute's inception), the White House (three Presidents), and in Congress (three Chairmen of the Senate Alcoholism and Drug Abuse Subcommittee), have resulted in a lack of continuity which impeded development of any consistency in policies, much less an over-all strategy backed by a consensus.

The transition in leadership from Chafetz to Noble in 1975–1976 resulted in an abrupt switch in the institute's approach to prevention: from Chafetz's "responsible drinking" concept, which eschewed control measures, to adoption of the public health model involving the scrutiny of advertising, labeling, legal age of purchase of alcoholic beverages and other controls. The shift in the institute's prevention policies was dramatized by a directive to purge NIAAA's literature of the words "responsible drinking" and variations thereof (28).

In the spring of 1977, Noble pursued the approach in the first draft of a five-year National Plan to Combat Alcohol Abuse and Alcoholism (29) when the stabilization of per capita consumption at the level of 2.7 gallons was spelled out as a major objective. The reaction by groups in the alcohol field was explosive and sparked a major constituency crisis which escalated when the institute's third report to Congress was released in the fall of 1978 (30). The document, which emphasized the negative consequences of alcohol use, was sharply assailed at a November, 1978, meeting of the National Coalition for Adequate Alcoholism Programs, a group of some 25 organizations in the field (31).

The transition from Chafetz to Noble coincided with another significant shift on Capitol Hill, with Sen. William D. Hathaway (D-Me.) succeeding the retired Harold Hughes as Chairman of the Senate Subcommittee on Alcoholism and Narcotics in January, 1975. While Noble instituted new approaches to prevention, Hathaway's reign on the Subcommittee featured a new critical look at the alcoholic beverage industry in terms of its advertising practices coupled with some abortive legislative steps toward dedicated taxation. In August, 1976, the Maine Democrat drafted a bill which would have increased the federal alcohol excise tax by 2.5% and dedicated the resultant revenues (about $113 million annually) to a trust fund in support of occupational alcoholism programs (32). As it turned out, Hathaway was persuaded to delete the excise tax provision from his proposal, when he introduced it as an amendment to the Tax Reform Act of 1976 (33). Although the Senate adopted the "empty" trust fund amendment by Hathaway, the provision was later dropped when the big tax bill went to conference with the House. Subsequently, Hathaway introduced a modified version of the trust fund bill, but it never advanced beyond the hearing stage. The Senator, who was defeated for reelection in 1978, was an outspoken critic of alcoholic beverage advertising and media treatment of drinking, and broke virgin ground for the Subcommittee with hearings in March, 1976, focusing on these issues (34).

Hathaway was succeeded as Chairman of the Senate Subcommittee by Sen. Donald W. Riegle (D-Mich.) who became an outspoken advocate of warning labels following passage in the Senate in May, 1979, of a health caution amendment by Sen. J. Strom Thurmond (R-S.C.) (35). Passage of the Thurmond amendment precipitated a series of developments, including a strong statement in support of warning labels by then HEW secretary Califano, a position which was reaffirmed in qualified form by NIAAA director DeLuca on behalf of Califano's successor, Patricia Roberts Harris, at a Senate hearing in September, 1979 (36).

Gerald L. Klerman, who became ADAMHA administrator in 1977, was in general agreement with Noble's views on prevention, and later became a highly articulate advocate of the public health model. At the same time, he supported the views of Food and Drug Administration Commissioner Donald Kennedy who, in the fall of 1977, urged the Treasury Department to start proceedings looking toward the imposition of a label, warning women about the fetal alcohol syndrome, on alcoholic beverages (37).

The succession of events commencing with Noble's scrapping of the responsible drinking approach in 1976 was accompanied by mounting opposition from segments of the alcohol field, including the alcoholic beverage industry associations, and gave rise to emotional cries of "Neo-Prohibitionism." The fight over the Thurmond warning label amendment became clouded by emotion. A staff report by the Senate Subcommittee on Alcoholism and Drug Abuse in September, 1979, became a cause célèbre within the National Coalition for Adequate Alcoholism Programs. The concerns of the Coalition were expressed in a January 2, 1980, letter to Riegle by chairman Leo Perlis, who referred to "not-so-very subtle innuendoes in your staff report linking opposition to warning labels to the influence of the alcoholic beverage industry." Perlis said the letter reflected the majority opinion of the Coalition's membership (38).

Many explanations can be offered to account for the depth of the schisms created by the warning label controversy which promises to continue into at least the early years of the 1980s. In one sense, the emergence of issues related to consumption, warning labels and advertising marked the beginnings of a federal focus on alcohol use and problems, as opposed to alcoholism. It has been said that the treatment and prevention constituencies should be separate: they are dealing with apples and oranges. One is concerned with alleviating the suffering of alcoholism; the other is concerned about the range of other problems attendant on the uses, misuse or "abuse" of alcohol (terms which generate controversy themselves). The fact that the alcoholism constituency was predominantly a treatment-focused constituency when the federal focus began to shift to the larger problems inherent in developing a strategy in prevention explains much of the political discord which accompanied the transition.

The lack of a coordinated, comprehensive federal policy gave rise to the idea of a National Commission on Alcoholism and Other Alcohol-Related Problems, a recommendation of the Liaison Task Panel on Alcoholism and Other Alcohol-Related Problems of the President's Commission on Mental Health in its final report submitted in April, 1978 (39). The universal support of the commission from the alcohol field and the strong backing of HEW and key Congressional figures led to the authorization of the Commission in early 1980 as part of the NIAAA renewal act (14). The unanimity of opinion that a high-level commission was needed to look at the issues was testament itself that the first 10 years of the Hughes Act had not developed a consensus in approaches to the nation's problems with alcohol (40).

NIAAA director DeLuca said, early in his term, that one of the major tasks of the 1980s will be for society to weigh the benefits of alcohol as a recreational and social beverage against the costs of its use in terms of not only alcoholism but a range of negative consequences, from traffic fatalities to lost production.

Viewed from the historical perspective, this statement from the lead federal alcoholism official represents a new posture of the federal government in its approach to beverage alcohol. The 21st Amendment (repeal of Prohibition) signalled a headlong retreat of the federal government from any attempt to control use of alcohol by the national government (leaving such matters up to the states). This, in part, explains why there was no federal role in alcoholism research, treatment and prevention in 1945, and, hence, no chapter comparable to this one in the first edition of *Alcohol, Science and Society*.

The Hughes Act resulted in the first concentrated governmental effort to alleviate the problem of alcoholism, and the first 10 years saw the development of a widespread treatment system. At the same time, it led to the first tentative steps by the national government toward an area viewed as out-of-bounds for almost half a century. In hindsight, it was a natural progression from the treating of casualties of a major public health problem to the examination of measures which might reduce the problem, including steps to control the availability of beverage alcohol (the "agent" in the public health model of prevention).

In fact, what has transpired was implicit in the nomenclature of the Hughes Act—the Comprehensive Alcohol Abuse and Alcoholism Prevention, Treatment and Rehabilitation Act, as well as the National Institute on Alcohol Abuse and Alcoholism, and now more explicitly in the National Commission on Alcoholism and Other Alcohol-Related Problems.

Alcoholism was the focus of the first period of the Hughes Act and the nature of the federal role in regard to alcohol problems was only beginning to take shape at the end of the 1970s. What the ultimate form of the federal role will be must await developments in the 1980s and beyond. It should make an interesting chapter in the third edition of *Alcohol, Science and Society*.

Epilogue

Since the writing of this chapter on the federal role, the political backdrop in the national alcoholism movement underwent a radical alteration with a new cast of conservative players on the Washington scene brought in on the electoral tide of November, 1980. The Reagan Administration translated what it regarded as a popular mandate to reduce the size of the federal government into a sweeping program of slashing, dismantling and consolidating a range of social and health programs—including alcoholism, drug abuse and mental health.

Constituency interests were confronted with a crisis of unparalleled dimensions in the history of the federal alcoholism effort when in March, 1981, the Administration proposed to strip NIAAA of its largest funding programs—project and formula grants—lump them into a health services block grant, and cut them by 25% before distribution to the states. If fully implemented, NIAAA would become a minor research facility with a truncated training component as the Administration proposed to repeal the core provisions of the Hughes Act.

With the precise outcome of the historic battle still in doubt in the spring of 1981, one result seemed inevitable: that the focus of the federal effort on the funding of treatment programs would be phased out rapidly, if not eliminated precipitously. What would have been a natural, evolutionary development appeared likely to become an accelerated process which threatened to disrupt and even dismantle the publicly funded treatment network that was established during the 1970s. The question was not so much as whether this development could be forestalled but whether the federal effort, as embodied by NIAAA, would be able to hold onto a significant role in the areas of research, prevention and training.

REFERENCES

1. COOPERATIVE COMMISSION ON THE STUDY OF ALCOHOLISM. Alcohol problems; a report to the nation. Prepared by T. F. A. Plaut. New York; Oxford University Press; 1967.
2. U.S. PRESIDENT'S COMMISSION ON LAW ENFORCEMENT AND ADMINISTRATION OF JUSTICE. TASK FORCE ON DRUNKENNESS. Task Force report; drunkenness—annotations, consultants' papers, and related material. Washington, D.C.; U.S. Govt Print. Off.; 1967.
3. 361 F. 2d 50 (D.C. Cir. 1966).
4. 356 F. 2d 761 (4th Cir. 1966).
5. 392 U.S. 514 (1968).
6. Public law 91-616.
7. Public law 89-564.
8. Public law 90-222.
9. Public law 91-177.
10. Public law 90-574.
11. Public law 91-211.
12. Over 200 projects are in operation under NIAAA's community action poverty program. ... Alcsm Rep. 1(No. 2): 6, 1972.
13. CHAFETZ, M. E. Alcoholism: people and process; a psychopolitical case study. Psychiat. Ann. 6(No. 3): 6–17, 1976.
14. Public law 93-282.
15. The new law, entitled the Comprehensive Alcohol Abuse and Alcoholism Prevention, Treatment and Rehabilitation Act Amendments of 1974. ... Alcsm Rep. 2(No. 15): 1–2, 1974.
16. A new organizational set-up for NIAAA. ... Alcsm Rep. 1(No. 2): 1–3, 1973.

17. A proposal by administrator Gerald Klerman to centralize initial review. . . . Alcsm Rep. **6**(No. 11): 1–5, 1978.

18. At AR's presstime, alcoholism constituency leaders were being told that NIAAA director. . . . Alcsm Rep. **6**(No. 13): 1, 1978.

19. Key elements of the alcohol initiatives announced May 1 by HEW secretary Califano. . . . Alcsm Rep. **7**(No. 14): 3, 1979.

20. U.S. district court Judge Thomas A. Flannery, ignoring last minute pleas. . . . Alcsm Rep. **2**(No. 9): 1–2, 1974.

21. Following is a roundup of testimony before the Senate Alcoholism and Narcotics Subcommittee. . . . Alcsm Rep. **4**(No. 9): 3–6, 1976.

22. U.S. CONGRESS. HOUSE. COMMITTEE ON INTERSTATE AND FOREIGN COMMERCE. Alcohol abuse and alcoholism amendments of 1976 to accompany H.R.12677. (Rep. No. 94-1092, 94th Congress, 2d session.) 2 vols. Washington, D.C.; U.S. Govt Print. Off.; 1976.

23. NIAAA told Congress that its handling of grants to national organizations. . . . Alcsm Rep. **5**(No. 6): 1–3, 1977.

24. Budgetary developments continued to dominate the national scene. . . . Alcsm Rep. **8**(No. 15): 1–2, 1980.

25. Suggestions of improprieties on the part of NIAAA officials. . . . Alcsm Rep. **3**(No. 2): 2–3, 1974.

26. Public law 96-180.

27. A significant shift in policy directions in funding treatment programs. . . . Alcsm Rep. **8**(No. 8): 4–5, 1980.

28. In a major departure from past policy in the area of prevention. . . . Alcsm Rep. **5**(No. 1): 1–2, 1976.

29. Holding the line at the present level of per capita alcohol consumption. . . . Alcsm Rep. **5**(No. 16): 1–3, 1977.

30. U.S. NATIONAL INSTITUTE ON ALCOHOL ABUSE AND ALCOHOLISM. Alcohol and health; third special report to the U.S. Congress from the Secretary of Health, Education and Welfare, June 1978; technical support document. Ernest P. Noble, editor. (DHEW Publ. No. ADM 79-832.) Washington, D.C.; U.S. Govt Print. Off.; 1979.

31. HEW's third special report to Congress on alcohol and health. . . . Alcsm Rep. **7**(No. 3): 5, 1978.

32. The Senate approved a proposal. . . . Alcsm Rep. **4**(No. 20): 2–3, 1976.

33. U.S. CONGRESS. SENATE. COMMITTEE ON FINANCE. Certain committee amendments to H.R. 10612; hearings before the Committee on Finance . . . 94th Congress, 2d session. Washington, D.C.; U.S. Govt Print. Off.; 1976.

34. U.S. CONGRESS. SENATE. COMMITTEE ON LABOR AND PUBLIC WELFARE. SUBCOMMITTEE ON ALCOHOLISM AND NARCOTICS. Media images of alcohol; the effect of advertising and other media on alcohol abuse, 1976. Hearings, 94th Congress, 2d session, March 8, 11, 1976. Washington, D.C.; U.S. Govt Print. Off.; 1976.

35. In a surprise move, the Senate approved an amendment by Sen. Strom Thurmond. . . . Alcsm Rep. **7**(No. 14): 6–7, 1979.

36. A wide division of views for and against health warning labels. . . . Alcsm Rep. **7**(No. 23): 2–5, 1979.
37. The head of HEW's Food and Drug Administration. . . . Alcsm Rep. **6**(No. 3): 3, 1977.
38. Criticism of a Senate Alcoholism and Drug Abuse Subcommittee staff report. . . . Alcsm. Rep. **8**(No. 6): 5–6, 1980.
39. Report of the Liaison Task Panel on alcohol-related problems. Pp. 2078–2102. In: U.S. PRESIDENT'S COMMISSION ON MENTAL HEALTH. Report to the President. Vol. 4. Appendix. Washington, D.C.; U.S. Govt Print. Off.; 1978.
40. The final report of the alcohol task panel. . . . Alcsm Rep. **6**(No. 13): 2–3, 1978.

24

Prevention: Rise, Decline and Renaissance

Joseph R. Gusfield

During the past decade there has been a renaissance of interest in preventing the incidence of alcohol problems, a result of a developing movement of change in alcohol policies. The 1978 report to Congress from the National Institute on Alcohol Abuse and Alcoholism (NIAAA) noted that: "Recent years have seen a considerable increase in interest in prevention programs and a growing recognition of the necessary differences of approach between prevention and treatment programs. Imaginative, realistic, and effective prevention programs are beginning to emerge" (1, *p. 131*).

In this chapter I will examine prevention as a perspective toward alcohol policies during the post-Repeal period and the past decades. The shifting imageries of alcohol problems and the explanatory theories associated with them are seen as social and cultural movements related to historical contexts of wider expanse than those of alcohol policies and alcohol studies.

Although I will refer to various studies and reviews of research[1] on prevention, this discussion is an analysis of how the revival of interest in prevention both contains and generates a new conceptualization of the problem of alcohol. Policies and programs are not only means toward the realization of given goals and ends, they are also embedded in assumptions, perceptions and theories about the reality of the problem, its nature and its sources. Such assumptions lead us to attend to some matters and to ignore others. In Kenneth Burke's apt phrase: "Every way of seeing is also a way of not seeing."

The Concept of Prevention

Definitions take their meaning from contrast as much as from simi-

[1]There are a number of excellent reviews of the literature in this field. The corpus of such studies or parts of it, can be obtained from: Wilkinson (2); Lau (3); Blane (4); Popham et al. (5); Gusfield (6); Wechsler (7) and especially Gomberg et al. (8).

larity. Prevention should be seen as a contrast to treatment; as an alternative, even conflicting orientation toward alcohol and health. For a variety of reasons, the changes in governing images of alcohol problems have given these terms significant importance as contradictory perspectives and policies. As contrasts they bear important affinities to more general movements in medicine and in politics.

I stress the contrast between prevention and treatment because conventionally studies have ignored the distinction and, consequently, denied certain significant political perceptions. Typically alcohol studies note three forms of prevention: primary, secondary and tertiary. Primary prevention includes actions to prevent agents from causing disease or injury. Secondary prevention is concerned with the early detection and treatment of alcoholics or potential alcoholics or problem drinkers. Tertiary is concerned with preventing further deterioration and reducing disability. Put in another fashion, tertiary prevention is aimed at existing "problem drinkers," secondary at beginning "problem drinkers" and persons with high risk to alcohol problems, and primary prevention at general populations with various levels of risk. The nomenclature is drawn from public health and is, especially in secondary and tertiary prevention, primarily oriented toward preventing disease and chronic conditions. The governing imagery is that of disease and the vocabulary that of epidemiology—host, agent, environment.

The most controversial measures, and the ones forming the cutting edge of discussion in recent years, have been those of primary prevention, especially legal and economic controls of the availability of alcohol. It is these that are forming new images of alcohol problems.

In generating a conception of prevention in this chapter, it is necessary to contrast, within a medical model, attempts at *treatment*, reduction of the *prevalence*, and attempts at *prevention*, or reduction of the *incidence*. Treatment achieves health by curing the unhealthy, prevention by eradicating or minimizing the occurrence of disease. The former is necessarily attentive to the sick—those who deviate from normal conditions. The latter is wider in its scope; it is oriented toward those who might become sick. Treatment is necessarily individualistic and remedial; prevention more collective and environmental.

Among alcohol programs, treatment-oriented policies involve efforts to rehabilitate individuals who suffer, are likely to suffer or are seen by others to be suffering from problems explainable by consumption of alcohol. Recovery centers, clinics, counselling, peer group intervention, case-finding, including early case-finding, and research on the effectiveness of different modes of treatment are all instances of treatment-oriented programs and measures. Measures which locate the source of problems as a condition of the person and seek his or her rehabilitation are included in my usage of treatment.

Prevention, as I use it here, is policy directed toward forestalling an occurrence. It operates more collectively than does treatment because the prevention concept directs attention toward environmental conditions, situational opportunities and restraints, normative standards and legal controls. It attends to events as well as to conditions of persons. Laws which regulate the minimum age of purchase of alcoholic beverages are instances of preventive measures. The laws affect all within an age category, whether they drink or are at risk of alcohol problems. Laws directed at driving while under the influence of alcohol (DWI) are concerned with forestalling events—accidents—rather than rehabilitation of individuals. Educational campaigns to promote moderate drinking or abstinence attempt to control behavior through information or persuasion without reference to specific persons or level of risk.

It is useful to distinguish between different types of prevention measures: (1) those which attempt to persuade people to do something or refrain from doing something, to deter drinking or behavior associated with it and (2) those which attempt to shape the situation or environment within which drinking or the problematic event can occur or the condition of the person develop (6). The first is persuasional. Although others may be, and often are, exposed to the form of persuasion, it is not directed at the abstainer, the non-troublesome drinker or those not at risk of being or creating a problem.

The second type is situational. It affects environmental conditions which make acts possible and provides alternatives to them or supplies opportunities and deterrences. Alcohol education in the public schools is an example of a persuasional form of prevention, intended to modify attitudes through information and personal clarification of values (4). Development of normative standards of moderate drinking through introducing alcohol in previously abstinent situations is an instance of situational measures (2). An illustration of a situational measure is the program introduced at Notre Dame University. Faced with increasing numbers of alcohol-related deaths and injuries involving students, buses were introduced to enable students to ride across the state line from Indiana to Michigan, where purchase of alcoholic beverages was legal at lower ages, thus avoiding the coupling of drinking with driving.

Generally, the decline of interest in prevention after Repeal meant a lessening of efforts toward the social control of drinking.

Treatment-orientation approaches utilized imagery of alcoholism as the problem of alcohol use and alcohol problems as the actions of deviant and diseased persons. The renaissance of prevention promises a return to an imagery of alcohol problems as problems of normal drinkers as well and of alcohol policies as implicated in the social control of drinking and drinking situations. The remainder of this chapter is an elaboration of this thesis and its implications.

The Fall of Prevention: The Alcoholism Movement

The year 1933 is a watershed in the history of alcohol policy in the U.S. Repeal of the Eighteenth Amendment meant the collapse of the major orientation toward alcohol problems of the 19th and early 20th centuries—the prevention of problems through limitations on the availability of the commodity. Reflecting on the history of the mid-1930s and the development of the Yale (now Rutgers) Center of Alcohol Studies, Mark Keller remarked, "of the four action objectives [research, education, treatment and prevention] promulgated by the Center of Alcohol Studies, three were actively pursued. The fourth— prevention—was given only lip service, as it has by everyone else ever since" (9 *p. 25*).

The neglect of prevention which Keller lamented is an accurate description of one consequence of the turn away from the definition of alcohol use as a morally suspect form of behavior and its redefinition as a matter of personal taste made problematic only by a group of people with a form of sickness. The change to the dominance of treatment policies is also a change from the dominance of drinking and drunkenness as social problems to alcoholism as the problem. For this reason, the activities of alcohol-concerned publics in the post-Repeal era have been called the alcoholism movement (10).

A major consensus about alcohol use developed in America during the 19th century. Alcohol came to be viewed as a dangerous commodity, a substance held responsible for accidents, crimes, and moral lapse (11). Alcohol use was no longer proper within the workplace (11–14).[2] This consensus is by no means a simple recognition of fact: It is deeply embedded in the changed character of the workday and the workplace accompanying the rise of industrial organizations. The Temperance Movement succeeded in defining alcohol as a commodity antithetical to industrial work and the workplace (13, 15). Sobriety and inebriety became important symbols of

[2]Also, GUSFIELD, J. R. Passage to play; a ritual theory of social drinking. Presented at the International Congress of Ethnological and Anthropological Sciences, Delhi, India, December 1978.

ethical and economic character. Abstinence was viewed as essential to the qualities of self-control needed in an industrial, market economy (12, 16). The use of alcohol was a moral as well as an economic problem, alcoholism was one of the problems of alcohol, and the alcohol problem was embedded in the total society as a matter of collective significance. The governing images of alcohol use were those of sin, stigma and choice. Law and religion could be called on to help in resolution.

By focusing attention on the alcoholic, the alcoholism movement conceptualized alcohol problems in a different fashion. Deviance and disease became the governing images. Once alcohol problems were seen primarily as those occurring to the drinker who had limited control over his or her drinking, a sharp distinction was made between the social drinker and the alcoholic. Not drinking but deviant drinking and deviant drinkers became the focus of attention.

The disease concept of alcoholism made questions of alcohol misuse those of medicine and health. Accordingly it was felt that scientific research could illuminate solutions as it can in matters of heart disease or diabetes. If the problems of alcohol pertain only to sick people suffering from an external agent—disease—then policies to be pursued are legitimately those of treatment and cure. The deviant is not sinful but sick (17). Alcohol problems are not collective, engaging many groups and levels of society; they are those of alcoholics and others interested in their welfare, a matter of sickness rather than morality. If alcoholism is an illness it is also a medical and scientific object. A technology of treatment should then be possible and a profession should be able to be trained to administer it.

The growth of the alcoholism movement has been manifest in the attention legislatures and voluntary associations have given to providing funds for treatment, to educating the general public, and to changing the public image of alcoholics from one of moral stigma to one of accepted illness. The success of Alcoholics Anonymous in capturing public attention has been a major force in directing public actions toward alcoholism.

The focus on alcoholism has not been a simple matter of logical deduction. A number of events and processes made the treatment emphasis in policy more feasible. First, the alcoholism movement depoliticized the alcohol problem. During the Temperance and Prohibition periods the alcohol question was a focus of significant political conflict not only between generalized culture patterns but between groups carrying differing orientations to the use of alcohol: Protestants vs Catholics; natives vs immigrants; rural vs urban (12). The medicalization of the alcohol problem defused this source of political tension. Alcohol disappeared from a high place on the agenda of public issues and political controversy. The styles of life supporting drinking or abstinence were not in conflict as the object of attention, the alcoholic, was deviant to both. Second, the alcoholic beverage industry did not find the alcoholism movement a source of great threat. Coming after a period in which the entire industry had been declared illegal, the emphasis on treatment was a welcome relief. Implicitly the industry and the movement shared a belief in the slogan of the manufacturers: "The fault is in the man and not in the bottle." From this perspective what is to be prevented is the incidence of alcoholism. Prevention came to mean early case-finding and education about how to detect alcoholism or early "problem drinking" and where to go for treatment. It was cleansed of restrictions on the behavior of "normal drinkers."

Logically, of course, prevention refers to the attempts to minimize or forestall the occurrence of an event or situation. From that consideration, emphasis on the control of alcohol availability can be directed at prevention of alcoholism from a health as well as a moral perspective. In its concern for treating alcoholics the alcoholism movement shied away from the political albatross of availability of alcohol as a strategy of prevention. It is quite appropriate to characterize the dominant policy of American society toward prevention of alcohol problems from 1933 to 1970 as one of efforts to persuade individuals with drinking problems to drink less or to undergo treatment for alcoholism.

Nevertheless, in several important ways the pre-Repeal conception of alcohol as a special commodity persisted in American society. Some primary prevention programs were an accepted legacy of prohibition. Following Repeal, all states restricted the sale of alcoholic beverages in one or another fashion, even though prohibition continued in only a few states (now in none) and in a small proportion of local jurisdictions, now comprising about 3 to 5% of the population.[3] Place and time and age of customer remain as major restrictions which distinguish the sale of alcohol from other commodities. Laws regulating the minimum age of purchase came into widespread existence after Repeal (18). Most states established Alcohol Beverage Control boards, ostensibly to promote temperance, but they regulated the market and licensed sellers, largely to avoid the undesirable effects of the old saloon (2, 19). Public drunkenness and

[3]ROOM, R., CAMERON, T., FRIEDMAN, J., SMITH, D. and WALLACK, L. Profile of policies and programmes for the prevention of alcohol-related problems. Second draft, University of California Social Research Group, 1979.

driving while intoxicated continued to be illegal. The DWI laws represent one of the clearest preventive approaches to an alcohol-related problem, that of automobile accidents, by attempting to persuade drivers to refrain from drinking or driving after drinking.

What happened during the period of the dominance of the alcoholism movement may be seen as the defusing of the political conflict around alcohol problems. The movement separated the general population and its behavior towards alcohol from that of alcoholics—a special type of deviant group which could not handle alcoholic beverages competently; people unlike most. An emphasis on treatment of the alcoholic defined policies as those which affected only the minority. The "special commodity" status of alcohol was never seriously attacked, however, but alcohol was removed from the agenda of major issues in American society and politics.

Implications of Alcoholism Imagery

Once alcohol problems are placed in the framework of disease and deviance and alcoholics in a special category, measures to alleviate the problem are directed at the special category. From this standpoint measures that might restrict the behavior of the general population are excessively confining or unneeded. Most people do not drink heavily and, even if they are drunk occasionally, they manage it competently. With the imagery of the alcoholism movement what is to be prevented is alcoholism. Thus prevention is either minimized as a strategy or attached to treatment as a form of early casefinding or education which will persuade those "suffering" from the disease to seek treatment.

The preoccupation with "alcoholism problems" has two significant implications: (1) it directs public attention to the alcoholic image as the description of all persons who experience difficulties imputed to alcohol use; and (2) it leads public officials to define all alcohol-related problems as those resulting from persons suffering from alcoholism. In both cases policies are aimed at reforming persons rather than preventing events and at alcoholism rather than instances of drunken behavior and its consequences.

The classic picture of the alcoholic as the Skid Row resident has undergone considerable revision in recent years. The concept of the "problem drinker" has been offered as a way of describing the variety and flexibility of alcohol problems as they affect individuals (20–22). It is doubtful, however, that the concept has had much impact on the general public which has subsumed "problem drinker" as another form of "alcoholic" (23, pp. 51–82).

The issues surrounding drinking and driving illustrate the problem of prevention conceived within the framework of alcoholism or within alternative frameworks. Much research has been devoted to the questions of whether a large segment of the DWI population is composed of, and a large proportion of alcohol-related driving events are produced by, alcoholics, problem drinkers or social drinkers. The debate has produced the usual rash of research studies and the usual shifts in "fashion" from the consensus of the late 1960s and early 1970s on the problem-drinking or alcoholismic character of many of those drivers to the more recent criticisms and qualifications of the thesis (1, *p. 87; 24–27*).

If DWI offenders are perceived as people who, as a class, have more than the "normal" problems of self-control with alcohol then legal measures, which hold the threat of punishment, can have little impact on deterring driving after drinking. In recent years many courts have inaugurated courses to persuade offenders of their alcohol problems and lead them into treatment. If driving while impaired by alcohol is seen as a problem of the alcoholic or problem drinker than it is only another form of the general problem of alcoholism. Policy then consists of case-finding: discovering the alcoholic and pre-alcoholic, convincing him or her to accept treatment, and providing treatment.

This singlistic view of alcohol problems can be contrasted with one which stresses the different forms of problems seen as alcohol-related. If the DWI issue is conceptualized as chiefly involving social drinkers or even occasional drinkers then it is an issue of the consequences of drunkeness or even just drinking; an event rather than a personal characteristic; indeed it is one so ubiquitous in American life that even though social drinkers may be deterred by one conviction, the pool of that population is so large as to constitute a continuous problem (23, 28, 29). DWI laws are thus limited means of preventing the event when the target group is alcoholics or problem drinkers but not when the target is the general population of drinkers (6, 26).

In a sense, the alcoholism movement has made it more difficult to perceive multiplicity in alcohol problems. By making the problem of alcohol that of alcoholics, it has minimized the consequences of drinking and drunkenness in the general population of drinkers and the sources of alcohol problems stemming from the general environment of institutional practices. It has also made the plurality of alcohol problems into a single one of characteristics of deviant drinkers. A wider spectrum of alcohol problems could recognize the differences between driving after drinking, diminished work productivity, marital strife and cirrhosis of the liver. Although they may be discovered in greater degree in problem drinkers they are by no means exclusively theirs. Events of drunkenness, for example, occur more frequently among problem drinkers but they may be only a small part of the total alcohol-related risk-taking events in the society. A recent study of U.S. Air Force personnel (30) found that although the 3½% of the persons defined as very heavy drinkers accounted for 26% of "drunk days," the remainder of drinkers accounted for 74% of the drunk days. A concentration on the person in the image of the alcoholic provides a different perspective from a concentration on the events.

The Renaissance of Prevention and Its Imagery

The sense of a new activism is apparent to many who have been working in the alcohol field for a number of years. The commission of conferences, symposia and working groups on problems of prevention is one sign of activity and interest.[4] Because of the renewed attention to measures that limit the availability of alcohol and other programs and policies to control the uses of alcohol, I refer to this as "the new temperance movement." Richard Bonnie refers to this as well as several other measures "to engineer changes in collective life style as an essential component of the new public health initiatives" as "the new paternalism" (32, *p. 39*). A leading publication, frequently cited in contemporary alcohol policy circles, indicates both the social control orientation of the movement and its setting in the imagery of public health. This report, the result of an international collaborative effort, bears the title *Alcohol Control Policies in Public Health Perspective* (33).

An important part of that report's orientation builds on research stemming from the work by a French statistician, Sully Ledermann, on consumption of alcohol (34) and that of the Addiction Research Foundation (35) on the association between rates of alcohol consumption and rates of alcohol problems. Briefly put, increases in rates of cirrhosis of the liver are correlated with increases in the

---

[4]See especially the report of the conference sponsored by the California Office of Alcoholism in 1974 (31). The NIAAA has convened several task-forces to prepare position papers and the Alcohol Drug and Mental Health Administration (ADAMHA) has been preparing a position paper on prevention (KLERMAN, G. ADAMHA prevention policy and review. Unpublished draft, 1980). Of particular significance is the commission by ADAMHA of a panel of the National Academy of Sciences on prevention, including a subgroup on alcoholism. Their report and the papers commissioned for it will be an important step in prevention policy. (That material is not yet available for attribution.)

national rate of consumption of alcohol and decreases are related to decreases in such rates.

The importance of this approach is that it shifts attention toward general national policies as means of preventing alcohol-related problems. Unlike the focus on treatment, the movement is concerned with limiting the availability of alcohol, including pricing mechanisms such as taxation as measures through which to effect problems by diminishing consumption. The prevention of liver cirrhosis and, in another facet of recent campaigns, the prevention of the fetal alcohol syndrome lead to measures applicable to the total population. From the perspective of this discussion what is significant about the distribution of consumption approach is less its scientific validity than its historical importance in raising control measures, including availability of alcoholic beverages, to a point of serious reconsideration.

What has emerged is a growing demand for approaches to alcohol problems through primary prevention rather than treatment or secondary and tertiary prevention. The "new temperance movement" is thus another form of health movement. The effort to educate the public about the fetal alcohol syndrome, for example, presents information in a context of pre-natal care (1, *pp. 39–45*). The concern for lowering national rates of consumption is clearest in a focus on the dangers of cirrhosis of the liver.

Prevention Measures and Their Effectiveness

In this section I will discuss studies of the scope and effectiveness of various types of measures, but the limits of such studies should be clearly understood. Reports of specific, isolated events, such as the effect of a strike among alcoholic beverage industry employees or a demonstration project to test the effect of advertising on diminishing consumption of alcohol abound in the literature. They should be sharply contrasted with studies of large-scale programs composed of many acts and actions, no one of which may be significant although the entirety builds into a climate of attitude and behavior control distinctly different from the sum of its parts.[5]

A second important caveat in the use of individual studies as support or negation of suggested policies is the difficulty in importing studies from another context. Differences in social structure, culture and institutions make it hazardous to reach conclusions about

[5] I am indebted to Nils Christie for pointing this out during the 1974 California conference on prevention.

what might happen in the U.S. based on results found in Sweden or vice-versa. Additionally, studies of one program or commodity cannot, without further study, be applied to another (36). Maccoby's study (37) suggests that the use of mass communications alone is ineffective in changing health habits relevant to heart problems but could be effective when accompanied by a community organization program. Whether the results are generalizable to alcohol consumption would not be known without further study. (The California Office of Alcoholism is currently undertaking such a study but results are not yet available.)

Since Prohibition represented a prevention program of major scope its effect on alcohol problems is of great concern. Did it lead to diminished consumption? Did it lead to decreased incidence of alcohol problems? Unfortunately the data are far from perfect. However, analysis of grains production, drunkenness arrests and rates of liver cirrhosis during Prohibition and post-Prohibition periods all oppose the popular view that the "noble experiment" was ineffective or led to increased drinking and drinking problems (38).[6] The increase in alcohol consumption that marked the period preceding Prohibition did not continue after Repeal: America became a nation of more drinkers, but of more moderate drinkers.

The Prohibition period suggests other effects, however. A sharp rise in the price of alcoholic beverages (available only on "black markets") and a shift to manufacturing spirits rather than beer diminished drinking among the poorer classes and increased the demand for spirits among the middle classes. Prohibition appears to have supported the long-run trends toward a more moderate drinking pattern in America. Its impact on basic sentiments toward alcohol policy seems slight (38). I have described above how after Repeal alcohol continued to be a special and controlled commodity, even more so in the case of laws regulating minimum age of purchase.

Prevention measures include far more than such sweeping and restrictive legislation as Prohibition entailed. Six types of measures undertaken or discussed in recent years are part of the emerging new impulse toward prevention and control. Again, it is important to distinguish persuasional from situational measures and those which attempt to change persons from those which attempt to control events.

1. Case-finding. One frequently encountered meaning of prevention is that of preventing alcoholism, problem drinking or liver

[6]Also, AARON, P. and MUSTO, D. Temperance and Prohibition in America; a historical overview. Unpublished report prepared for the panel on Alternative Policies for the Prevention of Alcohol Problems of the National Academy of Sciences, 1980.

cirrhosis by finding those whose drinking habits have already placed them in a population at greater risk than is normal. Such people can then be targets for persuasion into treatment or habit change. The programs of required classes on alcohol education for convicted DWI offenders is an example.

What is being prevented by case-finding? One ready answer is that alcoholism and problem drinking is being discovered before it becomes even more a source of suffering and trouble than already has occurred. In the instance of DWI offenders there is an additional rationale; a high proportion of offenders are also problem drinkers or alcoholics (26). From this viewpoint, the event can be controlled by changing individuals who are problem-drinking prone. Whether this rationale is correct or not, to date the results of such educational classes have not been salutary. Classes for offenders do not appear to have diminished the rates of rearrests among those exposed to them (39–41), although they may be more effective in reducing alcoholism.

Similarly, the shifts from jail as punishment for public drunkenness to decriminalization, detoxification and treatment may be useful to bring alcoholics into the treatment facilities but their effectiveness as mechanisms for removing drunks from shopping and entertainment areas is much in doubt (42).

Here, as in so much of the discussion of alcohol problems, the efficacy of specific measures depends greatly on how the problem is delineated.

2. *Education.* Faith in education and information is almost a cardinal principle of American life. There is a great, almost mystical faith in the power of advertising and communications media in the U.S. Education through school and especially through the mass media is often surrounded with an almost religious awe at the "power of the press." Providing information about the harmful effects of a behavior in a context of attitudinal approval is presumed to be an assured means of creating change. That such measures persist in the face of so many adverse studies testifies either to the great American belief in the wisdom of the individual or to a great predilection for cheap solutions. It is one of the clearest arguments for the persistence of magic in modern societies.

A growing disappointment with campaigns to control alcohol use through dissemination of information has led the advocates of prevention to policies of situational forms. Blane's influential review of educational measures (4) reached the conclusion that while educational programs may affect information and even attitudes, the impact on behavior is slight. Even Wilkinson, who advocates advertising and educational campaigns, sees them as only effective in a context of accompanying situational measures (2, *pp. 43 ff*).

3. *Legislation Deterring the Drinker.* The field of alcohol controls is filled with laws aimed at deterring the delinquent behavior of drinkers. The legislation against public drunkenness and driving while intoxicated represent two such familiar laws. They operate within a utilitarian philosophy of reward and punishment. Threats of fines, jail or loss of driver license are expected to deter the individual from the illegal acts (43). Increased punishment is expected to lead to increased conformity.

Such expectations have great limits. Efforts to increase law enforcement or to raise the punishment level have not proven effective as measures to diminish alcohol-related automobile deaths in the U.S. (23, 44–45), although they may have had temporary success in England (46–47). This is not to say that such laws may not be effective deterrents to even more public drunkenness or driving after drinking. No American state has attempted so radical an experiment in the interests of research as to repeal DWI laws. It seems doubtful, however, that increasing punishments to the individual will decrease the occurrence of the events.

4. *Situational Measures.* Legislation aimed at controlling the situation or environment of the drinker did not disappear with Repeal. As pointed out above, laws restricting sales to minors, zoning regulations, licensing requirements and other governmental acts have continued to define alcohol as a special commodity and to provide a less than fully permissive environment for the purchase of beer, wine and especially spirits. Much of the discussion since 1970 has focused on efforts to increase the range, intensity and salience of further measures to control the availability of alcoholic beverages to the general public or particular groups. The situational measures are applicable to all types of drinkers—occasional, moderate and heavy, problem-free, problem and alcoholismic.

A major point of debate within policy and research circles has been that of increasing taxation on alcoholic beverages (5, 35), a view partly supported by *Alcohol Control Policies in Public Health Perspective* (33). It represents a distinctively new approach to problems of alcohol control since Repeal. This policy follows from the work on national consumption patterns described above. Correlations between national consumption patterns and the incidence of liver cirrhosis appear to be convincing evidence that diminutions in total consumption affect problem drinkers and alcoholics as well as social drinkers. Both kinds of drinkers appear to be influenced by price as alcohol seems to be a more price-elastic commodity than has been assumed (33, *pp. 65–83*). Thus, with inflation (the price has declined relative to other commodities), with a general worldwide increase in consumption of alcohol, and with cirrhosis rates rising in many nations, the argument for further tax measures has gained greater support.[7]

The policy is more debatable in its effects on other alcohol problems, however. Schmidt and de Lint (35) posit a constant relation between cirrhosis rates and other alcohol problems but that ratio is disputable. Some recent evidence also suggests a positive relation between automobile deaths and alcohol prices.[8]

Measures to affect the availability of alcohol are easier to enforce than measures aimed at individual drinkers because they impinge on a relatively small and visible segment of the population—sellers and distributors. They represent the kind of legal sanctions most easily applied since little discretion is needed in determining dereliction and the sanctions are clear and easily imposed, as in loss of license (50).

The fact that such laws are readily observed does not necessarily imply their effectiveness in reaching desired goals of preventing or minimizing the occurrence of alcohol problems. Increasing or decreasing the numbers of retail outlets or hours of sale does not appear to have changed rates of consumption or alcohol problems significantly where such experiments have been attempted. Lowering or raising the age of permissible sale, however, does appear to have some effect in some places on alcohol-involved deaths from automobile crashes (5, *pp. 592–594; 51*).

The minimum-age laws are instructive. In the early 1970s, in the wake of lowering the voting age to 18, 18 American states lowered the minimum age of permissible sale to 18 or 19. Studies of the effects of this move, and of the subsequent raising of the age in 6 states, indicate that lowering the age was accompanied by an increase in youthful automobile accidents and deaths due to drinking, a generally increased amount of drinking and an increase in sales (7, 52). Results of studies investigating the effect of raising the age are not yet available. It should also be pointed out that in the leading study (52), while sharp changes occurred in Michigan and Maine, none were found in Vermont.

[7]See the exchange between Parker and Harman (48) and Schmidt and Popham (49).

[8]COOK, P. The effect of liquor taxes on drinking, cirrhosis and auto accidents. Unpublished report prepared for the panel on Alternative Policies for the Prevention of Alcohol Problems of the National Academy of Sciences, 1980.

The laws governing minimum age of purchase are significant in another sense. To be able to take away rights shortly after they have been given, as in the case of the 6 legislatures which subsequently raised the legal age of purchase by 1 to 3 years, indicates a high political capacity for control of alcohol problems. The alcoholic beverage industry does not appear to have been able to stem the legislative tide of such controls.

5. *Reconceptualization of Problems.* The policies discussed above direct public attention to drinking as the central cause of the problem and as the target of policies. Policies attempt to make the world and the person safe from drinking and drunkenness. Suppose that framework were reversed and instead policy-makers asked, How can we make the world and the person safe for drinking and drunkenness? As Ralph Nader said at the Congressional hearings on automobile safety, "Let us produce an automobile that will be safe on the assumption that it will be driven by fools and drunkards."

When a problem is defined as an alcohol problem that formulation directs attention toward drinking as the cause and preventing drinking as the solution. The "reality" of the problem is thus a capsulated form of theory and policy. It is possible to reconceptualize some such alcohol-related problems in ways which diminish the centrality of alcohol. For example, the Monday morning slump in industrial attendance and productivity (the modern equivalent of the "Saint Monday" in industrializing periods of history) is often seen as a problem resulting from weekend drinking. It can also be seen as a problem of scheduling work in such a fashion that it conflicts with the individual needs of workers. A number of firms, especially in Europe, have experimented with more individual work schedules as a means of making the work experience less alienative (53).

In a similar vein, issues of drinking and driving have placed emphasis on not drinking before driving or not drinking. No concerted experimentation, however, has been made into alternative modes of transportation; for example, a policy which charged taverns with responsibility for transporting its customers home. The invention of an automobile which cannot be started by an inebriated driver represents another strategy of preventing the problem (54). Perhaps we might seek a form of alcoholic beverage which was equally disinhibiting but affected motor capacities less strikingly.

Such reconceptualizations operate on a moral premise in which drinking is neutral. The aim is to provide an environment in which it is safe to drink and to be drunk. In the landmark Powell case,[9] the majority of the Supreme Court agreed with the dissenting opinion that alcoholics should not be punished for their condition but upheld Powell's conviction on a public drunkenness charge because his crime was not being drunk but being drunk in public. A possible "solution" to this boundary problem lies in providing places for homeless men to drink where they will not impinge on people such as merchants and shoppers. Some communities are beginning to experiment with "wet parks" which would accomplish such boundary-maintaining programs.[10]

6. *The Sociocultural Environment.* There has long existed an orientation that has explained alcohol problems as aspects of larger institutional and historical elements. Marxists may find Capitalism and problems of work at the root of personal strains and stresses that produce excessive drinking and alcoholism (55, 56). An emphasis on the general cultural and social conditions should not be dismissed but, by themselves, are too global for preventive strategies. Measures to deal with such deeply-seated institutional elements are beyond the ken of the fundamentally reformist orientation which dominates alcohol policies. Often they rest on nothing more than a series of assertions about the relation between background conditions and alcohol use. Perhaps Griffith Edwards has summed up what can be best said at present: "Since we are not able to manipulate personality and produce a race with no neurosis, the only realistic method of exerting a benign influence on prevalence of alcohol addiction is by control of environmental conditions of drinking" (57, *p. 424*).

Another orientation has been that of efforts at integrated drinking (2, 58). Here the relation between drinking norms and drinking problems has been the object of scrutiny. In one form or another, proponents of integrated drinking have argued that alcohol problems are accentuated in cultural settings in which drinking is proscribed. A more accepting attitude toward drinking would then provide for a learning process in which people are socialized into a controlled pattern of "responsible drinking." This view is bolstered by reference to research among American Jews and Italian–Americans which shows a low rate of alcoholism coupled with a high proportion of nonabstainers. However, as several scientists have

[9]*Powell v Texas*, 392 U.S. 514 (1968).

[10]Sacramento, California has experimented with such a park and it is under consideration in other cities.

pointed out, high rates of nonabstinence do not imply high use of alcohol on any one occasion. Nor is it the case that an increase in the moderate use of alcohol means a diminished use of alcohol excessively on other occasions (59).[11]

I have placed this discussion of the sociocultural context at the end of this section because it suggests an important dimension in preventive measures and programs. Research and policy contribute significantly to the sociocultural climate because they affect the conception of the object—alcohol—and of the alcohol problem. They aid in maintaining or developing new governing images. In turn, they are themselves a part of changing conceptions which bear on the images of alcohol problems and their resolution. They influence the development of the very sociocultural environments from which they are drawn.

In considering prevention as a movement, it is this function as an agent in the reconceptualization of alcohol problems that I have emphasized. In the final section I turn my attention to the larger social and cultural environments from which the new prevention movement draws its support and to which it in turn contributes.

Sources of New Alcohol Movements

The new turn toward prevention in the alcohol field draws on a number of sources and resources both inside the arena of alcohol concerns and outside in large social and cultural movements. The 1970s differ from the pre-Repeal eras in ways significant to an understanding of the changing imagery and programmatic interests relevant to alcohol problems. It is a cliché to say that through these decades America was changing; what is important is the significance of some of the changes in the development of prevention strategies and the governing images of alcohol.

Chief among the changes is the greatly enlarged role of the State, especially at the federal level, as a provider of human welfare. Increasingly, Americans have viewed the State as the central source of welfare, including the provision of medical services. In the late 1970s government—federal, state and local—was the source of two of every five dollars spent on medical services in the U.S. (60, p. 6).

The entry of the State into the health provision field coincided and interacted with two other important developments. One is the

[11]Also, ROOM, R. The effect of drinking laws on drinking behavior. Presented at the annual meeting of the Society for the Study of Social Problems, Denver, Colorado, August 1971.

greatly increased commitment to health in the U.S.: between 1960 and 1976 expenditures for medical care rose from 5.2 to 8.6% of the gross national product, and the 1981 estimate is 10.2.[12]

The second, perhaps related to the first, is the growing criticism of conventional medicine and the domination of treatment orientations to health, which is evolving into a movement toward preventive medicine (60, 62–64). The criticism stems from many sources; realization of the slim returns in health from large expenditures for new technology, an egalitarian impulse against expertise and expert authority, a demand for a more equitable distribution of medical services.

The State has become more sensitive to the costs of medicine as well as the quality (60, 65). The search for alternatives to clinical medicine is thus one possible solution to the costs crisis; so too is the rising importance of various self-help and holistic medicine programs. Campaigns to induce exercise or life-style changes emerge in a context of preventive health. The campaign against cigarette smoking, actively led by the Public Health Service, is a good example of one form of preventive medicine that has almost become a model for prevention activities.

It is noteworthy that in these activities, government officials and health researchers have played a leading role. The entry of this echelon of reformers, a salient feature of movements in contemporary America, is what Moynihan has dubbed as "the professionalization of reform" (66).

The alcohol field has reflected these trends. Since the rise of the alcoholism movement, alcohol problems have been approached largely as matters of health and medicine. The entry of the State into the treatment of alcoholism was a major trend following World War II, with treatment-oriented state agencies being established in 38 states during the period 1945–1953. With the inauguration of the NIAAA in 1970 recognition of alcohol problems as national and state concerns had come of age. In recent years it has given greater attention to prevention[4] and in 1973 established a Division of Prevention. Although the initial programs were directed toward education of youth, increasingly NIAAA has moved into community organization and toward consideration of other forms of primary prevention. Another direction, taken outside the health field, is the Alcohol Safety Action Programs conducted by the Department of Transportation to deter driving after drinking.

[12]Gibson and Mueller (61, *p. 4*), quoted by Starr (62, *p. 176*).

The Cultural Background of Prevention

The growing political importance of health as a public value may be a clue to subtler changes in the American culture that might generate political support of alcoholism prevention programs. The old constituencies that formed around the Temperance and Prohibition battles have lost vibrancy and structural support. Alcohol use is no longer a symbol which sharply divides religions, classes, or major sociological categories. While differences exist between varying styles of life that sanction or condone drinking and drunkenness, as I have argued elsewhere (12, *pp. 139–165*) these are no longer clearly linked to such institutional-communal bases as Protestant–Catholic, urban–rural or native–immigrant.

The emphasis on the workplace and on self-control which the 19th century made the justificatory values for temperance contained a dualism and a tension around which polar types emerged. In one orientation, life must be monistic; the same attitudes demanded in work are also demanded in leisure. Self-control, discipline and a mien of seriousness should pervade both. Alcohol is a constant threat and drunkenness an event of great danger without reward. In a diverse attitude, life is dualistic. Drinking and drunkenness are threats to work and life in its serious side, but leisure involves us in "time out": a clear differentiation from work and an acceptance of play.[13]

The emphasis on health as a value is, perhaps, part of a newer orientation to the self and to the relation between work and leisure. The widespread expression of interest in health and safety is seen in diet, nutrition and exercise programs, in the styles of life now enjoined in heart disease campaigns, in the various self-improvement and personal potential movements. Health is itself becoming a leisure-time pursuit as well as a prerequisite to work and play. The concern for a variety of illnesses, including not only cirrhosis but also heart disease, cancer, birth defects, general effects on metabolism, and the exposure to risk and illness which drunkenness as well as alcoholism entail, put alcohol in a significant place as a topic in discussions of appropriate, healthly, life styles.

The import of these concerns and the prevention measures they entail is that they too return to an emphasis on the drinker. It is rash to predict the future acceptance and growth of primary pre-

[13]In this analysis of "time out" behavior I draw on the work of MacAndrew and Edgerton (67), the study of bars by Cavan (68) and my discussion of ritual and drinking (cited in footnote 2). The analysis of competence and incompetence in drinking owes much to the ideas of Joseph Kotarba and is elaborated in our National Science Foundation report (69).

vention programs. Social science is filled with the ghosts of prophecies. By their very appearance, movements of change test the existence of the latent supports and generate new ones. The relative success of the cigarette smoking programs in changing the normative environment for smoking in the U.S. has given much encouragement to other campaigns to change deeply held habits.

What I am suggesting is that the prevention movement possesses a reconceptualization of alcohol problems and their solutions which represents a distinctively new element in the public attitudes and definitions. Unlike the era of the alcoholism movement, such programs necessarily represent controls over environmental conditions and personal choice. As such they become drawn into the arena of political conflict and struggle.

The authors of the original *Alcohol, Science and Society* hoped that Science might rescue us from the disputes and struggles of political conflict that Prohibition had entailed. As alcohol issues become involved in measures of primary prevention, law and public policy are drawn on as mechanisms of control. The higher the place of alcohol on the agendae of public attention, the more that detachment from politics may be dissolved. The issues of personal autonomy vs public health and safety may again be raised in the context of resolving alcohol problems.

REFERENCES

1. U.S. NATIONAL INSTITUTE ON ALCOHOL ABUSE AND ALCOHOLISM. Alcohol and health; third special report to the U.S. Congress from the Secretary of Health, Education and Welfare, June 1978; technical support document. Ernest P. Noble, editor. (DHEW Publ. No. ADM 79-832.) Washington, D.C.; U.S. Govt Print. Off.; 1979.
2. WILKINSON, R. L. The prevention of drinking problems; alcohol control and cultural influences. New York; Oxford University Press; 1970.
3. LAU, H.-H. Cost of alcoholic beverages as a determinant of alcohol consumption. Pp. 211–245. In: GIBBINS, R. J., ISRAEL, Y., KALANT, H., POPHAM, R. E., SCHMIDT, W. and SMART, R. G., eds. Recent advances in alcohol and drug problems. Vol. 2. New York; Wiley; 1975.
4. BLANE, H. Education and the prevention of alcoholism. Pp. 519–578. In: KISSIN, B. and BEGLEITER, H., eds. The biology of alcoholism. Vol. 4. Social aspects of alcoholism. New York; Plenum; 1976.
5. POPHAM, R. E., SCHMIDT, W. and DE LINT, J. The effects of legal restraint on drinking. Pp. 579–625. In: KISSIN, B. and BEGLEITER, H., eds. The biology of alcoholism. Vol. 4. Social aspects of alcoholism. New York; Plenum; 1976.
6. GUSFIELD, J. R. The prevention of drinking problems. Pp. 267–291.

In: FILSTEAD, W. J., ROSSI, J. and KELLER, M., eds. Alcohol and alcohol problems; new thinking and new directions. Cambridge, Mass.; Ballinger/Lippincott; 1976.

7. WECHSLER, H., ed. Minimum-drinking-age laws. Lexington, Mass.; Heath; 1980.

8. [GOMBERG, E., BRESLOW, L., HAMBURG, B. A. M. and NOBLE, E. P.] Prevention issues. Pp. 105–138. In: INSTITUTE OF MEDICINE. DIVISION OF HEALTH PROMOTION AND DISEASE PREVENTION. Alcoholism, alcohol abuse, and related problems; opportunities for research. Report of a study. Washington, D.C.; National Academy Press; 1980.

9. KELLER, M. Problems with alcohol; an historical perspective. Pp. 5–28. In: FILSTEAD, W. J., ROSSI, J. and KELLER, M., eds. Alcohol and alcohol problems; new thinking and new directions. Cambridge, Mass.; Ballinger/Lippincott; 1976.

10. ROOM, R. G. W. Governing images of alcohol and drug problems; the structure, sources, and sequels of conceptualizations of intractable problems. Ph.D. dissertation, University of California, Berkeley; 1978.

11. LEVINE, H. G. Demon of the middle class; self-control, liquor, and the ideology of temperance in 19th-century America. Ph.D. dissertation, University of California, Berkeley; 1978.

12. GUSFIELD, J. R. Symbolic crusade; status politics and the American temperance movement. Urbana; University of Illinois Press; 1963.

13. HARRISON, B. H. Drink and the Victorians; the temperance question in England, 1815–1872. Pittsburgh; University of Pittsburgh Press; 1971.

14. TYRRELL, I. R. Temperance and economic change in the Antebellum North. Pp. 45–67. In: BLOCKER, J. S., JR., ed. Alcohol, reform and society; the liquor issue in social context. (Contributions in American history, No. 83.) Westport, Conn.; Greenwood Press; 1979.

15. GUTMAN, H. Work, culture and society in industrializing America. New York; Vintage; 1967.

16. LEVINE, H. G. The discovery of addiction; changing conceptions of habitual drunkenness in America. J. Stud. Alc. 39: 143–174, 1978.

17. GUSFIELD, J. R. Moral passage; the symbolic process in public designations of deviance. Social Probl. 15: 175–188, 1967.

18. MOSHER, J. F. The history of youthful-drinking laws; implications for current policy. Pp. 11–38. In: WECHSLER, H. ed. Minimum-drinking-age laws. Lexington, Mass.; Heath; 1980.

19. MEDICINE IN THE PUBLIC INTEREST. A study in the actual effects of alcohol beverage control law. Vol. 1. Washington, D.C.; 1976.

20. CAHALAN, D., CISIN, I. H. and CROSSLEY, H. American drinking practices; a national study of drinking behavior and attitudes. (Rutgers Center of Alcohol Studies, Monogr. No. 6.) New Brunswick, N.J.; 1969.

21. CAHALAN, D. Problem drinkers; a national survey. San Francisco; Jossey-Bass; 1970.

22. CAHALAN, D. and ROOM, R. Problem drinking among American men. (Rutgers Center of Alcohol Studies, Monogr. No. 7.) New Brunswick, N.J.; 1974.

23. GUSFIELD, J. R. The culture of public problems; drinking-driving and the symbolic order. Chicago; University of Chicago Press; 1981.

24. U.S. DEPARTMENT OF TRANSPORATION. Alcohol and highway safety; a report to the Congress from the Secretary of Transportation. Washington, D.C.; 1968.

25. WALLER, J. A. and TURKEL, H. W. Alcoholism and traffic deaths. New Engl. J. Med. **275**: 532–536, 1966.

26. WALLER, J. A. Identification of problem drinking among drunken drivers. J. Amer. med. Ass. **200**: 114–120, 1967.

27. CAMERON, T. Alcohol and traffic. In: CALIFORNIA UNIVERSITY. SCHOOL OF PUBLIC HEALTH. SOCIAL RESEARCH GROUP. The epidemiological literature of alcohol, casualties and crime: systematic quantitative summaries; by Aarens, M. et al. 2 vol. (Rep. No. C-19.) Berkeley; 1977.

28. WOLFE, A. C. Characteristics of late-night, week-end drivers; results of the U.S. national roadside breath-testing survey and several local surveys. Pp. 41–49. In: ISRAELSTAM, S. and LAMBERT, S., eds. Alcohol, drugs and traffic safety; proceedings of the sixth international conference. Toronto; Addiction Research Foundation; 1975.

29. ZYLMAN, R. DWI enforcement programs: why are they not more effective? Accident Anal. Prev., Oxford **7**: 179–190, 1975.

30. POLICH, J. M. and ORVIS, B. R. Alcohol problems; patterns and prevalence in the U.S. Air Force. A project Air Force report prepared for the U.S. Air Force. (Report No. R-2308-AF.) Santa Monica; Rand Corporation; 1979.

31. ROOM, R. and SHEFFIELD, S., eds. The prevention of alcohol problems; report of a conference. Sacramento; California Health and Welfare Agency, Office of Alcoholism; 1974.

32. BONNIE, R. J. Discouraging unhealthy personal choices through government regulation; some thoughts about the minimum drinking age. Pp. 39–58. In: WECHSLER, H., ed. Minimum-drinking-age laws. Lexington, Mass.; Heath; 1980.

33. BRUUN, K., EDWARDS, G., LUMIO, M., MÄKELÄ, K., PAN, L., POPHAM, R. E., ROOM, R., SCHMIDT, W., SKOG, O.-J., SULKUNEN, P. and ÖSTERBERG, E. Alcohol control policies in public health perspective. (Finnish Foundation for Alcohol Studies, Vol. 25.) Helsinki; 1975.

34. LEDERMANN, S. Alcool, alcoolisme, alcoolisation. 2 vol. (Institut national d'études demographiques, travaux et documents.) Paris; Presses universitaires de France; 1956–1964.

35. SCHMIDT, W. and DE LINT, J. Estimating the prevalence of alcoholism from alcohol consumption and mortality data. Quart. J. Stud. Alc. **31**: 957–964, 1970.

36. KATZ, E. and LAZERSFELD, P. Personal influence; the part played by people in the flow of mass communications. Glencoe, Ill.; Free Press; 1955.

37. MACCOBY, N., FARQUHAR, J. W., WOOD, P. D. and ALEXANDER, J. Reducing the risk of cardiovascular disease; effects of a community based campaign on knowledge and behavior. J. community Hlth **3**: 100–114, 1977.

38. GUSFIELD, J. R. Prohibition; the impact of political utopianism. In: BRAEMAN, J., BREMMER, R. and BRODY, D., eds. Change and continuity in twentieth-century America: the 1920's. (Modern America, No. 2.) Columbus; Ohio State University Press; 1968.

39. NICHOLS, J. L., WEINSTEIN, E. B., ELLINGSTAD, V. S. and STRUCK-MAN-JOHNSON, D. L. The specific deterrent effect of ASAP education and rehabilitation programs. J. saf. Res., Chicago 10: 177–187, 1978.

40. WENDLING, A. and KOLODJI, B. An evaluation of the El Cajon drinking driver. San Diego; San Diego State University Department of Sociology; 1977.

41. MICHELSON, L. The effectiveness of an alcohol safety school in reducing recidivism of drinking drivers. J. Stud. Alc. 40: 1060–1074, 1979.

42. REGIER, M. C. Social policy in action; perspectives on the implementation of alcoholism reforms. Lexington, Mass.; Lexington Books; 1979.

43. ZIMRING, F. E. and HAWKINS, G. Deterrence; the legal threat in crime control. Chicago; University of Chicago Press; 1973.

44. U.S. NATIONAL HIGHWAY TRAFFIC SAFETY ADMINISTRATION. OFFICE OF ALCOHOL COUNTERMEASURES. Alcohol Safety Action Projects; evaluation of operations. Vol. 2. Detailed analysis. Ch. 3. Evaluation of the enforcement countermeasures activities. (DOT Publ. No. HS 800-874.) Washington, D.C.; 1974.

45. ZADOR, P. Statistical evaluation of the effectiveness of "Alcohol Safety Action Projects." Accident Anal. Prev., Oxford 8: 51–66, 1976.

46. ROSS, H. L. Law, science and accidents; the British Road Safety Act of 1967. J. legal Stud. 2 (No. 1): 1–73, 1973.

47. ROSS, H. L. Deterrence regained; the Cheshire constabulary's breathalyzer blitz. J. legal Stud. 6: 241–249, 1977.

48. PARKER, D. A. and HARMAN, M. S. The distribution of consumption model of prevention of alcohol problems; a critical assessment. J. Stud. Alc. 39: 377–399, 1978.

49. SCHMIDT, W. and POPHAM, R. E. The single distribution theory of alcohol consumption; a rejoinder to the critique of Parker and Harman. J. Stud. Alc. 39: 400–419, 1978.

50. HANDLER, J. Social movements and the legal system; a theory of law reform and social change. (Institute for Research on Poverty Ser.) New York; Academic Press; 1979.

51. WILLIAMS, A. F., RICH, R. F., ZADOR, P. L. and ROBERTSON, L. S. The legal minimum drinking age and fatal motor vehicle crashes. J. legal Stud. 4: 219–239, 1975.

52. DOUGLASS, R. L. The legal drinking age and traffic casualties; a special case of changing alcohol availability in a public health context. Pp. 93–132. In: WECHSLER, H., ed. Minimum-drinking-age laws. Lexington, Mass.; Heath; 1980.

53. UPJOHN INSTITUTE FOR EMPLOYMENT RESEARCH. Work in America. Cambridge, Mass.; MIT Press; 1973.

54. THOMPSON, R. R., TENNANT, J. A. and REPA, B. S. Vehicle-borne drunk driver countermeasures. Pp. 347–363. In: ISRAELSTAM, S. and LAM-

BERT, S., eds. Alcohol, drugs and traffic safety; proceedings of the sixth international conference. Toronto; Addiction Research Foundation; 1975.

55. STIVERS, R. A hair of the dog; Irish drinking and American stereotype. University Park; Pennsylvania State University Press; 1976.

56. STIVERS, R. Culture and alcoholism. Pp. 573–602. In: TARTER, R. E. and SUGERMAN, A. A., eds. Alcoholism; interdisciplinary approaches to an enduring problem. Reading, Mass.; Addison-Wesley; 1976.

57. EDWARDS, G. Public health implications of liquor control. Lancet **2**: 424–425, 1971.

58. FRANKEL, B. G. and WHITEHEAD, P. C. Sociological perspectives on drinking and damage. Pp. 13–43. In: BLOCKER, J. S., JR., ed. Alcohol, reform and society; the liquor issue in social context. (Contributions in American history, No. 83.) Westport, Conn.; Greenwood Press; 1979.

59. MÄKELÄ, K. Consumption level and cultural drinking patterns as determinants of alcohol problems. J. drug Issues **5**: 348–357, 1975.

60. GINZBERG, E. The limits of health reform. New York; Basic Books; 1977.

61. GIBSON, R. and MUELLER, M. National health expenditures, fiscal year 1976. Social Secur. Bull. **40** (No. 4): 3–22, 1977.

62. STARR, P. Medicine and the waning of professional sovereignty. Daedalus **107**: 175–193, 1978.

63. CARLSON, R. J. The end of medicine. New York; Wiley-Interscience; 1975.

64. KNOWLES, J. H., ed. Doing better and feeling worse. New York; Norton; 1977.

65. U.S. SURGEON GENERAL'S OFFICE. Healthy people; the Surgeon General's report on health promotion and disease prevention. (DHEW Publ. No. PHS 79-55071.) Washington, D.C.; U.S. Govt Print. Off.; 1979.

66. MOYNIHAN, D. P. The professionalization of reform. Public Interest, No. 1, pp. 6–10, 1965.

67. MACANDREW, C. and EDGERTON, B. Drunken comportment; a social explanation. Chicago; Alaine; 1969.

68. CAVAN, S. Liquor license; an ethnography of bar behavior. Chicago; Alaine; 1966.

69. GUSFIELD, J. R., KOTARBA, J. and RASMUSSEN, P. The world of the drinking driver. Washington, D.C.; National Science Foundation; 1979.

Subject Index

427

Author Index

437